Springer Monographs in Mathematics

Chjan Lim Joseph Nebus

Vorticity, Statistical Mechanics, and Monte Carlo Simulation

Chjan Lim
Department of Mathematical Sciences
Rensselaer Polytechnic Institute
Troy, NY 12180-3590
USA
limc@rpi.edu

Joseph Nebus
Department of Mathematics
National University of Singapore
Singapore
nebusj@rpi.edu

Mathematics Subject Classification (2000): 76-01, 76B47, 82-01, 82B26, 82B80

ISBN 978-1-4419-2247-2 e-ISBN 978-0-387-49431-9
Printed on acid-free paper.

9 8 7 6 5 4 3 2 1

springer.com

To Siew Leng and Sean – CCL

To Joseph Francis and Dr Mary Casey – JN

Preface

This book is meant for an audience of advanced undergraduates and graduate students taking courses on the statistical mechanics approach to turbulent flows and on stochastic simulations. It is also suitable for the self-study of professionals involved in the research and modelling of large scale stochastic fluid flows with a substantial vortical component.

Several related ideas motivate the approach in this book, namely, the application of equilibrium statistical mechanics to two-dimensional and 2.5-dimensional fluid flows in the spirit of Onsager [337], and Kraichnan [227], is taken to be a valid starting point, and the primary importance of non-linear convection effects combined with the gravitational and rotational properties of large scale stratified flows over the secondary effects of viscosity is assumed. The latter point is corroborated by the many successful studies of fluid viscosity which limit its effects to specific and narrow regions such as boundary layers, and to the initial and transient phases of the experiment such as in the Ekman layer and spin-up [154] [344].

The main point of applying equilibrium statistical methods to the problems in this book is underscored by the values of the Knudsen number $K = \lambda/l$ (where λ is the mean free path of the molecules of the fluid and l is the smallest relevant macroscopic length scale in the flow) in the body of two-dimensional and 2.5-dimensional large scale fluid flows treated here, namely $K < 10^{-6}$. We further elucidate this point by stressing the fact that in this book, the methods of statistical mechanics are applied not to the fluid as an ensemble of molecules but rather to the flow as an ensemble of vorticity parcels. Nonetheless, many of the techniques used in the statistical treatment of molecular thermodynamics, including the spin-lattice models pioneered in the study of magnetism in condensed matter physics, can be adapted for our primary purpose here.

Our approach of applying equilibrium statistical mechanics to vortical flows centers on the extremization of the free energy $F = U - TS$ where U is the internal energy and S is the entropy. Besides the standard application of Planck's theorem to thermal systems at positive temperatures, where one minimizes the free energy, we are also interested in vortex problems at negative temperatures, where one maximizes the free energy to obtain the thermodynamically stable statistical equilibria. This point is explored in a simple mean field theory for barotropic flows on a rotating sphere that relates for the first time positive and negative critical temperatures of phase transitions to the key variables of planetary spin, relative enstrophy and kinetic energy.

We note that at low enough positive temperatures T, the minimization of F can be profitably approximated by the easier ground state problem. We further note that the ground state problem by virtue of the minimization of augmented energy functionals, is directly related to steady-state flows of the associated Euler equations. Finally, these special steady-states are related back to decaying Navier-Stokes flows by the Principle of Selective Decay or Minimum Enstrophy, which states that the Dirichlet Quotient (defined as enstrophy over energy) in many damped two-dimensional viscous flows tends asymptotically to a minimum, achieved by the special steady-states.

Vortex statistics is special not only because negative temperatures occur at high energies (a curious phenomenon we will explain in detail), but because it is also characterized by the wide range of temperatures over which extremals of the free energy are close to the corresponding extremals of the internal energy. We will explore the physical reasons for these interesting phenomena in several archetypical examples of vortex dynamics. The most important of these problems are the crystalline or polyhedral equilibria of N point vortices on the sphere, the thermal equilibria of the Onsager vortex gas on the unbounded plane with respect to dynamical equilibria of the rotating two-dimensional Euler equations, and the thermal equilibria at negative temperatures of barotropic vortex dynamics on a rotating sphere.

The works, teachings and direct interactions of many scientists have contributed to this book; it is not possible to mention all their names. We acknowledge in gratitude, Denis Blackmore, Alexandre Chorin, Shui Nee Chow, Peter Constantin, Weinan E, Marty Golubitsky, Leslie Greengard, Tom Hou, Choy Heng Lai, Peter Lax, Andy Majda, Paul Newton, Don Saari, David Sattinger, Junping Shi, Lu Ting, and Lawrence Sirovich who taught one of us kinetic theory.

We would like to thank graduate students Tim Andersen and Xueru Ding for their contributions to graduate seminars at Rensselaer Polytechnic Institute where some of the ideas for this book were tested, and Syed M. Assad for past and continuing collaborations of which some have found their way into this book.

We would like to thank the Department of Mathematical Sciences at Rensselaer Polytechnic Institute and the Departments of Computational Science and of Mathematics at the National University of Singapore for their support of our work.

We thank Achi Dosanjh of Springer Science for initiating and completing the editorial process for this book.

We acknowledge the support of Dr. Chris Arney and Dr. Robert Launer, US Army Research Office and Dr. Anil Deane and Dr. Gary Johnson, US Department of Energy in the form of research grants during the research and writing phase.

Rensselaer Polytechnic Institute,
National University of Singapore,
June 2006

Chjan C Lim
Joseph Nebus

Contents

1

Introduction

1.1 Connecting Statistical Mechanics to Vortex Problems

The "unreasonable effectiveness of mathematics in the physical sciences," a phrase coined by Eugene Wigner[1], is often mentioned as an exotic property of mathematics. It is an expression of the wonder that models constructed from purely theoretical considerations and reasoned out can predict real systems with great precision. A modern instance of this wonder is found in the writings of Subrahmanyan Chandrasekhar[2]: "In my entire scientific life, the most shattering experience has been the realization that an exact solution of Einstein's equations of general relativity, discovered by Roy Kerr, provides the absolutely exact representation of untold numbers of massive black holes that populate the Universe."

The relationship between mathematics and reality was remarked on even by the Pythagoreans, but was probably not seen as noteworthy to the point of becoming cliché in the earliest era of mechanics. A prediction of the orbits of planets and comets based on Newtonian mechanics and gravitation, for example, could be excellent; but since the laws describing gravitation came from observations of planetary orbits that should be expected. If observation and theory did not agree the theory would not have been used.

It is probably in the laws of gasses that this effectiveness became distracting. The dynamics of a gas can in theory be developed by a Newtonian system, provided one knows to represent it as particle interactions – atomic theory, not generally accepted until the 19th century and still worth debate until Brownian motion was explained – and one knows the laws by which

[1] Eugene Paul Wigner, 1902 - 1995, introduced "parity" to quantum mechanics and discovered the curious properties of the strong nuclear force. [288]

[2] Subrahmanyan Chandrasekhar, 1910 - 1995, was a master of the mechanisms of stars, and predicted the existence of black holes from 1930. [288]

they interact – quantum theory – and one can handle the staggering number of variables needed. These were formidable challenges.

Attempted instead were models based on simple theoretical constraints and few details. Assuming simply that gas was made of particles and these particles moved independently provides the ideal gas law which had been noted and publicized by Robert Boyle, 1627-1691, and Jacques-Alexandre-César Charles, 1746-1823, and others centuries earlier. For such a simple model matching the general behavior of real gases was a surprise. Adding the assumptions atoms had a minimum size and some interaction allowed Johannes Diderik van der Waals, 1837-1923, to offer in 1873 a correction to the ideal gas law and an even better match to observation [25] [64].

Erwin Schrödinger[3] [389], in his compelling 1944 essay "What is life?", presented a famous argument for the apparent exactness and determinateness of macroscopic laws which are nonetheless based on physical laws of a statistical nature for the detailed components of a system. He argued the macroscopic law of bulk diffusion (which is clearly deterministic) is based on the completely and purely statistical phenomenon of random walks at the microstate level. The random walk of molecules is not only statistical in nature but it is also completely symmetrical: a molecule takes a jump to the right or left in say a one-dimensional model with equal probability. Yet its consequence at the macroscopic level is clearly asymmetric because the law of bulk diffusion has a clear direction: from high concentration to low concentration.

This is our textbook's inspiration. A simple model of fluid flow will be made from theoretical considerations. The model will be studied through several alternative strategies and adjusted to make it more natural.

We are more concerned with statistical equilibria than with dynamical equilibria. A dynamical equilibrium requires the components of a state to be in a spatially rigid and temporally stationary relationship with each other. This is too restrictive for the problems in this book. Statistical equilibria have stationary macroscopic variables which offer vastly more degrees of freedom in the fine details of the state. A rule of thumb for the appearance of seemingly exact macroscopic laws is that the macroscopic system must have large enough number N of microscopic components in order for the fluctuations of size $1/\sqrt{N}$ in the macroscopic variable to be small.

This book is our attempt to connect two main topics of asymptotic states in vortex flows and equilibrium statistical mechanics. While fully developed turbulence in a damped driven flow is a non-equilibrium phenomena, many powerful arguments (by Kolmogorov [223], [224], Oboukhov [330]) have been presented, asserting that for certain inertial ranges in the power spectrum of driven viscous flows, the methods of equilibrium statistical mechanics can be adopted. We will avoid such arguments and treat the phenomena of isolated

[3] Erwin Rudolf Josef Alexander Schrödinger, 1887 - 1961, got the inspiration for the wave form of quantum mechanics from a student's suggestion at a seminar Schrödinger gave on the electron theory of Louis de Broglie, 1892 - 1987. [288]

inviscid fluid turbulence within the context of equilibrium statistical mechanics.

The concept of negative temperature was introduced into vortex dynamics by Lars Onsager. Vortical systems in two dimensions and in 2.5 dimensions (which we will describe) support negative temperatures at high kinetic energies where the thermal equilibria are characterized by highly organized large-scale coherent structures. Thus, besides the standard application of Planck's theorem to thermal systems at positive temperatures, where one minimizes the free energy, we are also interested in vortex problems at negative temperatures, where one maximizes the free energy to obtain stable statistical equilibria.

In addition to the first common theme of Monte Carlo simulations of organized and of turbulent fluid flows in this book, a second theme is the relationship between dynamics and equilibrium statistical mechanics: the **extremals** (maxima and minima) of the energy determine the equilibria, but the extremals of the free energy give us the most probable states of the equilibrium statistics.

Vortex statistics has many noteworthy examples where the range of temperatures is quite large, over which extremals of the free energy are close to the corresponding extremals of the internal energy. We will explore the physical reasons for these interesting phenomena in several archetypical examples of vortex dynamics. The most important of these problems are the crystalline or polyhedral equilibria of N point vortices on the sphere, the thermal equilibria of the Onsager vortex gas on the unbounded plane with respect to dynamical equilibria of the rotating two-dimensional Euler equations, and the thermal equilibria at negative temperatures of barotropic vortex dynamics on a rotating sphere.

In the first case, N similar point vortices on a sphere, the Monte Carlo simulator running at positive temperatures achieves thermal equilibria which are very close to the polyhedral relative equilibria of the dynamical equations. These polyhedral crystalline states have extremely regular and uniform spatial separations, and thus minimize the interaction energy though without simultaneously maximizing the entropy. This situation provides one of the canonical ways in which minimizers of the free energy are close to the dynamical equilibria for a range of positive temperatures.

The second case concerns the unbounded Onsager point vortex gas, whose thermal equilibria of uniform vorticity distributions in a disk are close to the dynamic equilibria of the rotating two-dimensional Euler equations over a wide range of positive temperatures. The physical reason in this case is the same as the first, that is, the free energy minimizers are given by vortex states which minimize the internal energy.

The physical reason for the third case is that free energy maximizers corresponding to stable thermal equilibria at negative temperatures are achieved by vortex states with very low entropy. Unlike standard thermodynamic applications where entropy is maximized, the solid-body rotation flow states have the minimum entropy and maximum kinetic energy over allowed flow states with the same relative enstrophy.

A non-extensive continuum limit is allowed for two-dimensional flows in a fixed finite domain. The canonical examples for such flow domains are the fixed bounded regions on the plane, and finite but boundary-less domains such as the surface of the sphere or of the torus. Finite boundary-less domains are computationally convenient because boundary conditions are often complex. And among finite, boundary-less domains the problem of flows on the sphere is more important than flows on more topologically complex surfaces because of their applicability to atmospheric sciences, to which we will return.

Our principal focus is the of inviscid fluids. We justify this choice – which seems to exclude most of the fluids of the real world – on several grounds. The first is that often the viscosity of a fluid is a minor effect, and these slightly viscous fluids can be modelled by inviscid fluids, where we represent the interior of the flow by a fluid without viscosity, and add boundary layers in the regions where viscosity becomes relatively significant. More, even if we want to consider viscous fluids, we can still represent an interesting aspect of them – the non-linear convective aspects of the flows – by treating this portion of the flow as an inviscid fluid.

And furthermore much of what we can study in the thermal equilibrium of inviscid two-dimensional vortex dynamics (such as the minimizers of free energy functions) can be extended naturally to the ground states of augmented energy functionals, or to the steady states of two-dimensional Euler equations. These in turn are related, by the Principle of Selective Decay, also termed the Principle of Minimum Enstrophy, to the asymptotic flows of the decaying two-dimensional Navier[4]-Stokes[5] equations. The minimizers of the Dirichlet quotient, the ratio of enstrophy to energy, corresponds to the inviscid steady states we explicitly study [26].

1.2 Euler's Equation for Inviscid Fluid Flow

To write Leonhard Euler's[6] equation for fluid flow, we begin with the fluid velocity. Letting $\mathbf{u}$ stand for the velocity and ρ the density of the fluid, we

[4] Claude Louis Marie Henri Navier, 1785 - 1836, was in his day famous as a builder of bridges. He developed the first theory of suspension bridges, which had previously been empirical affairs. [288]

[5] George Gabriel Stokes, 1819 - 1903, besides the theory of fluid flow and the theorem about the integrals over surfaces, provided the first theory and the name for fluorescence. [288]

[6] It is almost impossible to overstate the contributions of Leonhard Euler, 1707 - 1783. The scope is suggested by the fact after his death it required fifty years to complete the publication of the backlog of his papers. He is credited with the modern uses of the symbols e, i, π, Σ, the finite differences Δt and $\Delta^2 t$, and $f(x)$ as a general symbol for a function. [288]

choose some fluid properties. We want the fluid to be incompressible, inviscid, and to experience no outside forces.

The obviously important properties of the fluid are the density at a time t and a point $\mathbf{r}$ – call that $\rho(t, \mathbf{r})$ – and the velocity, again a function of time and position. Call that $\mathbf{u}(t, \mathbf{r})$. We will build on three properties [88].

First is the conservation of mass. Suppose the fluid is incompressible, which is nearly correct for interesting fluids such as water at ordinary temperatures and pressures. Incompressibility demands a divergence of zero:

$$\nabla \cdot \mathbf{u} = 0 \tag{1.1}$$

A nonzero divergence over some region A corresponds to either a net loss or net gain of mass, so the fluid density is changing and the fluid is either expanding or compressing.

The next property is the conservation of momentum. The momentum inside region A, the total of mass times velocity, will be

$$\mathbf{p} = \int_A \rho \mathbf{u} dV \tag{1.2}$$

(with dV the differential volume within the region A). So the rate of change of the momentum in time will be

$$\frac{\partial}{\partial t}\mathbf{p} = \int_A \frac{\partial}{\partial t}(\rho \mathbf{u}) dV \tag{1.3}$$

Without external pressure, or gravity, or viscosity or intermolecular forces the momentum over the region A cannot change on the interior. Only on the surface can momentum enter or exit A:

$$\frac{\partial}{\partial t}\mathbf{p} = \int_{\partial A} (\rho \mathbf{u})\mathbf{u} \cdot \mathbf{dS} \tag{1.4}$$

with ∂A the surface of A and $\mathbf{dS}$ the differential element of area for that surface. Using Green's theorem[7], the integral is

$$\frac{\partial}{\partial t}\mathbf{p} = -\int_A \rho(\mathbf{u} \cdot \nabla)\mathbf{v} dV \tag{1.5}$$

If there is a force, which we will generalize by calling the pressure and denoting it as $P(\mathbf{r}, r)$, then momentum may enter or exit the region A, but again only through its surface. Even more particularly only the component of the force which is parallel to the outward unit normal vector $\hat{n}$ can affect the fluid, in or out. So the change in the momentum of the fluid caused by the pressure term P, is

[7] George Green, 1793 - 1841, besides his theorem connecting surface integrals to volume integrals, is credited with introducing the term "potential function" in the way we use it today. [288]

$$\frac{\partial}{\partial t}\mathbf{p} = \int_{\partial A} P\hat{n}d\mathbf{S} \tag{1.6}$$

$$= -\int_{A} \nabla P d\mathbf{V} \tag{1.7}$$

using again Green's theorem.

As we may have momentum gained or lost through either the fluid flow or through the pressure we add the two terms:

$$\frac{\partial}{\partial t}\mathbf{p} = -\int_{A} \rho(\mathbf{u}\cdot\nabla)\mathbf{v}dV - \int_{A} \nabla P d\mathbf{V} \tag{1.8}$$

Between equations 1.3 and 1.8 we have two representations of the derivative of momentum with respect to time. Setting them equal

$$\frac{\partial}{\partial t}\mathbf{p} = \int_{A} \frac{\partial}{\partial t}(\rho\mathbf{u})dV = \int_{A} \rho(\mathbf{u}\cdot\nabla)\mathbf{u}dV - \int_{A} \nabla P d\mathbf{V} \tag{1.9}$$

for all regions A. For the middle and right half of equation 1.9 to be equal independently of A requires the integrands be equal[8]:

$$\frac{\partial}{\partial t}(\rho\mathbf{u}) = -\rho(\mathbf{u}\cdot\nabla)\mathbf{u} - \nabla P \tag{1.10}$$

which is Euler's equation for inviscid, incompressible, unforced fluid flow. Assuming incompressibility makes ρ constant in time, so we may divide it out.

Having introduced the pressure, we will proceed now to drop it for nearly the entirety of the book, as we will find abundant interesting material even before adding pressure to the system. In this form and confined to one spatial dimension is often known as Burgers' equation[9], though we will keep a bit more freedom in space:

$$\frac{\partial}{\partial t}\mathbf{u} + \mathbf{u}\cdot\nabla u = 0 \tag{1.11}$$

Up to this point we have considered only two important physical properties. The third we will add in order to convert this equation into a form more suitable for treatment as a particle problem, which we will do in chapter 6. There we will also change our attention from the velocity of the fluid into the vorticity, that is, the curl of the velocity. This combination lets us recast the flow of an inviscid fluid as a statistical mechanics problem.

[8] Strictly speaking, they must be equal "almost everywhere" – the set of points that are exceptions must have measure zero. For example, finitely many exceptions are allowed.

[9] Johannes Martinus Burgers, 1895 - 1981. Burgers is known also for the Burgers dislocation, a method of describing the irregularities in a crystal. He was also expert on the study of polymers, and a sort of viscoelastic material is named for him.

Our roots in statistical mechanics and thermodynamics suggests a question: is there a temperature to a vortex dynamical system? Statistical mechanics defines the temperature of any system to be the derivative of energy with respect to the entropy. In the kinetic theory of gases this equals the physical heat. Although there is no physical heat in this problem, there is an energy and there is an entropy; therefore, it has a temperature.

This extension of temperature is not unique. Most physical models have an energy. The entropy of a model can be given through information-theoretical methods – as long as a system can contain information, it has an entropy. Therefore the idea of temperature can be applied to systems that have no resemblance at all to the gas particles the idea began with.

One fascinating consequence is that vortex systems can have a negative temperature. There are configurations for which adding more energy will decrease the entropy of the system. The derivative is then negative and the temperature is therefore less than zero. More remarkably these negative temperature states are extremely high-energy ones. These negative temperature states will receive considerable attention. (Vortex dynamics is not the only context in which negative temperatures arise. They can develop in systems in which a maximum possible energy exists. One noteworthy example is in describing the states of a laser.)

We will need to simplify our problem to be able to apply statistical mechanics methods to it. We want a large but finite number of particles or lattice sites which obey some interaction law. Our interests will lead us to rewrite the Euler equation from several perspectives. In one we will describe the vorticity of the fluid as a set of discrete "charged" particles which are free to move. In another we will construct a piecewise-continuous approximation to the vorticity based around a fixed set of mesh sites and allow the value of the function on these pieces to vary.

If we are interested in the "vortex gas" problem, placing a set of vortices of fixed strength and allowing them to move, then we could write it as a dynamical systems problem, with a Hamiltonian[10], a representation using the form of classical mechanics. With that we can use tools such as the Monte Carlo Metropolis Rule to explore this space and study the equilibrium statistical mechanics.

Unfortunately the Monte Carlo study of the vortex gas problem does not well handle vortices of positive and negative strengths mixed together. The Metropolis-Hastings rule will tend to make vortices of opposite sign cluster together. Similarly negative temperatures cannot be meaningfully applied; trying simply causes all like-signed vortices to cluster together. But as long as

[10] These functions were introduced by Sir William Rowan Hamilton, 1805 - 1865, and have become a fundamental approach to dynamical systems. Hamilton also discovered quaternions, famously carving the inspired equation $i^2 = j^2 = k^2 = ijk = -1$ into the stones of the Brougham Bridge. [288]

we are interested in a single sign and positive temperatures interesting work may be done.

In the lattice problem (our mesh sites may not be the regularly organized rows and columns of a proper lattice, but it is a fixed set of sites) we approximate the continuous vorticity field by a piecewise-continuous approximation. Changes in the fluid flow are represented by changes in the relative strengths of lattice sites. This approach resembles strongly a finite-elements study. This approach also well handles both positive and negative vorticities, and both positive and negative temperatures are meaningfully studied.

There is also a useful approach not based on points and site vorticities at all. Anyone who has studied enough differential equations has encountered Fourier decompositions of problems – supposing that the solution to a differential equation is the sum of sine and cosines of several periods, and finding the relative amplitudes of the different components. This sort of approach is called the spectral method. The analogy to identifying the components of a material by the intensities at different frequencies of the spectrum of light that has passed through the material is plain.

Through these approaches, we plan to show how analytical and computational mathematics complement one another. Analytic study of fluid flow provides a problem well-suited to numerical study. Numerical experiments will improve the understanding of old and will inspire new analysis. In combination we make both approaches stronger.

To end this introduction, we remember that the astronomical and cosmological examples alluded to above have a different scale of predictability than fluid phenomena such as flow turbulence and the weather. Astronomers have predicted solar and lunar eclipses to the second for centuries. But the weather cannot even now be predicted with anywhere near the same scale of accuracy. The accuracy in astronomical prediction is largely dependent on the exactness of the initial data at an earlier epoch. The inaccuracies in weather prediction persists in spite of greatly improved meteorological methods and instruments for measuring the state of an atmosphere. These are two very different realms of applied and computational mathematics, underscoring the theoretical and technical difficulties of the latter.

2

Probability

2.1 Introduction

In science fiction writer Stanley G Weinbaum's 1935 short story "The Lotus Eaters" the ultimate law of physics is declared to be the law of chance. Given the overwhelming power of statistical mechanics and of quantum mechanics this assessment is hard to dispute. The study of probability combines beautiful reasoning beginning from abstract first principles and describes the observable world with remarkable clarity. So before moving into Monte Carlo methods it is worthwhile reviewing the fundamentals of probability.

2.2 Basic Notions

Let the letter S represent a **sample space**, which is the set of all the possible outcomes of a particular experiment. An **event** U is a subset of the sample space: $U \subseteq S$. We typically think of an event as just one of the possible outcomes, but we will be interested in the probability of several outcomes occurring together, or of one outcome occurring given that another has.

We want to define is the **probability** $P\{E\}$ of the event E occurring. The first thought is to do this by a limit. Let the number $n(E)$ represent the number of times outcome E occurs in the first n repetitions of an experiment. Then

$$P\{E\} = \lim_{n \to \infty} \frac{n(E)}{n} \tag{2.1}$$

which is known as the **limiting frequency** of E.

As often happens, this intuitive definition requires the assumption of several difficult-to-justify axioms. The first troublesome axiom is that the limit described above exists. Another is the supposition that the limiting process will converge on the same limit every time a series of experiments is run. Another is that we can reliably estimate how many experiments are needed to establish this probability.

Given these problems we put aside the intuitive definition and build one around simpler axioms. Define a **probability space** as a pair (S, P) consisting of a sample space S made of events E, and a function P defined on the events E. An event is, to be precise, a specific subset of the sample space; we can build an intuitive feel for it by calling it a possible outcome of some process, using the ordinary English senses of the words. The function P satisfies these axioms [371] [426].

Axiom 1 *For each* $E \subseteq S$, $0 \leq P\{E\} \leq 1$.

That is, probabilities are defined to be between zero and one (inclusively). The greater the probability the greater the chance an event occurs.

Axiom 2 $P\{S\} = 1$.

This is a difficult axiom to dispute; the outcome of an experiment must be some element of the sample space S.

Axiom 3 *If* $E_1, E_2, E_3, \cdots, E_n$ *is any collection of mutually exclusive events, then* $P\{\cup_{i=1}^{n} E_i\} = \sum_{i=1}^{n} P\{E_i\}$.

That is, if it is not possible for several of a set of events to simultaneously occur, then the probability that exactly one of them does occur is the sum of the probabilities of any of them occurring.

Axiom 4 *If* E *and* F *are events and* α *a number between* 0 *and* 1, *then there exists an event* S *for which* $P\{S\} = \alpha$ *and for which* $P\{S \cap E\} = P\{S\}P\{E\}$ *and* $P\{S \cap F\} = P\{S\}P\{F\}$.

That is, whatever events we have we can assume the existence of other independent events which are as probable or improbable as desired.

From these axioms[1] we are able to build the studies of probability familiar to every student, as well as the properties used for statistical mechanics. Trusting that the ideas of independent events and of conditional probability are familiar we move on to random variables.

2.3 Random Variables and Distribution Functions

The sample space S is if nothing else a set. We can define a function from S to the real numbers. Such a real-valued function is known as a **random variable**.

[1] There is one more assumption we must make, if the sample space has infinitely many elements – we must assume that P is defined only for events which are measurable in the sample space. This restriction will not impair our work, but it is needed for the accurate analysis of infinitely large sets.

For example, if the experiment is tossing ten fair coins, the sample space is all 2^{10} possible ways the coins may fall, and a random variable of interest might be the number of heads which turn up. If the experiment is rolling a pair of dice, the sample space is the set of 36 combinations of outcomes of the dice; our random variable is the sum of the numbers on those dice.

These variables are interesting because the typical view of statistical mechanics is that detailed information of a system is not important – what we want to know are properties like energy or entropy, which are random variables by this definition. We will move towards first the distribution functions of a random variable X – the probability of X falling within a specified range – and the expectation value – the mean value of X.

Define the **cumulative distribution function** F of the random variable X to be probability that X is less than or equal to d:

$$F(d) = P\{X \leq d\} \tag{2.2}$$

this function satisfies several properties; among them [58] [371] [426]:

1. F is non-decreasing: if $a < b$, then $F(a) \leq F(b)$.
2. $\lim_{d \to \infty} F(d) = 1$.
3. $\lim_{d \to -\infty} F(d) = 0$.
4. F is continuous from the right: for any decreasing sequence d_n which converges to d, $\lim_{n \to \infty} F(d_n) = F(d)$. (Why is F not necessarily continuous?)

Define the **probability mass function** $p(x)$ of X to be

$$p(x) = P\{X = x\} \tag{2.3}$$

and satisfying the condition X must take on one of the values x_i:

$$\sum_{i=1}^{\infty} p(x_i) = 1 \tag{2.4}$$

The cumulative distribution function $F(c)$ we construct by letting

$$F(c) = \sum_{\text{all } x \leq c} p(x) \tag{2.5}$$

which it is easy to verify satisfies the above properties.

We are not interested only in discrete random variables. Continuous random variables such as the energy[2] of a system are handled similarly to the discrete case.

[2] The energy of a real dynamical system may be drawn from an interval. In the numerical simulations we do, we confine ourselves to the floating point number system and so will use what is "really" a discrete system. Generally though we will use the terminology of continuous random variables.

Let X be a random variable which may have any value from an uncountably infinite set of values. This X is a **continuous random variable** if there exists a function f, defined on all real numbers and non-negative, for which

$$P\{X \in A\} = \int_A f(x)dx \tag{2.6}$$

for every subset A of the reals. We also require of f that

$$\int_{-\infty}^{\infty} f(x)dx = 1 \tag{2.7}$$

as $\int_{-\infty}^{\infty} f(x)dx$ is the probability that X is some real number. This function f is the **probability density function** of X.

We may go from the probability density function to the cumulative distribution function by setting

$$F(c) = P\{X \in (-\infty, c]\} \tag{2.8}$$

$$= \int_{-\infty}^{c} f(x)dx \tag{2.9}$$

Equivalently,

$$\frac{d}{dc}F(c) = f(c) \tag{2.10}$$

and the probability of observing X to be within the range (c, d) is equal to

$$P\{X \in (-\infty, d]\} - P\{X \in (-\infty, c]\} = \int_{-\infty}^{d} f(x)dx - \int_{-\infty}^{c} f(x)dx \tag{2.11}$$

$$= \int_{c}^{d} f(x)dx \tag{2.12}$$

The probability density function is an approximation of the probability that, for any c, that X will be near c. Consider an interval of width ϵ centered around c:

$$P\{c - \frac{\epsilon}{2} \leq X \leq c + \frac{\epsilon}{2}\} = \int_{c-\frac{\epsilon}{2}}^{c+\frac{\epsilon}{2}} f(x)dx \tag{2.13}$$

$$\approx \epsilon f(c) \tag{2.14}$$

if f is continuous and ϵ sufficiently small.

2.4 Expectation Values and Averages

Given a random variable – the sum of a pair of dice, the distance a car may travel without needing refuelling, the kinetic energy of a gas – it is hard to avoid wondering what its average value is. The word "average" has multiple meanings; ours is a weighted arithmetic mean. Values of X more likely to occur should weigh more in the average than improbable ones. The weighted mean we finally study is called the expectation value of X.

Define the **expectation value** E of discrete variable X to be [371] [426]

$$E[X] = \sum_i x_i P\{X = x_i\} \tag{2.15}$$

and for a continuous variable X to be

$$E[X] = \int_{-\infty}^{\infty} x f(x) dx \tag{2.16}$$

(which are the same formulas used to find the center of mass for a set of discrete masses and for a continuous mass, a similarity reflected in the terms "probability mass" and "probability density" functions).

This expectation value is the "average" which matches our intuitive expectation: if we were to run a large number of experiments and measure X for each of them, the mean of these measured values of X will tend towards $E[X]$.

Given the real-valued function g we can define a new random variable $Y = g(X)$. If we know the probability mass or density function for X, do we know the expectation value for Y?

Proposition 1. *Suppose X is discrete; then*

$$E[Y] = E[g(X)] = \sum_i g(x_i) p(x_i) \tag{2.17}$$

Suppose X is continuous; then

$$E[Y] = E[g(X)] = \int_{-\infty}^{\infty} g(x) f(x) dx \tag{2.18}$$

Proof. Notice that there is for each y_j a set of points G_j, such that for any x_i in G_j we have $g(x_i) = y_j$. This lets us reorganize, in the discrete case, the sum over all the different values x_i into a sum over all the different values y_j of $g(x_i)$. So:

$$\sum_i g(x_i) p(x_i) = \sum_j \sum_{x_i \in G_j} g(x_i) p(x_i) \tag{2.19}$$

$$= \sum_j \sum_{x_i \in G_j} y_j p(x_i) \tag{2.20}$$

$$= \sum_j y_j \sum_{x_i \in G_j} p(x_i) \tag{2.21}$$

$$= \sum_j y_j P\{g(X) = y_j\} \tag{2.22}$$

$$= E[g(X)] \tag{2.23}$$

□

The proof of the continuous case is similar in inspiration, although to use the same outline requires the Lebesgue[3]-Stieltjes[4] definition of an integral. To avoid introducing that much supporting material we will instead narrow the focus so that g is a nonnegative function, and introduce a lemma.

Lemma 1. *For a nonnegative continuous random variable* Y,

$$E[Y] = \int_0^\infty P\{Y \geq y\}dy \tag{2.24}$$

Proof. Let f_Y be the probability density function for Y. Therefore $P\{Y > y\} = \int_y^\infty f_Y(x)dx$. And (notice the swapping the order of integration on the second line of this derivation)

$$\int_0^\infty P\{Y > y\}dy = \int_0^\infty \int_y^\infty f_Y(x)dxdy \tag{2.25}$$

$$= \int_0^\infty \left(\int_0^x dy\right) f_Y(x)dx \tag{2.26}$$

$$= \int_0^\infty x f_Y(x)dx \tag{2.27}$$

$$= E[Y] \tag{2.28}$$

Now we can complete the proof of the restricted form of Proposition 1. Let g be any continuous nonnegative function. For any y, let G_y be the set of x for which $g(x) > y$. We have

$$E[g(X)] = \int_0^\infty P\{g(X) > y\}dy \tag{2.29}$$

$$= \int_0^\infty \int_{G_y} f(x)dxdy \tag{2.30}$$

$$= \int_{G_0} \left(\int_0^{g(x)} dy\right) f(x)dx \tag{2.31}$$

$$= \int_{G_0} g(x)f(x)dx \tag{2.32}$$

[3] Henri Léon Lebesgue, 1875 - 1941, was one of the founders of measure theory, although he did not concentrate on that field. He found it too general a theory for his tastes and preferred to work on smaller, specific topics. [288]

[4] Thomas Jan Stieltjes, 1856 - 1894, besides extending the definition of the integral, is regarded as the founder of analysis of continued fractions, numbers defined by $1/(a+1/(b+1/(c+1/(\cdots))))$ for some integer sequence a, b, c, d, $\cdots$. His time as a student was spent reading Gauss and Jacobi, rather than attending lectures, which gave him a good background but made him fail his exams. [288]

□

The expectation value $E[g(X)]$ is in general *not* equal to $g(E[X])$. In fact the expectation value is – as one might guess from observing that it is either a sum or an integral – linear; $E[g(X)]$ will only equal $g(E[X])$ if the function g is of the form $a \times X + b$.

Proposition 2. *$E[aX + b] = aE[X] + b$ for constants a, b.*

Proof. When X is a discrete random variable,

$$E[aX + b] = \sum_i (ax_i + b)p(x_i) \tag{2.33}$$

$$= a\sum_i x_i p(x_i) + b\sum_i p(x_i) \tag{2.34}$$

$$= aE[X] + b \tag{2.35}$$

When X is a continuous random variable,

$$E[aX + b] = \int_{-\infty}^{\infty} (ax + b)f(x)dx \tag{2.36}$$

$$= a\int_{-\infty}^{\infty} xf(x)dx + b\int_{-\infty}^{\infty} f(x)dx \tag{2.37}$$

$$= aE[X] + b \tag{2.38}$$

□

2.5 Variance

Though the expectation value of a variable describes some of its behavior, it is not the only relevant quantity. A variable X which is always equal to zero has the same expectation value as a variable Y which is any integer between -5 and 5 with uniform probability. As a measure of how distributed the numbers are we can introduce (analogously to the moment of inertia of a set of mass) the **moment** of the variable X. For any n, the n**th moment** of X is

$$E[X^n] = \sum_i x_i^n p(x_i) \tag{2.39}$$

if X is discrete and

$$E[X^n] = \int_{-\infty}^{\infty} x_i^n f(x)dx \tag{2.40}$$

if X is continuous (and assuming the expectation value exists).

These moments are dependent on the values of X; if we define a new variable $Y = X + c$ for a constant c the moments of Y will be different from those of X though most would say the distribution of X is the same as that of Y. We can restore this "translation invariance" by using μ_X, the expectation value of X, as a reference point. Define the n**th central moment** of X to be

$$E[(X - \mu_X)^n] = \sum_I (x_i - \mu_X)^n p(x_i) \tag{2.41}$$

for the discrete case and

$$E[(X - \mu_X)^n] = \int_{-\infty}^{\infty} (x - \mu_X)^n f(x) dx \tag{2.42}$$

for the continuous (and again providing the integrals exist) [371].

The second central moment of X is known as the **variance**, $Var[X]$, and is quite often used. Its square root is known as the **standard deviation**, $\sigma[X]$. Typically the easiest way to calculate the variance is to use the second moment of X and the square of the first moment of X [371] [426].

Proposition 3. $Var[X] = E[X^2] - \mu_X^2$.

Proof. Since the variance of X is its second central moment we have

$$\begin{aligned} Var[X] &= E[(X - \mu_X)^2] && (2.43)\\ &= E[X^2 - 2\mu_X X + \mu_X^2] && (2.44)\\ &= E[X^2] - 2\mu_X E[X] + \mu_X^2 && (2.45)\\ &= E[X^2] - 2\mu_X^2 + \mu_X^2 && (2.46)\\ &= E[X^2] - \mu_X^2 && (2.47) \end{aligned}$$

As $\mu_X = E[X]$ then $Var[X] = E[X^2] - (E[X])^2$. The variance or the standard deviation most often describe whether the values of X are distributed close together (a small variance) or widely apart (a large variance).

□

Proposition 4. $Var[aX + b] = a^2 Var[X]$

Proof. Remembering proposition 2 and the definition of variance,

$$\begin{aligned} Var[aX + b] &= E[(aX + b - a\mu_X - b)^2] && (2.48)\\ &= E[(aX - a\mu_X)^2] && (2.49)\\ &= E[a^2(X - \mu_X)^2] && (2.50)\\ &= a^2 E[(X - \mu_X)^2] && (2.51)\\ &= a^2 Var[X] && (2.52) \end{aligned}$$

□

2.6 Multiple Variables and Independence

To this point we have considered only a single random variable X. It is almost inevitable we will want to measure several quantities in experiments, so we want to establish our probability tools for multiple variables.

Suppose that we have random variables X and Y. The **cumulative joint probability distribution function** is defined to be

$$F(c,d) = P\{X \leq c, Y \leq d\} \tag{2.53}$$
$$= P\{X \in (-\infty, c] \cap Y \in (-\infty, d]\} \tag{2.54}$$

We also define the **marginal distributions** of X and Y, which examine the probability for a single variable assuming the other is completely free. The marginal distribution of X is $F_X(c)$ and equals

$$F_X(c) = P\{X \leq c\} \tag{2.55}$$
$$= P\{X \leq c, Y < \infty\} \tag{2.56}$$
$$= \lim_{d \to \infty} P\{X \leq c, Y \leq d\} \tag{2.57}$$
$$= \lim_{d \to \infty} F(c,d) \tag{2.58}$$
$$= F(c,\infty) \tag{2.59}$$

and similarly

$$F_Y(d) = P\{Y \leq d\} \tag{2.60}$$
$$= P\{X < \infty, Y \leq d\} \tag{2.61}$$
$$= \lim_{c \to \infty} P\{X \leq c, Y \leq d\} \tag{2.62}$$
$$= \lim_{c \to \infty} F(c,d) \tag{2.63}$$
$$= F(\infty,d) \tag{2.64}$$

There is, for the case of discrete random variables, a **joint probability mass function** defined by

$$p(c,d) = P\{X = c, Y = d\} \tag{2.65}$$

with separate **probability mass functions** p_X and p_Y defined by

$$p_X(c) = P\{X = c\} \tag{2.66}$$
$$= \sum_d p(c,d) \tag{2.67}$$
$$p_Y(d) = P\{Y = d\} \tag{2.68}$$
$$= \sum_c p(c,d) \tag{2.69}$$

Similarly for continuous random variables the **joint probability density function** $f(x, y)$ is

$$P\{X \in C, Y \in D\} = \int_D \int_C f(x, y) dx dy \tag{2.70}$$

with separate **probability density functions** f_X and f_Y defined by

$$f_X(x) = \int_{-\infty}^{\infty} f(x, y) dy \tag{2.71}$$

$$f_Y(y) = \int_{-\infty}^{\infty} f(x, y) dx \tag{2.72}$$

We have mentioned independent random variables. We need now to give this idea a precise definition, as the introduction of several variables makes it less obvious independence could be taken for granted. The random variables X and Y are said to be **independent** if for any two subsets C and D of the probability space,

$$P\{X \in C, Y \in D\} = P\{X \in C\} P\{Y \in D\} \tag{2.73}$$

This is equivalent to showing that for all c and d,

$$P\{x \le c, y \le d\} = P\{x \le c\} P\{y \le d\} \tag{2.74}$$

Two variables are **dependent** if they are not independent. That is, they are independent if the events $E_C = \{X \in C\}$ and $E_D = \{Y \in D\}$ are independent.

Proposition 5. *If X, Y have joint probability density function $f(x, y)$ then they are independent if and only if*

$$f(c, d) = f_X(c) f_Y(d) \tag{2.75}$$

Another tool useful to examine independence is the **covariance**, a generalization of the variance of a single variable. For X and Y define the covariance $Cov[X, Y]$ to equal $E[(X - \mu_X)(Y - \mu_Y)]$, where $\mu_X = E[X]$ and $\mu_Y = E[Y]$.

Proposition 6. *If X and Y are independent then $Cov[X, Y] = 0$.*

Proof. We start with a lemma relating the covariance of X and Y to the expectation values of X, Y, and their product XY.

Lemma 2.

$$Cov[X, Y] = E[XY] - E[X] E[Y] \tag{2.76}$$

Proof.

$$Cov[X,Y] = E[XY - \mu_X Y - X\mu_Y + \mu_X\mu_Y] \quad (2.77)$$
$$= E[XY] - \mu_X E[Y] - \mu_Y E[X] + \mu_X\mu_Y \quad (2.78)$$
$$= E[XY] - \mu_X\mu_Y. \quad (2.79)$$

□

Returning to the proof of Proposition 5, we examine the expectation value of the product XY.

$$E[XY] = \sum_j \sum_i x_i y_i P\{X = x_i, Y = y_i\} \quad (2.80)$$
$$= \sum_j \sum_i x_i y_i P\{X = x_i\} P\{Y = y_i\} \quad (2.81)$$
$$= \left(\sum_j y_j P\{Y = y_j\}\right) \left(\sum_i x_i P\{X = x_i\}\right) \quad (2.82)$$
$$= E[Y]E[X] \quad (2.83)$$

So by Lemma 2,

$$Cov[X,Y] = 0. \quad (2.84)$$

□

2.7 Limiting Theorems

Next we look for assurance a finite number of experiments can measure something meaningful: will sufficiently many experiments approximate the "infinite" continuous limit? The **central limit theorem** shows the result of a sufficiently large number of independent, identically distributed random variables (sometimes abbreviated as "iid random variables") will approach the familiar Gaussian[5] bell curve or "normal" distribution of a random variable [58] [371] [426].

We begin with Markov's [6] Inequality. This is the first of two propositions which let us use only the mean and variance of a probability distribution to establish bounds on the probability of events.

[5] Johann Carl Friedrich Gauss, 1777 - 1855, is another person hard to exaggerate. He developed important principles for every field from probability to non-Euclidean geometry to mechanics; he lead a geodesic survey of the then-independent state of Hannover, and developed the least-squares approximation method in the hope of finding the asteroid Ceres, discovered 1 January 1801 by Giuseppe Piazzi, 1746-1826, and lost shortly after. His predictions were right. [288]

[6] Andrei Andreevich Markov, 1856 - 1922, besides Markov sequences and probability, studied continued fractions, which were pioneered by his teacher Pafnuty Lvovich Chebyshev. [92]

Proposition 7. *Let X be a nonnegative random variable and let a be any positive number. Then*

$$P\{X \geq a\} \leq \frac{E[X]}{a} \tag{2.85}$$

Proof. We will define a new variable I, which is equal to 1 whenever X is greater than or equal to a, and equal to 0 otherwise. (This sort of variable marking what subset satisfies some condition is called an **indicator function** or **characteristic function.**) So the expectation value $E[I]$ is equal to the probability that X is greater than or equal to a.

Since X is nonnegative, $\frac{X}{a} \geq 1 \geq I$. Therefore

$$E[I] \leq E\left[\frac{X}{a}\right] \tag{2.86}$$

$$E[I] \leq \frac{E[X]}{a} \tag{2.87}$$

$$P\{X \geq a\} \leq \frac{E[X]}{a} \tag{2.88}$$

□

Next is Chebyshev's[7] Inequality, which follows closely from Markov's Inequality.

Proposition 8. *Let X be a random variable with (finite) mean μ_X and variance σ_X^2. Then*

$$P\{|X - \mu_X| \geq \epsilon\} \leq \frac{\sigma_X^2}{\epsilon^2} \tag{2.89}$$

for any positive ϵ.

Proof. If X is a random variable, then $(X - \mu)^2$ is a nonnegative random variable. For the given ϵ let $a = \epsilon^2$; by Markov's Inequality we have

$$P\left\{(X - \mu)^2 \geq \epsilon^2\right\} \leq \frac{E[(X - \mu)^2]}{\epsilon^2} \tag{2.90}$$

By the definition of variance, $E[(X - \mu)^2]$ is equal to σ_X^2. And the condition $(X - \mu)^2 \geq \epsilon^2$ is equivalent to the condition $|X - \mu| \geq \epsilon$. So we have

[7] Pafnuty Lvovich Chebyshev, 1821 - 1894, was one of the great figures in the study of function approximation, and published the papers which gave a rigorous basis for using probability to study statistics of data sets. He was the first to prove Joseph Louis François Bertrand (1822-1900)'s conjecture that for every $n > 3$ there is at least one prime between n and $2n$, and almost proved the prime number theorem, that the number of primes less than n tends towards $\frac{n}{log(n)}$ as n grows. [288]

$$P\{|X-\mu|\geq\epsilon\}\leq\frac{\sigma_X^2}{\epsilon^2} \tag{2.91}$$

□

Though this inequality is too rough to be useful for estimating the probabilities of events in particular problems, its applies to random variables in general, and so is useful in the analysis of probability.

Next we introduce the **Weak Law of Large Numbers**.

Theorem 1. *Let $X_1, X_2, X_3, \cdots$, be a sequence of independent, identically distributed random variables, and let each of them have the same finite mean $\mu_X = E[X_i]$. For any $\epsilon > 0$ then*

$$\lim_{n\to\infty} P\left\{\left|\frac{X_1+X_2+X_3+\cdots+X_n}{n}-\mu_X\right|\geq\epsilon\right\}=0 \tag{2.92}$$

Proof. Let the variance of the variables X be σ_X^2. Assuming that σ_X^2 is finite, we know that

$$E\left[\frac{X_1+X_2+X_3+\cdots+X_n}{n}\right]=\mu_X \tag{2.93}$$

$$Var\left[\frac{X_1+X_2+X_3+\cdots+X_n}{n}\right]=\frac{\sigma_X^2}{n} \tag{2.94}$$

This fits exactly Chebyshev's Inequality, with $\frac{\sigma_X^2}{n}$ as the variance and $\frac{1}{n}(X_1+X_2+X_3+\cdots+X_n)$ as the random variable labelled X in the inequality.

□

Markov chains, which will become key in later chapters, were created in part due to Markov's desire to remove the requirement of independence from the requirements of the weak law of large numbers [58].

Now we come to the **Central Limit Theorem**, perhaps the most remarkable result of probability. It shows the collection of measurements of an independent variabe approximates towards the Gaussian distribution bell curve [371] [426].

Theorem 2. *Let $X_1, X_2, X_3, \cdots$, be a sequence of independent, identically distributed random variables, and let each of them have the same finite mean $\mu_X = E[X_i]$ and variance σ_X^2. Then*

$$\lim_{n\to\infty} P\left\{\frac{X_1+X_2+X_3+\cdots+X_n}{\sigma\sqrt{n}}\leq a\right\}=\frac{1}{\sqrt{2\pi}}\int_{-\infty}^{n} e^{-\frac{x^2}{2}}dx \tag{2.95}$$

That is, the distribution of X_i tends towards the standard normal distribution.

This theorem is most easily proven by introducing **moment generating function** of a probability distribution. For a discrete mass function it is

$$M(t) = \sum_i e^{tx_i} p(x_i) \tag{2.96}$$

and for a continuous probability density function is

$$M(t) = \int_{-\infty}^{\infty} e^{tx} f(x) dx \tag{2.97}$$

which is also equal to $E[e^{tX}]$.

For a Gaussian distribution bell curve with mean zero and variance one the moment generating function is $\exp\left(\frac{t^2}{2}\right)$. We will use this to prove the central limit theorem by supposing first that X_i has mean zero and variance one (this does not lose generality: if the mean is μ_X and the variance σ_X^2 define a new variable Y_i which equals $\frac{X_i - \mu_X}{\sigma_X}$). We show the moment generating function of $\frac{X_i}{\sqrt{n}}$ tends towards $e^{\frac{t^2}{2}}$.

$$M\left(\frac{t}{\sqrt{n}}\right) = E\left[exp\left\{\frac{t}{\sqrt{n}} X_i\right\}\right] \tag{2.98}$$

by construction. Let $L(t) = \log(M(t))$, which has the properties that

$$L(0) = 0 \tag{2.99}$$

$$L'(0) = \frac{M'(0)}{M(0)} = \mu = 0 \tag{2.100}$$

$$L''(0) = \frac{M(0)M''(0) - [M'(0)]^2}{[M(0)]^2} = E[X^2] = 1 \tag{2.101}$$

Look at the limit of $nL(\frac{t}{\sqrt{n}})$ as n goes to infinity.

$$\lim_{n\to\infty} \frac{L(\frac{t}{\sqrt{n}})}{n^{-1}} = \lim_{n\to\infty} \frac{-L'(\frac{t}{\sqrt{n}}) n^{-\frac{3}{2}} t}{-2n^{-2}} \tag{2.102}$$

$$= \lim_{n\to\infty} \left[\frac{L'(\frac{t}{\sqrt{n}}) t}{2n^{-\frac{1}{2}}}\right] \tag{2.103}$$

$$= \lim_{n\to\infty} \left[\frac{-L''(\frac{t}{\sqrt{n}}) n^{-\frac{3}{2}} t^2}{-2n^{-\frac{3}{2}}}\right] \tag{2.104}$$

$$= \lim_{n\to\infty} \left[L''(\frac{t}{\sqrt{n}}) \frac{t^2}{2}\right] \tag{2.105}$$

$$= \frac{t^2}{2} \tag{2.106}$$

So the distribution of $\frac{X_i}{\sqrt{n}}$ approaches that of the bell curve, with probability one.

□

The name "Weak Law of Large Numbers" (Theorem 1) suggests there should be a **Strong Law of Large Numbers**, and there is. It states that, with probability one, the arithmetic mean of a sequence of independent random variables converges to the mean of the distribution [371] [58].

Theorem 3. *Let X_1, X_2, X_3, $\cdots$, be a sequence of independent, identically distributed random variables, and let each of them have a finite mean $\mu_X = E[X_i]$. With probability one then*

$$P\left\{\left(\lim_{n\to\infty} \frac{X_1 + X_2 + X_3 + \cdots + X_n}{n}\right) = \mu_X\right\} = 1 \tag{2.107}$$

We will not prove exactly this Strong Law of Large Numbers; the proof can be dramatically simplified if we make the further assumption that the fourth moment of the random variables, $E[X^4]$, is finite. A proof not making this assumption is provided in Brémaud [58].

Proof. Suppose without a loss of generality that μ_X is zero. Let S_n be the sum of the first n values of X_i. We will work towards proving that as n grows large, $\frac{S_n}{n}$ goes to zero, and therefore the mean of X approaches μ_X.

Consider the expectation value of S_n^4:

$$\begin{aligned} E[S_n^4] = E[&(X_1 + X_2 + X_3 + \cdots + X_n) \\ &\times (X_1 + X_2 + X_3 + \cdots + X_n) \\ &\times (X_1 + X_2 + X_3 + \cdots + X_n) \\ &\times (X_1 + X_2 + X_3 + \cdots + X_n)] \end{aligned} \tag{2.108}$$

The expansion of the polynomial on the right hand side of equation 2.108 seems intimidating, but it is much simpler than it first appears: all of the terms in that polynomial are of one of the forms X_i^4, $X_i^3 X_j$, $X_i^2 X_j^2$, $X_i^2 X_j X_k$, or $X_i X_j, X_k, X_l$. As the expectation value is linear, the expectation value of this polynomial will be the sum of those of the component terms. Almost all of them drop out from the assumption that the mean μ_X, the expectation value $E[X]$, is zero:

$$E[X_i^3 X_j] = E[X_i^3]E[X_j] = 0 \tag{2.109}$$

$$E[X_i^2 X_j X_k] = E[X_i^2]E[X_j]E[X_k] = 0 \tag{2.110}$$

$$E[X_i X_j X_k X_l] = E[X_i]E[X_j]E[X_k]E[X_l] = 0 \tag{2.111}$$

Each pair of i and j will have six $X_i^2 X_j^2$ terms in the expansion; so for the sum of n terms there will be $6\binom{n}{2}$ terms. For each i, X_i^4 appears once in the expansion. So we can simplify equation 2.108 to

$$E[S_n^4] = nE[X_i^4] + 6\binom{n}{2}E[X_i^2 X_j^2] \tag{2.112}$$

$$E[S_n^4] = nE[X_i^4] + 3n(n-1)E[X_i^2]E[X_j^2] \tag{2.113}$$

(the last step justified by the independence of X_i and X_j).

Now the assumption the fourth moment is finite is used. Letting K equal $E[X_i^4]$ allows us to find an estimate on $E[X_i]^2$ as well, since the variance, which must be nonnegative, of X_i^2 will equal $E[(X_i^2)^2] - (E[X_i^2])^2$.

$$0 \leq E[X_i^4] - (E[X_i^2])^2 \tag{2.114}$$
$$(E[X_i^2])^2 \leq E[X_i^4] = K \tag{2.115}$$

So returning to equation 2.113 we have

$$E[S_n^4] \leq nK + 3n(n-1)K \tag{2.116}$$
$$E\left[\frac{S_n^4}{n^4}\right] \leq \frac{K}{n^3} + \frac{3K}{n^2} \tag{2.117}$$

We finish by considering the expectation value of a series whose terms are these numbers $\frac{S_n^4}{n^4}$: the numbers $\frac{S_n^4}{n^4}$ are with probability one going to zero.

$$E\left[\sum_{n=1}^{\infty} \frac{S_n^4}{n^4}\right] = \sum_{n=1}^{\infty} E\left[\frac{S_n^4}{n^4}\right] \tag{2.118}$$

which equation 2.117 shows is bounded. If there were a nonzero probability that $\sum_{n=1}^{\infty} \frac{S_n^4}{n^4}$ was infinitely large then its expectation value would be infinitely large. Thus with probability one, the sum $\sum_{n=1}^{\infty} \frac{S_n^4}{n^4}$ is finite. But for a series to converge its terms must go to zero. Therefore

$$\lim_{n \to \infty} \frac{S_n^4}{n^4} = 0 \tag{2.119}$$

and therefore $\frac{S_n}{n}$ must go to zero. Thus, with probability one,

$$\lim_{n \to \infty} \frac{X_1 + X_2 + X_3 + \cdots + X_n}{n} = \mu_X \tag{2.120}$$

□

Both the weak and the strong laws of large numbers state the "running average" of n values of X will be near the distribution mean of μ_X. The weak law allows the possibility the difference between that running average and the mean may grow arbitrarily large sometimes, or that the mean may be larger than some positive value ϵ an infinite number of times. The strong law shows the difference between the running average and μ_X will go to zero, and that the difference will be larger than ϵ at most finitely many times [371]. The argument by which Markov chain Monte Carlo algorithms estimate the equilibration of a system is an extension of this theorem [58].

2.8 Bayesian Probability

Essential also to our study of probability is the problem of **conditional probability**, the chance of one event given that another is assumed to hold. This is the foundation of Bayesian[8] statistics. The central question of this field is what is the probability of an event F, given the assumption an event G has occurred. The point to asking the question is to provide a quantitative answer to the process of **inference** – given that we have observed a sample with (for example) mean $E[x]$, what is the probability that the mean of the entire space μ_X, were it possible to sample, would be the same number [367]?

We begin with a proposition describing conditional probability [40]:

Proposition 9. *Given events E and F and a probability distribution $P(X)$, and the assumption that $P(F) \neq 0$, then the probability of event E given the assumption F occurs, denoted $P(E|F)$, is*

$$P(E|F) = \frac{P(E \cap F)}{P(F)} \tag{2.121}$$

It is useful to visualize the logic of the proof. Assuming each event is of equal probability, then the probability $P(F)$ is the area of the subset of space which satisfies event F, and $P(E \cap F)$ is the area of space which satisfies event E and F simultaneously. Thus obviously the probability of event E, given the assumption of event F, is the fraction of the space satisfying events F which also satisfies event E. In performing Monte Carlo experiments we will make assumptions like this about elements in phase space and the quantities we measure.

To prove this rigorously requires Axiom 4 [367]. Given events E and F there is some independent event S which has probability equal to $\frac{P(E \cap F)}{P(F)}$. But as event S is independent of both E and F, $P(S \cap F) = P(S)P(F) = P(E \cap F)$. So $S \cap F$ is as likely as $E \cap F$ is. Thus event S is just as probable as event E is, given F – and therefore $P(S|F)$ equals $P(E|F)$. Thus $P(E|F)P(F) = P(E \cap F)$ and the conclusion follows.

□

From this we can build some of the probability structure for conditional probability [40]:

Proposition 10. *For any events E, F, and G, assuming G is not the null set, and Ω is the space of all possible events*

1. $P(\emptyset|G) = 0 \leq P(E|G) \leq P(\Omega|G) = 1$
2. *If* $E \cap F \cap G \neq \emptyset$, *then* $P(E \cup F|G) = P(E|G) + P(F|G)$

---

[8] Thomas Bayes, 1702 - 1761, wrote important essays on probability, and the theory of fluxions, none of which were published under his own name during his lifetime. His notebooks include examinations of differential calculus, series, optics, celestial mechanics, and electricity. [288]

Only the second has a proof which is not obvious:

$$P(E \cup F|G) = \frac{P((E \cap G) \cup (F \cap G))}{P(G)} \tag{2.122}$$

$$= \frac{P(E \cap G)}{P(G)} + \frac{P(F \cap G)}{P(G)} = P(E|G) + P(F|G) \tag{2.123}$$

□

As corollary we may set up the additive structure [40]:

Proposition 11. *For all events F not the null set*

1. *If $\{E_j \cap F, j \in J\}$ is a finite collection of disjoint events then*

$$P\left(\bigcup_{j \in J} E_j|F\right) = \sum_{j \in J} P(E_j|F) \tag{2.124}$$

2. *For any event E, $P(E^c|F) = 1 - P(E|F)$.*

The first part follows from the first property of the preceding proposition; the second from letting E_2 be equal to E_1^c.

□

Together we have the most general statement of Bayes's Theorem [40]:

Theorem 4. *For any finite partition $\{E_j, j \in J\}$ of the probability space ω and for any non-empty event F*

$$P(E_k|F) = \frac{P(F|E_k)P(E_k)}{\sum_{j \in J} P(F|E_k)P(E_k)} \tag{2.125}$$

This follows from the fact

$$P(E_k|F) = \frac{P(E_k \cap F)}{P(F)} = \frac{P(F|E_K)P(E_k)}{P(F)} \tag{2.126}$$

and from the proposition regarding the additive structure by allowing $F = \cup_{j \in J} (F \cap E_j)$.

□

The above work has used discrete probabilities, but they can be extended (and, in fact, were originally done by Bayes [367]) to continuous variables, with integrals in place of summations.

2.9 Remarks

In this review we have established enough of the background of probability to construct Markov chain Monte Carlo algorithms. This is a technique in which we find the expectation value of a quantity over a probability space which is too large to be calculated in detail by finding a sampling – a selection of states called the Markov chain which has a probability distribution approximately that of the entire space. This sampled space then has roughly the same properties of any measured variable that the entire space does. In this way we are able to estimate expectation values (and variances) efficiently. More, Monte Carlo algorithms tend to be flexible and straightforward to implement, making them one of the cornerstones of statistical mechanics.

3

Statistical Mechanics

3.1 Introduction

One of the goals in this book is finding of an optimum state, an arrangement of site values subject to certain constraints. At first one may suppose these questions can be perfectly answered by analytic methods. Our energy function has only a scalar value, which is continuous and smooth (provided no two points have the same position). Its variables may be the strengths of each site, or they may be the positions of each site, but they are just an **n-tuple** (an ordered set, as in a vector) of real-valued coordinates. This is almost as well-behaved as one could hope for a function.

For a small number of particles maxima and minima can be found by simple Calculus methods. But when we want to examine dozens, hundreds, or thousands of particles analytic methods are no longer practical. Interestingly, in the limit of the infinite number of particles, the analytical approach becomes more productive, with the continuum limit offering more efficient tools again.

It the intermediate range, when there are too many particles to conveniently work with but not enough that the continuum can be considered, where computational methods work best. Many of the statistical tools we can use – particularly the Markov chain Monte Carlo method – are simple to program. Typically one does not need to calculate derivatives or gradients for the energy function as well – evaluating the original function suffices – which makes it easier and faster to compute new approximate solutions [353].

3.2 Ensembles

An **ensemble** is a collection of the representations of one model. For example, one model of a fluid is the collection of fixed mesh sites and the vorticities at these sites. The ensemble is the collection of all the different ways vorticities can be assigned the sites [161]. The ensemble can be a broad collection representing, for example, all the possible vorticity assignments that can be given, or it can be narrower, such as the collection of vorticity assignments with a chosen energy.

The collection of systems that have a single energy is described as the **microcanonical ensemble**. When a variety of different energies are allowed, this is called **canonical ensemble** [186]. We will often look at problems which are canonical in one property (such as energy) but microcanonical in another (such as enstrophy). There is one terminologically confusing point here: holding the number of sites in the problem constant is considered canonical.

The **grand canonical ensemble** is the collection one gets by letting all measured quantities vary, including the number of sites. All of the microcanonical and canonical ensembles are within the grand canonical ensemble, but that is usually a larger set than is convenient to work with for these problems [161] [186].

3.3 Partition Functions

Suppose our system is a collection of mesh sites and strengths for each site. What is the most likely state to find the system in? In the process of calculating the probabilities of each state we will construct the partition function.

A few assumptions are needed. First we suppose that any two states that have the same energy have the same probability of being found. Next we suppose that separate systems are independent: if one has duplicate systems, the probabilities of any given state are the same for the first as for the second [186]. We introduce these twin systems to simplify our derivation. We look at the chance that the first is in the state A and the second is in another state B. Let the energy of a state X be written as E_X.

Let $P_1(E)$ be the probability that the first system is in a state with energy E, and let $P_2(E)$ the probability that the second system is in a state with energy E. Then let $P_{1+2}(F)$ be the probability that the first and the second systems are in states that have a total energy F.

The probability that the first system is in A is $P_1(E_A)$ and the probability the second system is in B is $P_2(E_B)$. The probability that the first system is in A and the second is in B has to be $P_{1+2}(E_{A+B})$ – which must equal $P_{1+2}(E_A + E_B)$.

There is another construction: the probability of having one system in state A is also $P_1(E_A)$ and the probability of having the other in state B is $P_2(E_B)$. Since by assumption the two states are independent the chance of having A in one and B in the other is $P_1(E_A) \times P_2(E_B)$. Between the constructions we have

$$P_{1+2}(E_A + E_B) = P_1(E_A) \times P_2(E_B) \tag{3.1}$$

To find how probability depends on energy we differentiate 3.1 two separate times, first with respect to E_A and second with respect to E_B:

$$P'_{1+2}(E_A + E_B) = P'_1(E_A)P_2(E_B) \tag{3.2}$$
$$P'_{1+2}(E_A + E_B) = P_1(E_A)P'_2(E_B) \tag{3.3}$$

and so the right hand sides are equal,

$$P'_1(E_A)P_2(E_B) = P_1(E_A)P'_2(E_B) \tag{3.4}$$

which can be separated, putting all the E_A terms on one side and the E_B terms on the other. The result is

$$\frac{P'_1(E_A)}{P_1(E_A)} = \frac{P'_2(E_B)}{P_2(E_B)} \tag{3.5}$$

The two sides of 3.5 depend on different variables, E_A and E_B – so to be equal for all E_A and E_B both sides are equal to a constant independent of either variable. Call this variable $-\beta$, so

$$\frac{dP_1(E_A)}{dE_A} = -\beta P_q(E_A) \tag{3.6}$$
$$P_q(E_A) = Ce^{-\beta E_A} \tag{3.7}$$

with C some constant of integration which will be identified shortly.

The constant β does not depend on the system. In the study of gas particles β has a useful physical interpretation; it is equal to $\frac{1}{k_b T}$, where T is the temperature of the gas and k_b is the Boltzmann constant[1] [186]. For this reason β is named the inverse temperature, even though in our work the idea of temperature does not correspond to measuring heat.

We return to the expression $P_1(E_A) = Ce^{-\beta E_A}$. The sum of the probabilities for being in each state, for all possible states, must be 1, which lets us find C.

$$\sum_A P(E_A) = \sum_A Ce^{-\beta E_A} \tag{3.8}$$
$$= C\sum_A e^{-\beta E_A} = 1 \tag{3.9}$$

(since C does not depend on the state A), and we have to conclude

$$\frac{1}{C} = \sum_A e^{-\beta E_A} \tag{3.10}$$

This summation – the sum over all the states of a function proportional to the likelihood of finding the system in this state – is called the **partition**

[1] Named for Ludwig Boltzmann, 1844 - 1906, one of the most important figures of statistical mechanics. He connected the ideas of entropy and the dynamics of gases. The equation $S = k\ logW$, with S the entropy, W the number of configurations for a given state, and k his constant, is inscribed on his tombstone. [25] [288]

function. It is typically written as Z, from the first letter of Zustandssumme, which was Max Karl Ernst Ludwig Planck's[2] name for this summation, the "state sum" [64]. (The name "partition function" is credited to Charles George Darwin [3] and Sir Ralph Howard Fowler[4] [42] [161].)

$$Z = \sum_{A} e^{-\beta E_A} \tag{3.11}$$

contains considerable information. In a classical thermodynamics problem, one can use the partition function to calculate the internal energy, the **Helmholtz**[5] **free energy**, the entropy, and a variety of other interesting properties.

(The writing of equation 3.11 assumes a summation will accurately take the sum over all possible states. If there is a continuum of possible values for the energy then one has to move to an integral over them [161] [186].)

3.4 Constraints and Chemical Potentials

From the normalized probability function one can find the most probable state. But we are often interested in the most probable state subject to some additional constraint, such as states with a specified value of circulation.

The derivation of section 3.3 can be repeated with several properties studied. There is also another way of viewing the partition function with constraints, based on the method of Lagrange multipliers[6], which will be reviewed later in the next section.

Suppose the probability of finding a system in the state A is dependent on two quantities, E_A and Ω_A, and that as above we have two systems, 1 and 2. The probability of system 1 being in state A is $P(E_A, \Omega_A)$. And similarly the

[2] 1858 - 1947. Planck was literally the inventor of the quantum, introduced to explain the radiation of a blackbody. [25]

[3] 1845 - 1912. Darwin was the son of the "Origin of Species" author, and was the first to try to explain the origins of the Earth and Moon through the now-discarded hypothesis that the Moon was ripped from the nascent Earth by centrifugal force. [288]

[4] 1889 - 1944. Fowler's greatest work is either explaining white dwarf stars using the quantum mechanics of a degenerate gas, or in introducing Paul Adrien Maurice Dirac to quantum mechanics. [288]

[5] Hermann von Helmholtz, 1821 - 1894, is another key figure in thermodynamics. In 1847 he presented the proof that made the Conservation of Energy a key principle of physics and chemistry. He was also the first person to measure the speed of a nerve impulse, in response to his biology teacher's statement that it was something science could never measure. [25]

[6] Joseph-Louis Lagrange, 1736 - 1813, did much to establish modern calculus and physics notation, differential equations and the beginnings of abstract algebra. He was also one of the committee which designed the metric system. [25]

probability of system 2 being in state B is $P(E_B, \Omega_B)$. Finally the probability of system 1 being in state A while system 2 is in state B will be

$$P_{1+2}(E_A + E_B, \Omega_A + \Omega_B) = P_1(E_A, \Omega_A)P_2(E_B, \Omega_B) \tag{3.12}$$

First differentiate 3.12 with respect to E_A, and then differentiate it with respect to E_B, which must both be equal:

$$\frac{\partial}{\partial E}P_{1+2}(E_A + E_B, \Omega_A + \Omega_B) = \frac{\partial}{\partial E}P_1(E_A, \Omega_A)P_2(E_B, \Omega_B) \tag{3.13}$$

$$\frac{\partial}{\partial E}P_{1+2}(E_A + E_B, \Omega_A + \Omega_B) = P_1(E_A, \Omega_A)\frac{\partial}{\partial E}P_2(E_B, \Omega_B) \tag{3.14}$$

and setting the two right halves equal to one another provides

$$\frac{\frac{\partial}{\partial E}P_1(E_A, \Omega_A)}{P_1(E_A, \Omega_A)} = \frac{\frac{\partial}{\partial E}P_2(E_B, \Omega_B)}{P_2(E_B, \Omega_B)} \tag{3.15}$$

From this, the left hand side and the right hand side of the above equation involve different variables, and so must both be equal to the same constant $-\beta$. The dependence of P on energy must be exponential, with some further function of Ω yet to be determined:

$$P(E, \Omega) = e^{-\beta E} f(\Omega) \tag{3.16}$$

Determining $f(\Omega)$ can be done by looking at the derivatives of P with respect to Ω. Taking the derivative of 3.12 with respect to Ω_A, and also with respect to Ω_B, provides

$$P'_{1+2}(E_A + E_B, \Omega_A + \Omega_B) = \frac{\partial}{\partial E}P_1(E_A, \Omega_A)P_2(E_B, \Omega_B) \tag{3.17}$$

$$P'_{1+2}(E_A + E_B, \Omega_A + \Omega_B) = P_1(E_A, \Omega_A)\frac{\partial}{\partial E}P_2(E_B, \Omega_B) \tag{3.18}$$

and by setting both right halves equal to one another we get

$$\frac{\frac{\partial}{\partial \Omega}P_1(E_A, \Omega_A)}{P_1(E_A, \Omega_A)} = \frac{\frac{\partial}{\partial \Omega}P_2(E_B, \Omega_B)}{P_2(E_B, \Omega_B)} \tag{3.19}$$

This time setting the left hand side and the right hand side separately equal to the constant $-\mu$, we have

$$\frac{\partial}{\partial \Omega}P_1(E_A, \Omega_A) = -\mu P_1(E_A, \Omega_A) \tag{3.20}$$

$$P_1(E_A, \Omega_A) = Cg(E_A)e^{-\mu \Omega_A} \tag{3.21}$$

for some function $g(E)$. Between equations 3.16 and 3.21 the probability of system 1 being in state A must be

$$P_1(E_A, \Omega_A) = Ce^{-\beta E_A - \mu\Omega_A} \tag{3.22}$$

The partition function with an additional constrain will look like

$$Z = \sum_A e^{-\beta E(A) - \mu\Omega(A)} \tag{3.23}$$

μ is called the **chemical potential**, which reflects its origin in the study of gases, where these constraints correspond to interactions among particles. More constraints can be added by adding more potentials and constraints, as above – if the constraining functions are $\Omega(A)$ and $\Xi(A)$, and two constants (both called chemical potentials) μ_1 and μ_2 are added, then one has

$$Z = \sum_A e^{-\beta E(A) - \mu_1\Omega(A) - \mu_2\Xi(A)} \tag{3.24}$$

3.5 Partition Functions by Lagrange Multipliers

The Gibbsian[7] picture of statistical thermodynamics consists of N systems which form a virtual ensemble, with l distinct **states**, and energy levels e_1, e_2, $\cdots$, e_l. The active variables are the **occupation number**, a_1, a_2, $\cdots$, a_l, which represent the number of virtual systems in each state. We assume that the total number of virtual systems is a constant, which provides us with our first constraint:

$$\sum_{j=1}^{l} a_j = N \tag{3.25}$$

The total energy of the ensemble is assumed to be fixed as well. This yields our second constraint:

$$\sum_{j=1}^{l} a_j e_j = E \tag{3.26}$$

Let the vector $\mathbf{a} = (a_1, a_2, a_3, \cdots a_l)$ describe the occupation numbers of a system. Define $\Omega(\mathbf{a})$ to be the number of ways that the virtual systems can occupy the l states. Then

$$\Omega(\mathbf{a}) = \frac{N!}{\prod_{j=1}^{l} a_j!} \tag{3.27}$$

[7] Josiah Willard Gibbs, 1839 - 1903, was perhaps the finest theoretical physicist the United States ever produced. His work gave theoretical a basis to thermodynamics, the kinetic theory of gases and ultimately quantum mechanics. He advanced vectors as well, causing them to become the standard tool to represent multidimensional problems. [288]

We will compute the occupation vector **a** which maximizes

$$\log(\Omega(\mathbf{a})) = \log(N!) - \sum_{j=1}^{l} (a_j \log(a_j) - a_j) \tag{3.28}$$

subject to the constraints of equations 3.25 and 3.26. The problem of maximizing a quantity subject to constraints suggests the use of Lagrange multipliers.

Using the Lagrange multiplier approach, we define the augmented functional

$$F(\mathbf{a}) \equiv \log(\Omega(\mathbf{a})) - \lambda \sum_{j=1}^{l} a_j - \beta \sum_{j=1}^{l} a_j e_j \tag{3.29}$$

This converts the problem of a constrained optimization of $\log(\Omega(\mathbf{a}))$ into the unconstrained optimization of $F(\mathbf{a})$. The variables are still the value of each component a_j, so that the maximum will be found when

$$\frac{\partial}{\partial a_j} F(\mathbf{a}) = -\log(a_j) - \lambda - \beta e_j \tag{3.30}$$

$$= 0 \tag{3.31}$$

for $j = 1, 2, 3, \cdots, l$ and from which we find the maximum a_j^0 to be

$$a_j^0 = e^{-\lambda} e^{-\beta e_j} \tag{3.32}$$

for each j.

We can then profitably view the maximum occupation vector – and, so, the equilibrium occupation vector – $\mathbf{a}_0 = (a_1^0, a_2^0, a_3^0, \cdots, a_l^0)$ as the probability distribution of a single system over l energy levels, that is,

$$P(j) = \frac{a_j^0}{N} = \frac{e^{-\lambda - \beta e_j}}{\sum_{k=1}^{l} e^{-\lambda} e^{-\beta e_k}} \tag{3.33}$$

which is recognizably equation 3.11.

3.6 Microstate and Macrostates

Consider a system of three equally strong bar magnets A, B, and C, within a uniform magnetic field, locked so they may either point "up" – with their north pointing towards the magnetic field's north – or "down" – with their north pointing towards the magnetic field's south. The energy of these magnets will depend on the configuration of the magnet values.

One state may be to have A up, B down, and C down. Is this the same state as having A down, B up, and C down? Or of having A and B down and

C up? By the set-up of the problem these should all have the same energy, and so the same likelihood of appearing. Are they all the same state? Is the important description of state j that "A is up, B and C are down" or is it that "one is up, the other two down"? This will have an effect on what the expected value of the energy is. Rather than continue to allow the word "state" to be applied to several ideas, we will introduce two new names.

The **microstate** is the first description of the state above, the idea that we want to describe "magnet A up, magnets B and C down". It is the description of the system by listing each of its component values.

The **macrostate** is the second description, in which we just describe total properties but refrain from explicitly saying how the system arranges itself to have those properties. It is the description of the system by listing the aggregate totals of the entire system.

A microstate description implies a macrostate: the detailed information about the entire system can be used to derive the more abstracted description. If we know the velocity and mass of each of a collection of gas particles we can calculate with certainty the kinetic energy of the gas. A macrostate will (typically) have multiple microstates that fit within its description, however, so that we cannot derive "the" detailed information from its more abstracted description. The difference between the two is like two ways of representing census data: the macrostate representation of a neighborhood's population reports just how many people are in each house. The microstate representation reports which people are in which house.

For the more complicated sort of system which we find interesting, consider a lattice with M sites, and with a collection of N particles, which satisfy the requirement

$$1 \ll M \ll N \tag{3.34}$$

(The symbol $\ll$ meaning "is considerably less than".) For simplicity, just as is done in deriving the ideal gas law, we do not at the moment discuss the more realistic case in which particles interact. Instead we will assume that each site j has an energy e_j and a " circulation" Γ_j, and the occupation of one site does not affect the dynamics of another.

Suppose further that the system is in contact with a heat bath, a reservoir of heat at temperature $T = V_\beta$, and also a " circulation" bath with the chemical potential μ.

The energy of a microstate $\mathbf{a} \in \mathbf{A}$ is

$$E(\mathbf{a}) = E(\mathbf{A}) = \sum_{j=1}^{M} a_j e_j \tag{3.35}$$

and its circulation is

$$\Gamma(\mathbf{a}) = \Gamma(\mathbf{A}) = \sum_{j=1}^{M} a_j \Gamma_j \tag{3.36}$$

We also know of the constraint on the number of particles

$$\sum_{j=1}^{M} a_j = N \tag{3.37}$$

The **degeneracy** of a macrostate is the number of microstates $\mathbf{a} \in \mathbf{A}$

$$\Upsilon(\mathbf{A}) = \frac{N!}{\prod_{j=1}^{M} a_j!} \tag{3.38}$$

That is, it represents how many ways to distribute the fine details so that the summary picture is unchanged. The total number of microstates summed over all possible macrostates is

$$|D| = \frac{N!}{(N-M)!} \tag{3.39}$$

where D is the set of all of the microstates. The probability of any one particular macrostate $\mathbf{A}$, assuming each of the microstates $\mathbf{a} \in D$ are equally likely, will be

$$P(\mathbf{A}) = \Upsilon(\mathbf{A}) \frac{(N-M)!}{N!} \tag{3.40}$$

$$= \frac{(N-M)!}{\prod a_j!} \tag{3.41}$$

Now to add a bit more realism we assume the probability of a microstate $\mathbf{a} \in \mathbf{A}$ will depend not just on the energy of the macrostate $E(\mathbf{A})$ but also on its circulation $\Gamma(\mathbf{A})$. Therefore

$$P(\mathbf{a}) = k \exp\left(-\beta E(\mathbf{A}) - \mu \Gamma(\mathbf{A})\right) \tag{3.42}$$

where k is defined so that

$$\frac{1}{k} = Z = \sum_{\mathbf{a} \in D} \exp\left(-\beta E(\mathbf{a}) - \mu \Gamma(\mathbf{a})\right) \tag{3.43}$$

and is a normalization factor.

The probability of the macrostate $\mathbf{A}$ is

$$P(\mathbf{A}) = \Upsilon(\mathbf{A}) P(\mathbf{a}) \tag{3.44}$$

for any microstate $\mathbf{a} \in \mathbf{A}$, since the probability $P(\mathbf{a})$ is fixed by the values $E(\mathbf{A})$ and $\Gamma(\mathbf{A})$. Therefore

$$P(\mathbf{A}) = \frac{N!}{\prod a_j!} \frac{\exp\left(-\beta E(\mathbf{A}) - \mu \Gamma(\mathbf{A})\right)}{\sum_{\mathbf{a} \in D} \exp\left(-\beta E(\mathbf{a}) - \mu \Gamma(\mathbf{a})\right)} \tag{3.45}$$

$$= \frac{\exp\left(-\beta E(\mathbf{A}) - \mu \Gamma(\mathbf{A}) \Upsilon(\mathbf{A})\right)}{\sum_{\mathbf{A}'} \Upsilon(\mathbf{A}') \exp\left(-\beta E(\mathbf{A}') - \mu \Gamma(\mathbf{A}')\right)} \tag{3.46}$$

which states that the canonical partition function of equation 3.43 can be viewed also as a sum over all the distinct macrostates $\mathbf{A}'$.

Given that we know the probability of the macrostate $\mathbf{A}$ the obvious follow-up question is what might the most probable macrostate $\mathbf{A}_0$, given the distribution of equation 3.46? The problem is equivalent to maximizing

$$\log\left(P(\mathbf{A})\right) = \log\left(\Upsilon(\mathbf{A})\right) + \log\left(k\right) - \beta E(\mathbf{A}) - \mu \Gamma(\mathbf{A}) \tag{3.47}$$

by setting the partial derivatives with respect to each of the variables – each of them a_j still – equal to zero:

$$0 = \frac{\partial}{\partial a_j} P(\mathbf{A}) = -\log\left(a_j\right) - \beta e_j - \mu \Gamma_j \tag{3.48}$$

which eventually provides us

$$a_j^0 = \exp\left(-\beta e_j - \mu \Gamma_j\right) \tag{3.49}$$

The most probable macrostate $\mathbf{A}_0 = (a_1^0, a_2^0, a_3^0, \cdots, a_M^0)$ under the above equilibrium canonical setting satisfying equations 3.43, 3.46, and 3.49 can be viewed as a probability distribution itself. The probability of finding a particle in site j is

$$p(j) = \frac{a_j^0}{N} \tag{3.50}$$

for $j = 1, 2, 3, \cdots, M$.

It is not difficult to show that under microcanonical constraints on the energy and/or the " circulation" the result is the same in this free, or ideal, case.

3.7 Expectation Values

More interesting meaning can be found from the partition function and the Lagrangian multipliers. Consider the derivative of $log(Z)$ as from equation 3.24 taken with respect to the inverse temperature β:

$$\frac{\partial log(Z)}{\partial \beta} = \frac{1}{Z}\frac{\partial}{\partial \beta}\sum_A \exp\left(-\beta E_A - \mu_1 \Omega(A) - \mu_2 \Xi(A)\right) \tag{3.51}$$

$$= \frac{1}{Z}\sum_A \frac{\partial}{\partial \beta} \exp\left(-\beta E_A - \mu_1 \Omega(A) - \mu_2 \Xi(A)\right) \tag{3.52}$$

$$= \frac{1}{Z}\sum_A -E_A \exp\left(-\beta E_A - \mu_1 \Omega(A) - \mu_2 \Xi(A)\right) \tag{3.53}$$

$$= -\frac{1}{Z}\sum_A E_A \exp\left(-\beta E_A - \mu_1 \Omega(A) - \mu_2 \Xi(A)\right) \tag{3.54}$$

Since $\exp(-\beta E_A - \mu_1 \Omega(A) - \mu_2 \Xi(A))$ is proportional to the likelihood of state A appearing, then the summation of E_A multiplied by the probability of state A occurring is the expectation value of the energy. Partial differentiation with respect to a chemical potential μ_1 or μ_2 gives the expectation value for the other constrained quantities Ω or Ξ. Similarly differentiating Z with respect to the energy E yields the constant β, and differentiating with respect to another constrained quantity gives the corresponding chemical potential μ.

Relatively few states contributing much to the expectation value. Many states will have so large an energy their probability of occurring is too small, or have so few other states with nearby energies they cannot contribute much to the average. So if a large enough number of states can be generated, with probabilities as the partition function describe, then one can measure the energy (or another interesting quantity) at each of those generated states. The average over this sliver of available states can be close enough to the actual average over all possible states. Generating that representative set of states is the subject of the next chapter, on Monte Carlo.

3.8 Thermodynamics from Z

The discussion in sections 3.6 and 3.7 signifies the importance of the partition function

$$Z = \sum_{\text{microstates } \mathbf{a}} \exp(-\beta E(\mathbf{a}) - \mu \Gamma(\mathbf{a})) \tag{3.55}$$

where we have symbolically retained the " circulation" $\Gamma(\mathbf{a})$ as a placeholder for any number of canonical constraints. We have no particular interest in which quantity this is, only that it is something conserved canonically.

The expectation value of the energy E is now given by

$$\langle E \rangle = \sum_{\mathbf{a}} E(\mathbf{a}) \frac{\exp(-\beta E(\mathbf{a}) - \mu \Gamma(\mathbf{a}))}{Z} \tag{3.56}$$

$$= -\frac{\partial}{\partial \beta} \log(Z) \tag{3.57}$$

The mean values of other thermodynamic quantities represented in the partition function – such as the circulation – are found similarly:

$$\langle \Gamma \rangle = \sum_{\mathbf{a}} \Gamma(\mathbf{a}) \frac{\exp(-\beta E(\mathbf{a}) - \mu \Gamma(\mathbf{a}))}{Z} \tag{3.58}$$

$$= \frac{\partial}{\partial \mu} \log(Z) \tag{3.59}$$

Equations 3.57 and 3.59 leave little doubt that the quantity $\log(Z)$ plays an important role in thermodynamics, and deserves a special label:

$$F \equiv \log(Z) \tag{3.60}$$

F is now regarded as a function of all the Lagrange multipliers, β and μ and any others in the system, as well as the fixed parameters of the system such as e_j and Γ_j seen in the section 3.7.

Returning to the powerful Gibbsian picture from section 3.5 and recalling in particular equation 3.33, the occupation numbers

$$a_j^0 = \exp(-\lambda)\exp(-\beta e_j) \tag{3.61}$$

then the partition function

$$Z = \sum_{j=1}^{m} \exp(-\lambda)\exp(-\beta e_j) = N \tag{3.62}$$

is now a sum over m states, each with energy e_j. So then

$$a_j^0 = -\frac{N}{\beta}\frac{\partial}{\partial e_j}\log(Z) \tag{3.63}$$

$$= \frac{N}{Z}\exp(-\lambda)\exp(-\beta e_j) \tag{3.64}$$

and the energy per system is

$$\frac{E}{N} = -\frac{\partial}{\partial \beta}\log(Z) \tag{3.65}$$

In terms of the parameters N, E, and e_j for $j = 1, 2, 3, \cdots, m$, and the Lagrange multipliers β and λ, then the function $F = \log(Z)$ can be written in a differential form, as

$$dF = \frac{\partial F}{\partial \beta}d\beta + \frac{\partial F}{\partial \lambda}d\lambda + \sum_{j=1}^{m}\frac{\partial F}{\partial e_j}de_j \tag{3.66}$$

$$= -\frac{E}{N}d\beta - d\lambda - \frac{\beta}{N}\sum_{j=1}^{m} a_j^0 de_j \tag{3.67}$$

Thus,

$$d\left(F + \frac{E}{N}\beta + \lambda\right) = \beta\left(d\left(\frac{E}{N}\right) - \frac{1}{N}\sum a_j^0 de_j\right) \tag{3.68}$$

the right-hand side of which is interpreted as the β-weighted differential of the sum

$$U + \left(-\frac{1}{N}\sum a_j^0 e_j\right) \equiv \frac{1}{N}\left(E - \sum a_j^0 e_j\right) \tag{3.69}$$

where U is the average internal energy per system, and where $-\frac{1}{N}\sum a_j^0 de_j$ is the average work done by a system when the virtual ensemble of N systems is reversibly moved from energy levels e_j to $e_j + de_j$ (for each $j = 1, 2, 3, \cdots, m$) in contact with a heat bath.

By the first law of thermodynamics, the right-hand side of equation 3.67 is therefore equal to β times the heat gain per system, and so equation 3.67 becomes

$$d\left(\frac{F}{\beta} + U\right) = dQ \tag{3.70}$$

after dropping the constant λ from the left-hand side of equation 3.67. From here, it is easy to see that β is the inverse temperature $\frac{1}{k_B}T$ of the virtual ensemble (with k_b the Boltzmann constant).

Furthermore, the quantity $\frac{F}{\beta} + U$ is itself a perfect differential, and thus,

$$\frac{F}{\beta} + U \equiv \frac{\log(Z)}{\beta} + U = Q + Constant \tag{3.71}$$

Dropping this constant, we will label this quantity

$$\frac{1}{\beta}S' \equiv \frac{1}{\beta}F + U = Q \tag{3.72}$$

we see that it is closely related to

$$S = k_B \log(Z) + \frac{U}{T} \tag{3.73}$$

From the additivity of S' when two systems are combined, it follows that S must be the entropy. The Helmholtz free energy of a system is given by

$$F_H \equiv -k_B T \log(Z) = -\frac{\log(Z)}{\beta} \tag{3.74}$$

$$= U - TS \tag{3.75}$$

So the quantity F in equation 3.60 is thus the important quantity F_H, which enters into the renowned theorem due to Planck:

Theorem 5. ***Planck's Theorem: The minimization of free energy.*** *A thermodynamic system is in stable statistical equilibrium if and only if it minimizes F_H.*

This theorem will be applied to the significant problem of the Onsager[8] point vortex gas for the whole plane in a later chapter. To underscore the importance of F_H, we give a last application in this section to a thermodynamic system where volume V is one of the parameters. Then the pressure is

$$P = \frac{\partial F_H}{\partial V} \tag{3.76}$$

[8] Lars Onsager, 1903 - 1976, won the 1968 Nobel Prize in Chemistry for developing a general theory of irreversible chemical processes, that is, the field of non-statistical equilibrium. [287]

3.9 Fluctuations

We discuss the meaning of a most probable state in the context of a Boltzmann microcanonical ensemble. In particular, we show that this macrostate has overwhelmingly large probability. By showing this we are able to trust that our numerical methods for finding a statistical equilibrium will reach this most probable state. We do this by introducing the notion of thermal fluctuations.

Recall the Lagrange multiplier derivation of the most probable macrostate, the state vector

$$\mathbf{A}_0 = \left(a_1^0, a_2^0, a_3^0, \cdots, a_M^0\right) \tag{3.77}$$

where

$$a_k^0 = K \exp\left(\beta e_k\right) \tag{3.78}$$

This is known as the Maxwell[9]-Boltzmann distribution.

Consider a neighboring macrostate, however, called $\mathbf{A}$, defined to be

$$\mathbf{A} = \mathbf{A}_0 + \mathbf{\Delta} \tag{3.79}$$

$$\mathbf{\Delta} = (\Delta_1, \Delta_2, \Delta_3, \cdots, \Delta_M) \tag{3.80}$$

$$\sum_{k=1}^{M} \Delta_k = 0 \tag{3.81}$$

$$\sum_{k=1}^{M} e_k \Delta_k = 0 \tag{3.82}$$

where Δ_k is the perturbation in the occupation number a_k, that is, $\Delta_k \geq 0$ or $\Delta_k < 0$. The Maxwell-Boltzmann distribution $\mathbf{A}_0$ maximized the number of microstates in $\mathbf{A}_0$. So we will show that this maximum is exponentially large relative to the number $|\mathbf{A}|$ of microstates in the neighboring state $\mathbf{A}$.

Proposition 12. $|\mathbf{A}| \to |\mathbf{A}_0| \prod_{k=1}^{M} \exp\left(-\frac{1}{2}\frac{\Delta_k^0}{a_k^0}\right)$ *as* $\Delta_k \to 0$.

Proof. We will use the Taylor[10] polynomial expansion of equation 3.80. From it we find

[9] James Clerk Maxwell, 1831 - 1879, is renowned for his equations of electromagnetism. He also proved the rings of Saturn had to be composed of small particles rather than be solid rings or discs, and brought deserved publicity – and needed explanations – to the statistical mechanics work of Josiah Willard Gibbs. [288]

[10] Brooke Taylor, 1685 - 1731, was not the first to develop a Taylor polynomial. He did discover the method of integration by parts, and advanced the study of perspective, including inventing the term "linear perspective" and examining whether one could determine, based on a picture, where the artist's eye must have been. [288]

$$\log |\mathbf{A}| = \log |\mathbf{A}_0| + \sum_{k=1}^{M} \left(\frac{\partial}{\partial a_k} \log |\mathbf{A}| \right) |_{a_k^0} \Delta_k$$
$$+ \frac{1}{2} \sum_{k,l}^{M} \left(\frac{\partial^2}{\partial a_k \partial a_l} \log |\mathbf{A}| \right) |_{a_k^0} \Delta_k \Delta_l + h.o.t. \tag{3.83}$$

where *h.o.t.* stands for the third-order and higher terms in the polynomial expansion which we immediately suppose will be sufficiently small to ignore.

The first-order partial derivatives $\left(\frac{\partial}{\partial a_k} \log |\mathbf{A}| \right) |_{a_k^0}$ are equal to zero by the definition of $\mathbf{A}_0$ as a maximum. The second-order partial derivatives become

$$\left(\frac{\partial^2}{\partial a_k \partial a_l} \log |\mathbf{A}| \right) = \frac{\partial^2}{\partial a_k \partial a_l} \left(\log (N!) - \sum_{k=1}^{M} a_k (\log (a_k) - 1) \right) \tag{3.84}$$
$$= -\frac{\delta_{k,l}}{a_k} \tag{3.85}$$

after using Stirling's formula[11]. Then

$$\log |\mathbf{A}| = \log |\mathbf{A}_0| - \frac{1}{2} \sum_{k=1}^{M} \frac{\Delta_k^2}{a_k^0} + O(\Delta_k^3) \tag{3.86}$$

where the last term represents all the higher-order terms from the polynomial expansion.

For sufficiently small perturbations Δ_k away from the Maxwell-Boltzmann macrostate $\mathbf{A}_0$, the above proposition tells us that not only is $\mathbf{A}_0$ the maximizer of $|\mathbf{A}|$ but, more importantly, that

$$\frac{|\mathbf{A}|}{|\mathbf{A}_0|} = \prod_{k=1}^{M} \exp \left(-\frac{1}{2} \frac{\Delta_k^2}{a_k^0} \right) \tag{3.87}$$

which is exponentially small in Δ_k^2 with a rate constant $\frac{1}{2a_k}$. Therefore for $N > M \gg 1$, we deduce $a_k^0 \approx \frac{N}{M} > 1$, that the rate constant in equation 3.87 is on the order $O(\frac{M}{N}) < 1$, which is not extremely small. In this case, equation 3.87 implies that $|\mathbf{A}|$ is exponentially small compared to $|\mathbf{A}_0|$.

We can say more, and do by introducing the notion of **fluctuations**, or **variance**, per-site. Define the fluctuation of a_k to be

$$\overline{(a_k - \overline{a}_k)^2} = \overline{a}_k^2 - (\overline{a}_k)^2 \tag{3.88}$$

---

[11] James Stirling, 1692 - 1770, presented his approximation for the factorial in Proposition 28, Example 2, of 1730's *Methodus Differentialis.* He studied the shape of the Earth, the better ventilation of mine shafts, and directed the construction of new locks on the river Clyde which made the port at Glasgow more navigable and aided the city's rise as a commercial power. [288]

where $\overline{a}_k$ is the expectation value of a_k. By virtue of equation 3.87 we substitute a_k^0 in for $\overline{a}_k$ in equation 3.88 to get

$$\overline{\Delta_k^2} = \overline{\left(a_k - a_k^0\right)^2} = \overline{a}_k^2 - \left(a_k^0\right)^2 \tag{3.89}$$

Therefore equation 3.87 implies that

$$\overline{\Delta_k^2} = a_k^0, k = 1, 2, 3, \cdots, M \tag{3.90}$$

The above results give us a heuristic justification for the physical significance of the most probable, the Maxwell-Boltzmann, macrostate. A proof making use of contour integrals was developed by Darwin and Fowler [101] to which we refer readers who would like a rigorous demonstration.

Since equation 3.87 can be taken to be a Gaussian distribution probability measure for the random vector $\boldsymbol{\Delta}$, one with zero means $\overline{\Delta}_k = 0$, then we can attribute physical meaning to equation 3.90. In this case the fluctuation in occupation numbers – in densities – at each site k is equal to the expected and the most probable occupation number a_k^0.

Recalling the most probable macrostate $\mathbf{A}_0$ is itself directly related to a probability distribution of N particles over M cells, then we remember the probability of finding a particle in cell k is $\frac{a_k^0}{N}$. So the probability density fluctuations $\frac{\Delta_k^2}{N} = \frac{a_k^0}{N} \approx O\left(\frac{1}{M}\right)$ is therefore small for $M \gg 1$. Not only will we have a most probable state, but if there are sufficiently many particles and sufficiently many macrostates the fluctuations around that most probable state will be negligible.

3.10 Applications

The power of the Gibbsian picture is showcased below by applying the results in section 3.8 to two different statistics, namely (i) to the Fermi[12]-Dirac[13] statistics and (ii) to the classical or Boltzmann statistics. Using the same notation as in sections 3.6, 3.7, and 3.8, we let the energy levels or states of one particle be denoted by e_j, the energy of a gas of particles be

$$E(\mathbf{A}) = \sum a_j e_j \tag{3.91}$$

[12] Enrico Fermi, 1901 - 1954, discovered the statistics of particles that obey the Pauli exclusion principle and developed the theory of beta decay, explaining radioactivity and leading to the development of atomic bombs and nuclear power. [92]

[13] Paul Adrien Maurice Dirac, 1902 - 1984, created much of the notation and many of the terms describing quantum mechanics; it was he who noted the similarity between Heisenberg's commutators and Poisson brackets, identified the Heisenberg Uncertainty Principle as representing noncommutative operators, and predicted antimatter from theoretical considerations. Always averse to publicity he considered refusing the Nobel Prize in physics he won (with Schrödinger) until he was advised that would bring him more publicity. [288]

where

$$\sum a_j = N \tag{3.92}$$

and $(a_1, a_2, a_3, \cdots, a_m)$ are the occupation numbers of each energy state.

Thus the partition function is a sum over macrostates $\mathbf{A}_J$,

$$Z_n = \sum_{\mathbf{A}} \exp\left(-\beta E(\mathbf{A})\right) \tag{3.93}$$

So far equation 3.93 is applicable to both Bose[14]-Einstein[15] and Fermi-Dirac statistics. To specialize it to the Fermi-Dirac statistics, we enforce the additional condition

$$a_j \in \{0, 1\} \tag{3.94}$$

which is, of course, the Pauli[16] exclusion principle. If we write

$$z_j = \exp\left(-\beta e_j\right) \tag{3.95}$$

then we put

$$Z_n = \sum_{\mathbf{A}} \left(\prod_j z_j^{a_j} \right) \tag{3.96}$$

and we will prove in theorem 6 that

$$Z_N = \prod_j \left(1 + z_j\right) \tag{3.97}$$

when equation 3.94 holds, without imposing equation 3.92.

[14] Satyendranath Bose, 1894 - 1974, derived the equations describing blackbody radiation from Boltzmann statistics, making the argument for the existence of quanta of light much more compelling. Besides his research contributions he is noted for improving the educational system in India. [288]

[15] Albert Einstein, 1879 - 1955, is renowned for special and general relativity, and for his work in quantum mechanics. However, his work in statistical mechanics was no less important; among his five outstanding papers of 1905 are two explaining Brownian motion as a statistical phenomena and proving that atoms and molecules must exist, despite the many and apparently contradictory traits given the particles before quantum mechanics explained them. [288] [64]

[16] Wolfgang Ernst Pauli, 1900 - 1958, received his middle name from his father's wish to honor Ernst Mach, 1838 - 1916. In school he found the regular curriculum boring, and clandestinely read Einstein's papers in class. He predicted the neutron from the need to preserve several conservation laws, and noted the need for a fourth quantum number for electrons, the spin. [288]

Theorem 6. *If the conditions of equations 3.91, 3.93, 3.94, and 3.95 hold, then we can calculate the partition function as*

$$Z_N = \prod_{j=1}^{m} (1 + z_j) \tag{3.98}$$

Proof. From equation 3.95,

$$\begin{aligned} Z_N &= \sum_{\mathbf{A}} \left(\prod_{j=1}^{m} z_j a^j \right) && (3.99) \\ &= \sum_{a_1=0}^{1} \sum_{a_2=0}^{1} \cdots \sum_{a_m=0}^{1} \left(\prod_{j=1}^{m} z_j a^j \right) && (3.100) \\ &= \sum_{a_1=0}^{1} z_1^{a_1} \sum_{a_2=0}^{1} z_2^{a_2} \cdots \sum_{a_m=0}^{1} z_m^{a_m} && (3.101) \\ &= \prod_{j=1}^{m} (1 + z_j) && (3.102) \end{aligned}$$

□

On adding equation 3.92, the problem of evaluating Z_N is more difficult as it now reflects the selection of all terms where the occupation numbers are of order N in the variables z_j. The Residue Theorem accomplishes this if we first set

$$Z(\eta) \equiv \prod_{j=1}^{m} (1 + \eta z_j) \tag{3.103}$$

for the complex-valued variable η and then write

$$Z_N = \frac{1}{2\pi i} \oint_C \eta^{-(N+1)} Z(\eta) d\eta \tag{3.104}$$

where C is a simple curve[17] in the complex plane which surrounds exactly one singularity, namely, the one where $\eta = 0$.

The **Method of Steepest Descent,** also referred to as the **Saddle Point Method**, can then be used to explain 3.104, once we write it in a more transparent form. We take a moment to explain the method of steepest descent, following the derivation given by Arfken [24]:

Suppose we have the analytic functions $f(z)$ and $g(z)$ of the complex-valued variable z, and a variable s which we take to be real. We examine the asymptotic behavior of the integral of the form

[17] A curve is simple if it does not cross itself.

$$I(s) = \int_C g(z) \exp\left(s f(z)\right) dz \tag{3.105}$$

where the contour of integration C is either closed or lets $f(z)$ approach $-\infty$ and the integrand vanishes at the limits.

Now, representing $f(z)$ as $u(x,y) + iv(x,y)$ the integral is

$$I(s) = \int_C g(z) e^{su(x,y)} e^{isv(x,y)} dz \tag{3.106}$$

If s is a positive number the value of the integrand grows large when $u(x,y)$ is positive, and small when $u(x,y)$ is negative. The consequence of this is that if s is large, then most of the contribution to the value of the integral $I(s)$ comes from the region around the maximum of $u(x,y)$.

In the vicinity of the maximum $z_0 = (x_0, y_0)$ of u, we may suppose the imaginary component $v(x,y)$ is nearly the constant $v_0 = v(x_0, y_0)$. Therefore we can approximate the integral as

$$I(s) = e^{isv_0} \int_C g(z) e^{su(x,y)} dz \tag{3.107}$$

Furthermore, at the maximum,

$$\frac{\partial u}{\partial x} = \frac{\partial u}{\partial y} = 0 \tag{3.108}$$

$$\frac{df(z)}{dz}|_{z_0} = 0 \tag{3.109}$$

Since the function $f(z)$ is analytic, yet the integral 3.105 is bounded, it cannot have a global maximum or minimum. Since the derivative at the maximum of the real part is zero then z_0 must be a saddle point. In the vicinity of z_0 there are a set of curves where $u(x,y)$ is a constant and where $v(x,y)$ is a constant; those form a set of curves locally an approximately orthogonal set of coordinates. So the curve where $v(x,y) = v_0$ is the line of steepest descent from a saddle point out to any closed loop around z_0, giving the method its name, which is more often seen in the problem of finding minima of functions.

Around the saddle point we can use the Taylor series to approximate $f(z)$, as

$$f(z) = f(z_0) + \frac{1}{2}(z - z_0)^2 f''(z_0) + \cdots \tag{3.110}$$

The term $\frac{1}{2}(z - z_0)^2 f''(z_0)$ is real (the imaginary component is constant on the contour of steepest descent), and negative (along the contour the real component is decreasing). If then $f''(z_0)$ is not zero, then we may parametrize the contour with the new variable t, so that

$$f(z) - f(z_0) \approx \frac{1}{2}(z - z_0)^2 f''(z_0) \equiv -\frac{1}{2s} t^2 \tag{3.111}$$

If we use a polar coordinate representation of $z - z_0$ as $r\exp(i\theta)$ then

$$t^2 = -sf''(z_0)r^2\exp(2i\theta) \tag{3.112}$$
$$t = \pm r\sqrt{|sf''(z_0)|} \tag{3.113}$$

With this parametrization we can write the equation 3.107 as

$$I(s) \approx g(z_0)e^{sf(z_0)} \int_{-\infty}^{\infty} \exp\left(-\frac{1}{2}t^2\right) \frac{dz}{dt} dt \tag{3.114}$$

with

$$\frac{dz}{dt} = \left(\frac{dt}{dz}\right)^{-1} = \sqrt{|sf''(z_0)|}\exp(\mathrm{i}\theta) \tag{3.115}$$

so that the integral becomes

$$I(s) \approx \frac{g(z_0)\exp(sf(z_0))\exp(\mathrm{i}\theta)}{\sqrt{sf''(z_0)}} \int_{-\infty}^{\infty} \exp\left(-\frac{1}{2}t^2\right) dt \tag{3.116}$$
$$\approx \frac{\sqrt{2\pi}g(z_0)\exp(sf(z_0))\exp(i\theta)}{\sqrt{sf''(z_0)}} \tag{3.117}$$

The parameter θ is the phase of the contour of steepest descent as it passes through the saddle point, which assumes there is only one saddle point. If there are multiple saddle points passed through by the curve, one must add the term 3.117 as evaluated at each of the saddle points.

Through this involved derivation we have an analytic tool for evaluating a partition function and finding the configurations which maximize the Gibbs factor – which, as these are the statistical equilibria of the systems modelled, are exactly the configurations we want to find.

Returning now to the explanation of 3.104, we rewrite the partition function in a more useful form.

$$Z_N = \frac{1}{2\pi\mathrm{i}} \oint_C \exp(f(\eta))\, d\eta \tag{3.118}$$
$$= \frac{1}{2\pi\mathrm{i}} \oint_C d\eta \exp\left(f(\overline{\eta}) + f'(\overline{\eta})(\eta - \overline{\eta}) + \frac{1}{2}f''(\overline{\eta})(\eta - \overline{\eta})^2 + h.o.t.\right) \tag{3.119}$$

with $h.o.t.$ again higher order terms. That is, we simply write the Taylor series expansion of $f(\eta)$ around a point $\overline{\eta}$.

If the point $\overline{\eta}$ satisfies the restrictions

$$f'(\overline{\eta}) = 0 \tag{3.120}$$

and

$$f''((\bar{\eta})) \gg 1 \tag{3.121}$$

then equation 3.119 becomes

$$Z_N = \frac{\exp(f(\overline{\eta}))}{2\pi i} \oint_C d\eta \exp\left(-\frac{1}{2} f''(\overline{\beta})(\eta - \overline{\eta})^2 + h.o.t.\right) \tag{3.122}$$

$$= \frac{\exp(f(\overline{\eta}))}{\sqrt{2\pi f''(\overline{\eta})}} + h.o.t. \tag{3.123}$$

where $\overline{\eta}$ is, therefore, the saddle point of the function $f(\eta)$.

In order to solve equation 3.120, and thus find the saddle point, and so complete the derivation of Z_N, we compute

$$f'(\overline{\eta}) = -\frac{N+1}{\overline{\eta}} + \frac{d}{d\eta} \log(Z(\eta)) = 0 \tag{3.124}$$

The remainder term in equation 3.123 is vanishingly small as N increases to an infinitely large value. We show this result in the following theorem.

Theorem 7. *Providing equations 3.91, 3.92, and 3.94 hold, then the partition function Z_N of equation 3.93 tends, as N increases to ∞, to*

$$Z_N = \frac{\exp(f(\overline{\eta}))}{\sqrt{2\pi f''(\overline{\eta})}} \tag{3.125}$$

where $\overline{\eta}$ solves the saddle point condition of equation 3.124 and $f''(\overline{\eta}) \gg 1$.

The Helmholtz free energy is

$$F_H \equiv -\frac{1}{\beta} \log(Z_N) \tag{3.126}$$

$$= -\frac{1}{\beta}\left(f(\overline{\eta}) - \frac{1}{2}\log(f''(\overline{\eta}))\right) \tag{3.127}$$

$$= -\frac{1}{\beta}\left(\log(Z(\overline{\eta})) - (N+1)\log(\overline{\eta}) - \frac{1}{2}\log(f''(\overline{\eta}))\right) \tag{3.128}$$

4

The Monte Carlo Approach

4.1 Introduction

What will we find if we measure the energy of a system? The focus of this chapter is finding the expectation value of any random variable – energy or another property – by Monte Carlo methods.

Our first thought is a measurement should find the most probable energy. We know the probability of state j appearing, from the partition function:

$$P(E_j) = \frac{\exp(-\beta E_j)}{\sum_s \exp(-\beta E_s)} \tag{4.1}$$

The denominator is a constant, so we can maximize $P(E)$ by minimizing βE. If we assume β is positive then we find the maximum probability by minimizing the energy.

This solution is a bit too simple. For a start, it needs a more exact definition of what it is to be in a state; do we mean the microstate or the macrostate? This will have an effect on what the most expectation value of the energy is.

We will need an idea of how to fairly represent all the states of a system, and what we mean by those states. Then we need to handle the possibilities analytically, and how to draw out a representative slate for a numerical estimate. The tool we will use to create this representative sample is a Monte Carlo algorithm.

4.2 Microstates and Macrostates

In section 3.6 we described the characteristics that distinguish between microstates and macrostates, with the example of several magnets in a uniform magnetic field. What is provided in equation 4.1 is a $P(E_j)$ which is the probability of microstate j appearing. The summation in the denominator is over all microstates. This is generally not the probability that the system will have the energy E_j – that is the probability of the *macrostate* of energy E_j.

If many microstates have the same energy, then even though the chance of any one of them could be tiny, the chance that the system will be in *some* one of them can be great. The number of different states with the same energy is the degeneracy. Representing it as $\Upsilon(E)$, then the chance of observing the energy E in a system is

$$P_l(E) = \Upsilon(E) \times P(E) \tag{4.2}$$

and its maximum – the most likely state to be observed – will typically not be at the minimum βE microstate[1].

The degeneracy $\Upsilon(E)$ will usually start from 1 – typically there is a unique minimum-energy configuration – and increase as the energy increases. We can demonstrate this with another example made of bar magnets, each of equal strength, in a uniform magnetic field; in this case, ten such magnets. Each magnet has two equilibrium states, one up and one down. The down state is lower-energy than the up state, so the lowest possible energy state has all down. The degeneracy in this case is $\Upsilon(E_0) = 1$.

With enough energy E_b to flip one of the magnets up, we can choose any of the ten magnets and flip it over: the degeneracy is $\Upsilon(E_1) = 10$. With twice E_b we can choose any two of the ten magnets to flip over: the degeneracy is $\Upsilon(E_2) = \binom{10}{2} = 45$. At $3E_b$ any three may be flipped and the degeneracy is $\Upsilon(E_3) = \binom{10}{3} = 120$.

The degeneracy increases until the system has five times the base energy. At $8E_b$ the degeneracy is $\Upsilon(E_8) = \binom{10}{8}$, which has to equal $\binom{10}{2}$: choosing which eight to flip up is equivalent to choosing which two to leave down. At $9E_b$ the degeneracy is $\Upsilon(E_9) = \binom{10}{9} = \binom{10}{1} = 10$. And if there is ten (or more) times the energy needed to flip one magnet then the degeneracy is $\Upsilon(E_{10}) = 1$: every magnet must be flipped up.

When a system has a maximum possible energy, the degeneracy $\Upsilon(E)$ declines as the energy approaches that maximum. This is the phenomenon of negative temperatures: β the derivative of entropy with respect to energy is negative.

An analytically simple enough system can have its partition function and degeneracy exactly calculated. But few systems are simple enough to be understood this way; for example, the three-dimensional Ising[2] model is not yet analytically solved. Still we could find it numerically: for all the N possible states for the system, calculate their energies, and take the expectation value.

[1] The historically-minded reader will find this familiar to the reasoning that explained the radiation of a blackbody, by which Planck explained the "ultraviolet catastrophe". The principles are analogous.

[2] Ernst Ising, 1900 - 1998, examined ferromagnetism as his thesis problem, but did not appreciate for years the importance it had in the statistical mechanics and materials science communities. He was trapped in Luxembourg when Germany invaded in 1940, and was among many put to work dismantling the Maginot Line. [183]

$$\langle E \rangle = \frac{1}{N} \sum_{s} E(s) \tag{4.3}$$

This algorithm is too basic to use: its impracticality can be quickly shown by calculating the number of different configurations of the ten-magnet system. However, the essence of numerical simulation is that we do not need to calculate all of them. The law of large numbers indicates that the system will almost certainly be in one peak of most probable configurations, and anything too far from that peak may be ignored without much error.

This begs the question. If we knew the most probable states we would not need our estimate of the most probable states. We need a representative sampling of states, the expectation value for which equals the expectation value for all possible states. How do we create this representative sample? We begin, as many numerical methods do, by taking a wild guess.

4.3 Markov Chains

This initial guess serves the same role the first guess in Newton-Raphson iterations does, the arbitrary start of a process which eventually reach the desired solution. From our starting state we will generate a Markov chain of new states, which will cluster around the most probable states.

Any system has a set of all its possible states. We will explore that set by a stochastic process, one which relies on a certain randomness in how it will evolve. For example, if the system starts in the state A, perhaps it has a probability of 0.50 of being moved to state B, and has a probability of 0.30 of staying in A, and a 0.20 probability of moving to another state C. And then once it finishes that step, the system has new probabilities of the process moving it to A, or to B, or to C or another state, and so on.

In a Markov chain the probability of moving to another state depends only on the current state, and not on any other factors. We represent this with a **transition matrix**, a square matrix of size N – where N is the number of possible states – and in which the entry at row j, column k is the probability of the system moving from the state j to the state k in one iteration. The sum of the terms in any one column must be 1 – the system must be in some state at the end of the iteration. Each entry in this transition matrix must be nonnegative, and be at most 1.

Let $\mathbf{x}_0$ be the **probability vector**, a vector with N dimensions. Each state is a different dimension, and the value of element j in the vector is the probability the system is in state j. Each term is between zero and one inclusive, and the sum of all components is one. With the transition matrix M representing the effect of running our stochastic process, the probability of being in the various possible states after the first several iterations is

$$\mathbf{x}_1 = M\mathbf{x}_0 \tag{4.4}$$

$$\mathbf{x}_2 = M\mathbf{x}_1 = M^2\mathbf{x}_0 \tag{4.5}$$

$$\mathbf{x}_3 = M\mathbf{x}_2 = M^2\mathbf{x}_1 = M^3\mathbf{x}_0 \tag{4.6}$$

$$\vdots$$

$$\mathbf{x}_{i+1} = M\mathbf{x}_i = M^i\mathbf{x}_0 \tag{4.7}$$

The probability of getting from microstate j to microstate k in a single step is $M_{j,k}$. The chance of making it in two steps is $M^2_{j,k}$, and in n steps of $M^n_{j,k}$. The chance of ever getting from$j$ to k is $\sum_{n=1}^{\infty} M^n_{j,k}$. State j is called a **recurrent state** if $\sum_{n=1}^{\infty} M^n_{j,j}$ equals one – that is, if the system run long enough has probability one of returning. If $M^n_{j,j}$ is nonzero only when n is a whole multiple of some integer p, then state j has the **period** p; if p is 1 the state is **aperiodic**. If it is always possible, eventually, to get from any j to any k and back again – there are some p and q so that $M^p_{j,k} > 0$ and $M^q_{k,j} > 0$ – then M is an **irreducible** Markov chain. An irreducible aperiodic chain will justify Markov chain Monte Carlo in section 4.7.

After a great many iterations, the system settles to a statistical equilibrium, with the components no longer significantly changing. This equilibrium is the vector for which

$$\mathbf{x} = \lim_{i \to \infty} M^i\mathbf{x}_0 \tag{4.8}$$

which ought to be reached independently of whatever the initial vector $\mathbf{x}_0$ is.

Another interpretation of the statistical equilibrium vector is that it is the eigenvector of M with eigenvalue 1. If the vector $\mathbf{x}$ from 4.8 exists, then

$$\mathbf{x} = M\mathbf{x} \tag{4.9}$$

Note that the method of picking an arbitrary starting point and repeating matrix multiplication is one numerical method for finding a matrix's largest eigenvalue and its eigenvector.

The random walk is probably the best-known example of a Markov chain (though the self-avoiding random walk is not – its moves depend on its previous states), and we can view our Markov chains as a form of random walk through all possible states. Assuming that the phase space of possible states of a system does not have any traps it should be possible to get from any one state to any other state eventually.

Using these iterative or eigenvector methods seems to make the problem worse. Now we need not just all the states but the chance of transferring between them – and finding eigenvalues and eigenvectors of large, mostly nonzero, matrices is extremely computationally costly. We need to reduce this overhead.

4.4 Detailed Balance

We know something about the statistical equilibrium of a system. In equilibrium the probability of being in the state j has to be

$$\pi(j) = \frac{1}{Z} \exp(-\beta E_j) \tag{4.10}$$

with Z the partition function and a normalizing factor, so that $\sum_j \pi(j) = 1$.

Detailed balance, a principle holding for any time-reversible system, holds that at statistical equilibrium the rate at which any process occurs equals the rate at which its inverse occurs. The chance of the transition from state A to B equals the chance of the transition from B to A occurring. In chemical reactions – which may offer an intuitive guide to the property – it means the rate at which the components combine to form the product equals the rate at which some of the product breaks back down to its components.

Let $P_{A,B}$ represent the probability of state A evolving to B. The chance of observing A turning to B is the product $\pi(A) \times P_{A,B}$ – the probability we began in A *and* changed to B. To be in detailed balance means for all A and B

$$\pi(A) P_{A,B} = \pi(B) P_{B,A} \tag{4.11}$$

From equation 4.11 and from the probability of states occurring from equation 4.10 we can find the probability of observing the transition from A to B, in terms of the probability of observing the transition from B to A. The challenge is then to construct a Markov chain of states which reach this detailed balance.

$$\pi(A) P_{A,B} = \pi(B) P_{B,A} \tag{4.12}$$

$$P_{A,B} = P_{B,A} \frac{\pi(B)}{\pi(A)} \tag{4.13}$$

$$P_{A,B} = P_{B,A} \frac{\frac{1}{Z}\exp(-\beta E(B))}{\frac{1}{Z}\exp(-\beta E(A)))} \tag{4.14}$$

$$P_{A,B} = P_{B,A} \exp(-\beta(E(B) - E(A))) \tag{4.15}$$

4.5 The Metropolis Rule

The term **Monte Carlo** describes a collection of probability-based methods. The name is meant to evoke gambling: any one event is unpredictable, but the averages over many events are certain. The use of statistical methods to find exact results goes back centuries at least. Likely the best-known and startling example is Buffon's needle problem[3], in which short needles are dropped across

[3] Georges Compte de Buffon, 1707 - 1788, besides proposing and solving the needle problem, was a naturalist and proposed that one could define two animals to be of the same species if their offspring were fertile. [288]

an array of parallel straight lines. The chance of a needle intersecting one of the lines is proportional to the width of the lines, the length of the needle, and π, which provides an experimental (but inefficient) way to calculate π.

Modern Monte Carlo studies problems from numerical quadrature, to random walks, to polymer and crystal growth, to neutral network growth and decay. Some techniques allow the solution of differential equations by these methods. Monte Carlo methods – and the name[4] – came about after 1944, in the effort to simulate the diffusion of neutrons in fissionable materials [166].

There are many techniques, but the typical approach starts with an arbitrary state. We experimentally adjust that solution, making small changes at random. Steps that improve the solution are accepted, and steps which worsen it are rejected with a probability that depends on how much worse the change would be. This process continues until the detailed balance is met.

The Metropolis Rule for Monte Carlo, introduced by Nicholas Metropolis, A W Rosenbluth, M N Rosenbluth, A H Teller, and Edward Teller in a 1953 paper[5], "Equations of state calculations by fast computing machines," (J. Chem. Phys. 21, 1087-1092) is one of the great algorithms of the 20th century.

The algorithm is powerful and flexible; it can be used for problems from the absorption of neutrons by atomic nucleii to the growth of crystals to the travelling salesman problem. To customize it to the needs of the vortex problems we address here we fill in only a few details. What we will use is the Hastings[6] rule, and so this particular method is often called the Metropolis-Hastings algorithm.

We begin with state A. Typically Monte Carlo programs will try to change as few components as possible, for example by moving one particle. The desire to change as few variables as possible is a computational convenience. We need to calculate the difference in energy (and other quantities) between the new state and the old, and fewer changes make those calculations faster.

---

[4] Nicholas Constantine Metropolis, 1915 - 1999, and Stanislaw Ulm, 1909 - 1986, are credited with choosing the name. [288]

[5] Nicholas Metropolis is one of the founders of electronic computing, and developed the MANIAC I and II computers – the acronyms standing for 'Mathematical And Numerical Integrator And Computer,' which he hoped would end the intrusion of acronyms into every aspect of computing. He was more successful giving names to the synthetic chemical elements technetium and astatine. [439] Marshall Nicholas Rosenbluth, 1927 - 2003, began his career studying the scattering of electrons in nucleii. His thesis advisor – Dr Edward Teller – invited him to the hydrogen bomb project; he worked on the problem of fusion power much of his life. [63] Edward Teller, 1908 - 2003, was one of the pioneers of the fission and fusion bombs, and his lobbying efforts helped create the Lawrence Livermore National Laboratory for thermonuclear research. [286]

[6] W Keith Hastings, born 1930, began his career in financial mathematics but switched to statistical mechanics when he found his particular problem declared dead at a mathematics conference. [436]

So we find a modified state B. We then determine whether to accept or reject it. The probability of moving from state A to state B is $\exp(-\beta(E_B - E_A))$, so we find $\Delta E = E_B - E_A$ and our inverse temperature β. Draw a random number r from a uniform distribution on the interval $[0, 1]$; if $r < \exp(-\beta\Delta E)$ then the experiment is accepted. Repeat the process until an equilibrium is reached. (Note that if $\beta\Delta E < 0$ then the change is always accepted. The interpretation that in this case the probability of moving from A to B is greater than one is quietly ignored; the probability cannot be more than one.)

That this rule obeys the detailed balance principle is clear: the chance of the process moving the system from state A to state B in one iterate is $\pi(A)\exp(\beta(E(B) - E(A)))$.

The thermodynamic origins of statistical mechanics give us a metaphor, of placing the system into a heat bath at inverse temperature β. The Metropolis-Hastings algorithm simulates what happens if a system is given access to an unlimited bath from which to draw, or into which to dump, heat. For the system of gas particles, the average kinetic energy at detailed balance equals to the average kinetic energy at the temperature $T = \frac{1}{k_B\beta}$ (with k_B the Boltzmann constant). Reaching a statistical equilibrium may be referred to as reaching thermal equilibrium.

There are other criteria that may be applied: the detailed balance is satisfied if experiments are approved whenever the r drawn from $[0, 1]$ is less than $[\exp(\beta\Delta E) + 1]^{-1}$ instead. While this alternate acceptance criteria will accept and reject slightly different states than the above rule does, they will produce a Markov chain with similar properties.

The Markov chain settles around a few states around the peak where the product of the probability and degeneracy reaches a maximum. Assuming β is positive, then if we start from a state with higher than the most-probable energy we will see any states lowering energy approved, while few states increasing it are permitted. Thus we get a chain of states with, usually, decreasing energy whenever we are above the most-probable energy.

If we begin from below the most-probable energy, while the Metropolis-Hastings process would try to decrease the energy, the degeneracy of these lower-energy states is small enough the algorithm cannot find many. More moves increasing the energy are approved. The only energy at which the number of moves increasing the energy will equal the number decreasing is that most-probable energy state – which is another way of saying the system settles at the energy where detailed balance is satisfied.

4.6 Multiple Canonical Constraints

We have looked at systems in which several quantities, such as energy and circulation, affect the probability of a microstate appearing. When the probability depended only on its energy, its probability of appearing at a particular

β was $\frac{1}{Z}\exp(-\beta E)$. With several quantities, E and Ω as in section 3.4, we had an inverse temperature β and a chemical potential μ, and the probability of microstate A was

$$\pi(A) = \frac{\exp\left(-\beta E_A - \mu\Omega_A\right)}{\sum_j \exp\left(-\beta E_j - \mu\Omega_j\right)} \tag{4.16}$$

with Z the partition function, the sum of $\exp(-\beta E(j) - \mu\Omega(j))$ over all possible microstates j.

The kinetic theory of gases defines the **enthalpy** of a system to be the sum of the energy and of the pressure times the volume. We will appropriate the name enthalpy and modify it to be

$$H_A = E_A + \frac{\mu}{\beta}\Omega_A \tag{4.17}$$

with Ω_A the new canonically constrained quantity. The Metropolis-Hastings algorithm we rewrite with enthalpy in place of energy.

Given the current microstate j, we generate a new microstate k, and calculate the change in enthalpy $\Delta H = \Delta E + \frac{\mu}{\beta}\Delta\Omega$. We draw a random number r uniformly from the interval $(0, 1)$ and accept the move whenever

$$r \leq \exp\left(-\beta\Delta H\right) = \exp\left(-\beta\Delta E - \mu\Delta\Omega\right) \tag{4.18}$$

and reject the move otherwise.

(Some books and articles introduce enthalpy as $H = E + \mu\Omega$, which is a different scaling of μ. Whether to use this definition or that of 4.17 is a personal preference.)

This enthalpy can be extended. Each new conserved quantity Ξ requires its own chemical potential, but the algorithm remains the same, with

$$H_A = E_A + \frac{\mu_1}{\beta}\Omega_A + \frac{\mu_2}{\beta}\Xi_A \tag{4.19}$$

and the decision to accept or reject a proposed move being based on whether or not the randomly drawn number r satisfies

$$r \leq \exp\left(-\beta\Delta H\right) = \exp\left(-\beta\Delta E - \mu_1\Delta\Omega - \mu_2\Delta\Xi\right) \tag{4.20}$$

As with the metaphor of Metropolis-Hastings simulation as placing the system in a heat bath at inverse temperature β, we can regard multiple canonical constraints as giving a system access to several baths from which to draw or into which to dump energy, enstrophy, or whatever is interesting.

4.7 Ensemble Averages

We use the Markov chain of Metropolis-Hastings produced sequences to find an average of the property x. With M such states, and $x(j)$ the property measured at state number j, the mean value of x is

$$\bar{x} = \frac{1}{M} \sum_{j=1}^{M} x(j) \tag{4.21}$$

This number will approximate the expectation value of x, the average over all possible states. Given the probability of any state j occurring at inverse temperature β is $\frac{1}{Z} \exp(-\beta H(j))$ then this expectation value, if there are a discrete set of N possible states, will be

$$\langle x \rangle = \frac{\sum_{j=1}^{N} x(j) \exp\left(-\beta H(j)\right)}{\sum_{j=1}^{N} \exp\left(-\beta H(j)\right)} \tag{4.22}$$

or, if there is a continuum of possible states,

$$\langle x \rangle = \frac{\int x(\mathbf{s}) \exp\left(-\beta H(\mathbf{s})\right) d\mathbf{s}}{\int \exp\left(-\beta H(\mathbf{s})\right) d\mathbf{s}} \tag{4.23}$$

The **ergodic hypothesis** says if one takes a single system and constructs a long enough chain of microstates, then the averages of any measured quantity taken over these microstates will approximate the average the quantity would have over all of phase space. The fraction of "time" spent in any macrostate consisting of a certain energy range will be proportional to the fraction of the volume of phase space that is in that energy range macrostate.

Consider an irreducible aperiodic Markov chain. This sequence of states can explore all phase space without becoming trapped forever in one region – it is irreducible, so the chance of getting from one microstate to another is never zero. It can concentrate on the most probable microstates – an aperiodic chain may repeat its position. A long enough chain should explore phase space and spend approximately as much "time" – have as many links – in each macrostate as the whole phase space does.

Given an irreducible aperiodic Markov chain, $\pi(j)$ the probability the current microstate is j, and $M_{j,k}$ the probability of moving to k then

$$\pi(k) = \sum_{j} M_{j,k} \pi(j) \tag{4.24}$$

when the steady state distribution has been found, and the summation is taken over all microstates. (If we have a continuum of states, rather than a discrete set, this becomes an integral.) If we satisfy this (and the requirements $\pi(k) > 0$ and $\sum_j \pi(k) = 1$) then

$$\lim_{n \to \infty} M_{j,k}^{n} = \pi(k) \tag{4.25}$$

The choice of j is irrelevant: the statistical equilibrium does not depend on the initial state. We can find an equilibrium even by multiplying M by itself repeatedly. Every column of M^n approaches the equilibrium (if it exists).

The approach of Metropolis et al – well-presented by J. M. Hammersley and D. C. Handscomb [166] – is to build a transition matrix P with elements $P_{j,k}$ that will eventually satisfy the equilibrium distribution. Assume $P_{j,k} > 0$ for all j and k, and that P is a regular transition matrix: $\sum_k P_{j,k} = 1$ and $P_{j,k} = P_{k,j}$. We know the relative probability $\frac{\pi j}{\pi k}$ at equilibrium: $\exp(-\beta(H_j - H_k))$. From this we build the elements of the transition matrix M.

Define $M_{j,k}$ by the rule:

$$M_{j,k} = \left\{ \begin{array}{ll} P_{j,k}\frac{\pi(k)}{\pi(j)} & \text{if } \frac{\pi(k)}{\pi(j)} < 1 \\ P_{j,k} & \text{if } \frac{\pi(k)}{\pi(j)} \geq 1 \end{array} \right\} \text{if } j \neq k \tag{4.26}$$

$$M_{j,j} = P_{j,j} + \sum_k{}' P_{j,k}(1 - \frac{\pi(k)}{\pi(j)}) \tag{4.27}$$

where $\sum_k'$ means a summation over all states k for which $\frac{\pi(k)}{\pi(j)} \geq 1$. As one last bit of notation let $\sum_k''$ represent the summation over all $k \neq j$ for which $\frac{\pi(k)}{\pi(j)} \geq 1$, which will let us conveniently find $M_{j,k}$:

$$\sum_j M_{j,k} = P_{j,j} + \sum_k{}' P_{j,k}\left(1 - \frac{\pi(k)}{\pi(j)}\right)$$

$$+ \sum_k{}' P_{j,k}\frac{\pi(k)}{\pi(j)} + \sum_k{}'' P_{j,k} \tag{4.28}$$

$$= P_{j,j} + \sum_k{}' P_{j,k} + \sum_k{}'' P_{j,k} \tag{4.29}$$

$$= P_{j,j} + \sum_{k \neq j} P_{j,k} \tag{4.30}$$

$$= \sum_j P_{j,k} \tag{4.31}$$

$$= 1 \tag{4.32}$$

which means the matrix M is a regular matrix with nonzero terms as demanded. We have remaining only to show that equation 4.24 holds and the Markov chain Monte Carlo will be fully justified.

The assumption of detailed balance means the chance of observing a move from j to k is equal to the chance of observing the reverse: $\pi(j)M_{j,k} = \pi(k)M_{k,j}$. We claim this is satisfied by this matrix. Suppose for the states j and k, $\pi(j) = \pi(k)$. From equation 4.26 then

$$M_{j,k} = P_{j,k} = P_{k,j} = M_{k,j} \tag{4.33}$$

$$\pi(j)M_{j,k} = \pi(k)M_{k,j} \tag{4.34}$$

which satisfies the balance if $\pi(j) = \pi(k)$. If they are not equal, then – taking without loss of generality – suppose $\pi(k) < \pi(j)$. Then, again from 4.26 and our assumption that $P_{j,k} = P_{k,j}$ we have

$$M_{j,k} = P_{j,k}\frac{\pi(k)}{\pi(j)} = P_{k,j}\frac{\pi(k)}{\pi(j)} = M_{k,j}\frac{\pi(k)}{\pi(j)} \tag{4.35}$$

$$\pi(j)M_{j,k} = \pi(k)M_{k,j} \tag{4.36}$$

and a similar argument will hold if $\pi(j) < \pi(k)$. Finally,

$$\sum_j \pi(j)M_{j,k} = \sum_j \pi(k)M_{k,j} \tag{4.37}$$

$$= \pi(k)\sum_j M_{k,j} \tag{4.38}$$

$$= \pi(k) \times 1 \tag{4.39}$$

which is equation 4.24.

And this explains finally the method of the Metropolis-Hastings rule. From any microstate j some new microstate k is picked. We accept or reject that move, with probability $\frac{\pi(k)}{\pi(j)}$, which number[7] can be calculated knowing only microstates j and k. The resulting chain of states are distributed as the entire phase space is. Ensemble averages over a long enough chain will approximate averages over the whole of phase space [166].

We have the question of how long is a long enough chain. The expectation value of the difference between the ensemble average and the Markov chain average for N states is proportional to $\sqrt{\frac{1}{N}}$. What that proportion is depends on a constant called the **correlation time**, a measure of how many attempted changes have to be made before we have two independent states. Its value is not obvious before calculations are done, though. Worse, **critical slowing down** occurs: the correlation time grows longer if β is close to the inverse temperature of a phase transition. We will return to this phenomenon.

By tracking the expectation values of the quantity $f(i)$ and of the quantity $f^2(i)$ we can estimate the correlation time and from that the error margin of any measurement. If $f(N)$ is the value of the measured quantity after N links and $f(N+t)$ is its value after t more links are created, then

$$\frac{\langle f(N+t)f(N)\rangle}{\langle f^2\rangle - \langle f\rangle^2} \sim \exp\left(-\frac{t}{\tau}\right) \tag{4.40}$$

with τ the correlation time. The estimate of the error in f after N links is

$$\Delta f \sim \sqrt{\langle f^2\rangle - \langle f\rangle^2}\sqrt{\frac{\tau}{N}} \tag{4.41}$$

[7] Properly, the probability is $\min\left(1, \frac{\pi(k)}{\pi(j)}\right)$.

[166] [353]. $\sqrt{\langle f^2 \rangle - \langle f \rangle^2}$ is sometimes called the **spread** of f.

Often good practice is running several examples for as long a time as possible, with the energy (and other interesting properties) measured often, to examine their evolution. This provides some feel for the correlation time, and how long simulations need to be for fluctuations to grow sufficiently small.

4.8 Random Number Generation

Given our need for them, it is worth reviewing how to find random sequences of numbers. This assumes we know what a random sequence of numbers is; most people would say "9553702163" looks random, but it is from the digits of a famous number and can be created by a simple formula.

Typically to judge whether a sequence of numbers is "random" we consider whether the distribution of generated numbers matches an expected distribution (typically uniform or Gaussian distributions), and whether there are correlations between numbers (whether smaller strings repeat). A common but not infallible rough test is to use a file compression algorithm on the list; lists that do not significantly compress are quite likely random.

Taken literally this rules out every software-based random number generator. A computer can return only finitely many possible numbers, so there are only finitely many ways to arrange them and eventually the sequence must repeat from the beginning. This time to repeat is the **period** of the generator. We accept this, though, by using algorithms with periods much greater than the number of random numbers we will use.

As a pseudo-random number generator follows deterministic rules, in principle it generates the same sequence every run. This is convenient while debugging code which relies on the random number generator. To avoid repetition, however, pseudo-random number generators need a **seed**, an arbitrarily chosen starting point. A popular seed is the system time, generally the number of seconds (or milliseconds or microseconds) since an epochal date.

As we implement them, pseudo-random number generators are functions. Let x_0 be the seed or some default starting number, and build the sequence of numbers $x_1, x_2, x_3, \cdots$ by letting $x_{j+1} = f(x_j)$ for all positive integers j. (The function may depend on several terms, a function of x_j, x_{j-1}, x_{j-2}, $\cdots$, x_{j-N}.) The most common generators depend on modular arithmetic.

4.8.1 Linear Congruence Generators

D H Lehmer proposed in 1948 the linear congruence generator [142]. It uses three reference numbers – the **multiplier** a, the **increment** b, and the **modulus** M. Numbers are generated by the sequence

$$x_{j+1} = (ax_j + b) \, mod M \tag{4.42}$$

Linear congruence generators are simple and popular, but have some weaknesses. Plotting pairs of coordinates (x_j, x_{j+1}) find the points tend to lie along several lines; and the generator has a period of at most M.

A variant sets the increment $b = 0$, which, with setting the modulus M to be a power of two, makes the generator more computationally efficient, but trims the period greatly. With this modulus of 2^k, the maximum period is $\frac{M}{4}$, when the multiplier a is congruent to 3 or 5 modulus 8. A common modulus and multiplier are $M = 2^{31} - 1$ and $a = 7^5$; the modulus is one of the Mersenne primes. This provides a period on the order of 10^9.

4.8.2 Multiple Recursive and Fibonacci Generators

An extension to the congruential generator is to generate x_{j+1} by using the last k values. Given several multipliers a_0, a_1, a_2, $\cdots$ a_k one creates the sequence

$$x_{j+1} = (a_0 x_j + a_1 x_{j-1} + a_2 x_{j-2} + \cdots + a_k x_{j-k}) mod M \tag{4.43}$$

These were proposed by R C Tausworthe in 1965 to generate a random sequence by using modulus two. Donald Knuth extended the idea in 1981 to use any prime modulus. Ideally, they are capable of reaching a period of $M^k - 1$, though this is reached only for certain multipliers a_i.

The Fibonacci[8] generator is inspired by the Fibonacci sequence, in which $x_{j+1} = x_j + x_{j-1}$. With two offsets k and l, the lagged Fibonacci generator is

$$x_{j+1} = (x_{j-k} + x_{j-l}) mod M \tag{4.44}$$

For a prime M and for $l > k$, the period can be as large as $M^l - 1$, which requires good choices of k and l as well the first l values of x_j. [142]. A generalized Fibonacci generator does not rely on modular arithmetic; it takes any binary operation $f(x, y)$ and generates terms by

$$x_{j+1} = f(x_{j-k}, x_{j-l}) \tag{4.45}$$

Its performance is heavily dependent on the operation. It is tempting to choose a non-linear f. The benefits of that are unproven; only linear and the later-discussed inverse congruential generators have much theory justifying them.

4.8.3 Add-with-Carry and Subtract-with-Carry Generators

Add-with-carry and subtract-with-carry generators rely on two numbers, the **carry** c_j – which typically depends on the term generated – and the modular

[8] Leonardo Pisano, nicknamed Fibonacci, 1170c - 1250c, was most influential in popularizing Arabic numerals in Europe. He developed considerable amounts of number theory, as well as arithmetic useful to the merchants and surveyors of Italy. [288]

base M; they also call on previously generated numbers, from k and from l terms before the current value. The relationship to Fibonacci generators and linear congruential generators is plain. The sequence generated is

$$x_{j+1} = (x_{j-k} + x_{j-l} + c_j) mod M \tag{4.46}$$

with the rule that $c_1 = 0$, and $c_j = 0$ if $x_{j-k}+x_{j-l}+c_{j-1} < M$; c_j is otherwise 1. This generator is in some ways equivalent to using a linear congruential generator with a much larger prime modulus, leaving it still vulnerable to pairs of values ending up on a few lines.

The similar subtract-with-carry generator uses

$$x_{j+1} = (x_{j-k} - x_{j-l} - c_j) mod M \tag{4.47}$$

with the carry c_j equal to zero for c_1 and if $x_{j-k} - x_{j-l} - c_{j-1} < M$, but equal to 1 otherwise [142].

4.8.4 Inverse Congruential Generators

Given that one way a random number generator fails is producing sequences that fall on lines, it seems the obvious fix is to use a non-linear algorithm to generate x_j. But non-linear algorithms are generally not proven to be better than the linear-based methods.

An exception is the multiplicative inverse of a number modulo M. Given the multiplier a, the carry c, and a prime number M let

$$x_{j+1} = \left(a x_j{}^{-1} + c\right) mod M \tag{4.48}$$

where $x_j{}^{-1}$ is the multiplicative inverse of x_j in the integers mod M. Despite its good properties – it appears to be free of pairs of values falling on lines – finding mutliplicative inverses in modulo arithmetic is a challenge, which can make it difficult to implement. One modified inverse congruential generator with excellent performance, including a period M when $M = 2^p$ (and $p \geq 4$), is

$$x_{j+1} = j(a x_j + c)^{-1} mod M \tag{4.49}$$

with $a = 2 mod 4$ and $b = 1 mod 2$ [142].

Another generalization using a modulus M that is a power of two is suggested by Kato, Wu, and Yanagihara and combines the inverse and the ordinary linear congruence generators

$$x_{j+1} = (a x_j^{-1} + b x_j + c) mod M \tag{4.50}$$

though it is not clear this is better than the plain inverse congruential generator.

4.8.5 Combined Generators

In the search for better random number generators one considers using the output of several random number sequences and in some way combining them. There are several approaches which do this. One example, easy to code and with a long period, is a triple linear congruential generator described by Wichmann and Hill in 1982, which uses three sequences x_j, y_j, and z_j to generate a longer sequence u_i. This rule is

$$x_{j+1} = 171x_j mod 30269 \tag{4.51}$$

$$y_{j+1} = 172y_j mod 30307 \tag{4.52}$$

$$z_{j+1} = 170z_j mod 30323 \tag{4.53}$$

$$u_{j+1} = \left(\frac{x_j}{30269} + \frac{y_j}{30307} + \frac{z_j}{30323}\right) mod 1 \tag{4.54}$$

It produces u_j values between 0 and 1 with a period around 10^{12}.

A more general algorithm, proposed by L'Ecuyer in 1988, uses the terms of k different multiplicative congruential generators x^1, x^2, x^3, $\cdots$ x^k, each with prime moduli M_j and for which all the $(M_j - 1)/2$ are relatively prime. Assuming M_1 to be the largest of them all, the current terms in the sequences x_i^1, x_i^2, $\cdots$ x_j^k make the next element in the combined sequence x_{j+1} by

$$x_{j+1} = \sum_{l=1}^{k} (-1)^{l-1} x_j^l mod(M_1 - 1) \tag{4.55}$$

or normalize it by using the above x_{j+1}, creating the elements u_{j+1} by

$$u_{j+1} = \left\{ \begin{array}{ll} \frac{x_{j+1}}{M_1} & \text{if } x_{j+1} > 0 \\ \frac{M_1 - 1}{M_1} & \text{if } x_{j+1} = 0 \end{array} \right\} \tag{4.56}$$

so that u_j floats between 0 and 1. One specific generator suggested was

$$x_{j+1} = 40014x_j mod 2147483563 \tag{4.57}$$

$$y_{j+1} = 40692y_j mod 2147483399 \tag{4.58}$$

$$z_{j+1} = x_{j+1} - y_{i+j} \tag{4.59}$$

If $z_{j+1} < 1$ then take $z_{j+1} = z_{j+1} + 2147483562$. Finally,

$$u_{j+1} = 4.656613 z_{j+1} \times 10^{-10} \tag{4.60}$$

which produces u_j between 0 and 1 with a period on the order of 10^{18} [142].

None of these algorithms is so complicated as to be unprogrammable. However, for serious work one usually uses a random number generator such as from the Linpack package of algorithms, or from Donald Knuth's codes. These are more likely to be implemented efficiently, debugged, and tested for quality. More, serious programming efforts should accept several different random number generators, and results not relied on unless they prove comparable with different generators.

5

Spectral Methods

5.1 Introduction

In this chapter we introduce spectral methods. The principle is familiar to anyone who has used Fourier[1] series to solve a differential equation: we represent the unknown but presumed-to-exist solution as the summation of a (possibly infinite) number of constants times base functions, and convert the problem of solving a differential equation into solving a set of equations for the coefficients.

In order to understand the context it is necessary to review functional analysis, and particularly the study of operators, which act on functions in the way that functions act on real numbers. We are all quite familiar with two operators, the derivative and the integral operators. These take as input functions and return functions as output. We can build from this extend the notion of vector spaces and linear algebra into treatments of wide classes of functions.

Many of the concepts familiar to linear algebra extend naturally into these new fields. The result is an ability to manipulate functions in much the way we can manipulate vectors and matrices, and many of the statistical mechanics we will want to study become therefore manageable.

5.2 Inner Products

We are all familiar with the **vector space**, a collection of elements, generally called **vectors** but often given a name suitable to the specific problem, which

[1] Jean Baptiste Joseph, Baron de Fourier, 1768 - 1830, developed his series in the attempt to explain heat as a fluid permeating space. He narrowly escaped execution for political offenses during the post-Revolutionary Reign of Terror, and joined Napoleon Bonapartes army in Egypt in 1798. There he administrated the political and scientific organisations Napoleon set up while trapped in Egypt by the British Navy. [288]

is a set closed under vector addition and scalar multiplication. We will choose elements from the set of functions satisfying particular conditions.

An **operator** is a mapping between two **function spaces** (the collection of real-valued continuous functions defined on some reference interval), a **domain** and a **range**. The domain and range function spaces do not need to have the same number of dimensions, or be spaces of functions defined over the same intervals, or may even both be the real or the complex numbers. For example, linear combinations of derivatives with respect to different variables is an operator on any function of those variables (so we could rewrite differential equations in the form of an operator on the unknown function equalling a forcing function, which is often a useful form). When the range is the real or the complex numbers, the operator is often called a **functional**. An example of a functional is the Hamiltonian which maps the $4N$ dimensional space of site strengths and positions in space to the scalar energy.

The first tool needed for functional analysis is the **inner product**, which generalizes the dot product between vectors. The properties of an inner product are that for any three functions f, g, and h, and for a scalar α, then

$$\langle f+g,h\rangle = \langle f,h\rangle + \langle g,h\rangle \tag{5.1}$$

$$\langle \alpha f,g\rangle = \alpha \langle f,g\rangle \tag{5.2}$$

$$\langle \alpha f,g\rangle = \alpha \langle f,g\rangle \tag{5.3}$$

$$\langle f,g\rangle = \langle g,f\rangle \tag{5.4}$$

$$\langle f,f\rangle \geq 0 \tag{5.5}$$

with the note that $\langle f,f\rangle = 0$ if and only if f is identically zero.

From an inner product we define the **norm** of a function, corresponding exactly to the norm of a vector.

$$\|f\| = \sqrt{\langle f,f\rangle} \tag{5.6}$$

Two functions f and g are **orthogonal** if and only if their inner product is zero.

Assume we have a vector space whose elements are functions, which has an inner product and therefore a norm. If the vector space is **complete** – the limit of any convergent sequence of functions from the space is itself in the set – then we have a **Hilbert**[2] **space**. The Hilbert space $L^2(D)$ is defined to be the set of all functions, which are themselves defined over the set D, and which are square-integrable: for any $f \in L^2(D)$ the integral

$$\int_D f^2(\mathbf{x})d\mathbf{x} < \infty \tag{5.7}$$

[2] David Hilbert, 1862 - 1943, a founder of functional analysis and supporter of the axiomatization of mathematics, is noted for a set of 23 problems given to the Second International Congress of Mathematicians in Paris in 1900, and developed the equations of general relativity contemporaneously to Einstein. [288]

What elements fit in this space depend on the set D over which functions are defined (and this set may be finite or infinite, based on the real number line, the complex plane, or an n-dimensional real-number space, or more abstract sets). The inner product for the space $L^2(D)$ is then

$$\langle f, g \rangle = \int_D f(\mathbf{x}) g(\mathbf{x}) d\mathbf{x} \tag{5.8}$$

(which is analogous to the dot product of vectors).

Given an operator L and any two functions f and g in a Hilbert space, there is an operator known as the **adjoint** $L^\dagger$ defined by the relationship

$$\langle Lf, g \rangle = \langle f, L^\dagger g \rangle \tag{5.9}$$

which is similar to the definition of the adjoint matrix. Just as there are self-adjoint matrices there are **self-adjoint operators** L when $L = L^\dagger$. Such self-adjoint operators, provided they also satisfy the requirement that

$$\langle f^*, Lg \rangle \equiv \langle g, Lf^* \rangle \tag{5.10}$$

(where f^* is the complex conjugate of the function f), are known as **Hermitian[3] operators.**

An operator is **linear** if it meets exactly the same conditions which make a function linear: for any functions f and g, and for a scalar α, the operator L is linear if

$$L(f + g) = L(f) + L(g) \tag{5.11}$$
$$L(\alpha f) = \alpha L(f) \tag{5.12}$$

We also have for an operator L the **kernel**, which is a function $G(\mathbf{x}, \mathbf{y})$ so that, for any function f in the relevant Hilbert space,

$$y = \int_D G(\mathbf{x}, \mathbf{y}) f(\mathbf{y}) d\mathbf{y} \tag{5.13}$$

is a function which solves the equation

$$Ly(\mathbf{x}) + f(\mathbf{x}) = 0. \tag{5.14}$$

We have already seen kernels introduced as the fundamental solutions, representing the solution of (in our case) the inverse Laplace[4]-Beltrami[5] operator on a given domain.

[3] Charles Hermite, 1822 - 1901, made substantial discoveries in elliptic functions and in quadratic forms, and thus provided the logical connections between number theory, theta functions, and transformations of Abelian functions. He was the first to prove that e is a transcendental number, and is best known for the polynomials, matrices, and interpolation methods now named for him. [288]

[4] Pierre-Simon de Laplace, 1749 - 1827, was one of the committee which designed the metric system; working with Antoine-Laurent Lavoisier, 1743 - 1794, he helped demonstrate that respiration and combustion are forms of the same process. [25]

[5] Eugenio Beltrami, 1835 - 1900, studied non-Euclidean geometry and the properties of elasticity, and developed mechanical interpretations of Maxwell's equations.

Generally operators do not commute, something obvious from experience with the derivative and integral operators. If an operator L commutes with its own adjoint $L^\dagger$, then L is a **normal** operator.

5.3 Basis Functions

Just as with matrices and vectors there are **eigenfunctions** for operators, functions ϕ for which

$$L\phi = \lambda w \phi \tag{5.15}$$

where L is an operator, λ is a scalar (real or complex), and w is a real-valued weighting function, not everywhere zero, dependent on the operator. For a self-adjoint operator L we have three important physical properties, which are the results of the **Spectral Theorem**, and which are exactly analogous to the way the eigenvalues and eigenvectors of a matrix will form a basis for a Euclidean space.

First is that the eigenvalues of L are real. Suppose we have the eigenfunctions ϕ_j and ϕ_k with corresponding eigenvalues λ_j and λ_k. Then

$$L\phi_j - \lambda_j w \phi_j = 0 \tag{5.16}$$
$$L\phi_k - \lambda_k w \phi_k = 0 \tag{5.17}$$

The complex conjugate of the second equation is

$$L\phi_k^* - \lambda_k^* w \phi_k^* = 0 \tag{5.18}$$

We multiply equation 5.16 by ϕ_k^* and equation 5.18 by ϕ_j, and subtract one from the other, providing:

$$\phi_k^* L\phi_j - \phi_j - L\phi_k^* = \left(\lambda_k^* - \lambda_j\right) w \phi_j \phi_k^* \tag{5.19}$$

which, integrated over D, provides

$$\int_d \phi_k^* L\phi_j d\mathbf{x} - \int_D \phi_j - L\phi_k^* d\mathbf{x} = \left(\lambda_k^* - \lambda_j\right) \int_D w \phi_j \phi_k^* d\mathbf{x} \tag{5.20}$$

the left-hand side of which is zero, by the self-adjoint nature of the operator L. Therefore

$$\left(\lambda_k^* - \lambda_j\right) \int_D w \phi_j \phi_k^* d\mathbf{x} = 0 \tag{5.21}$$

His generalization of the Laplace operator is noted in the name of the operator here; he also recovered from obscurity the work of Giovanni Girolamo Saccheri, 1667 - 1733, who had proven the existence of non-Euclidean geometries but did not recognize it. [288]

But if $j = k$ the integral cannot be zero, unless ϕ_j is; therefore

$$\lambda_j^* = \lambda_j \tag{5.22}$$

and therefore the eigenvalue is real.

The next property is that the eigenfunctions of the self-adjoint operator are orthogonal. Whenever $j \neq k$, if $\lambda_j \neq k$, then

$$\langle \phi_j, \phi_k \rangle = 0 \tag{5.23}$$

This follows from the assumption that $(\lambda_j^* - \lambda_k) \neq 0$ in equation 5.21. Orthogonality is not guaranteed in the case that two distinct eigenfunctions have the same eigenvalue; this is known as a **degenerate** case. However, such eigenfunctions can be made orthogonal, such as by the Gram[6]-Schmidt[7] orthogonalization process. We omit the details because for the operators in which we are interested degenerate cases do not appear.

The final important property is completeness: the set of eigenfunctions ϕ_j for an operator form a basis set for a Hilbert space, exactly as the set of eigenvectors for a matrix form a basis set for a vector space. We would like to show that any function $F(\mathbf{x})$ in the Hilbert space may be approximated by

$$F(x) = \sum_{j=0}^{n} a_j \phi_j(\mathbf{x}) \tag{5.24}$$

for the right coefficients a_j, although – just as with Fourier series – what we are actually able to demonstrate is that the mean square error between the original function and our eigenfunction approximation may be made arbitrarily small. The function $F(x)$ and the infinite series will be equal **almost everywhere**, which means the elements in the domain which are exceptions to the two being equal form a set of measure zero – that is, there are either a finite set of exceptions, or if there are an infinite set of exceptions they fill no space.

What we know then is that the mean square error will satisfy

$$\lim_{N \to \infty} \int_D \left[F(x) - \sum_{j=0}^{N} a_j \phi_j(\mathbf{x}) \right]^2 w(\mathbf{x}) d\mathbf{x} = 0 \tag{5.25}$$

when we have found the coefficients to be

$$a_j = \int_D F(\mathbf{x}) \phi_j(\mathbf{x}) w(\mathbf{x}) d\mathbf{x} \tag{5.26}$$

[6] Jorgen Pedersen Gram, 1850 - 1916, worked in numeral applied and pure mathematical fields, and found a recurring interest in the modelling of forest growth. [288]

[7] Erhard Schmidt, 1876 - 1959, developed ideas which lead to the discovery of Hilbert spaces, creating functional analysis; he also gave the definition of a norm as $\|z\| = \sqrt{< z, z^* >}$. [288]

which follows exactly as the definition of coefficients for the Fourier series does. These coefficients similarly minimize the mean square error if we use a finite series approximation to the function $F(x)$ rather than the infinite approximation.

The **spectrum** $\sigma(L)$ of an operator L is the set of values r for which there is no inverse to the operator

$$S_r = L - rI \tag{5.27}$$

where I is the identity operator. The spectrum is obviously analogous to the eigenvalues of a matrix, and can similarly used to characterize an operator.

This presentation almost certainly reminds one of an example of the spectral decomposition of a function – namely, the Fourier series approximating a function. In the Fourier series the functions are those periodic with a period T (or, equivalently, and without loss of generality, functions defined on the interval $[0, T]$ for functions of a single variable; the extension to more variables is obvious). The basis functions are the sine and cosine functions with appropriate wavelengths, $\sin\left(\frac{2\pi k}{T}t\right)$ and $\cos\left(\frac{2\pi k}{T}t\right)$ for the integers k and variable t. In this case the operator with the appropriate eigenfunctions is simply the second derivative with respect to t. The coefficients found for equation 5.24 are exactly those on expects from the Fourier series.

5.4 Minimizers

Methods of minimizing functionals are obviously useful. We need some such method to find an energy minimum or to calculate a partition function. So we will set up some useful tools.

Definition 1. *A function f with domain some topological space*[8] *X and range the real numbers is called lower semicontinuous if and only if the sequence x_j converging to x implies that $f(x) \leq \lim_{j\to\infty} f(x_j)$.*

We use this definition right away to argue a theorem about minimizers:

Theorem 8. *If a lower semicontinuous function f defined on a compact space*[9] *X is bounded from below, then it takes on its minimum value.*

[8] A set X and a collection of open subsets T of elements of X is a **topological space** by definition when the null set is in T; the set X is in T; all the intersections of finitely many sets from T are in T; and all the unions of finitely many sets from T are in T.

[9] A compact space is a topological space for which every set X that is the union of any number of open sets can be covered by a finite number of those sets. For example, any closed and bounded subset of $\Re^n$ is a compact space.

Proof. Let $A = \inf f(x) > -\infty$, since f is bounded below. Now let $\{x_j\}_{j=1}^{\infty}$ be a minimizing sequence in the compact space X, that is,

$$f(x_j) \to A \text{ as } j \to \infty. \tag{5.28}$$

Since X is compact, there is a subsequence $\{x_{n_k}\}_{k=1}^{\infty}$ of $\{x_j\}_{j=1}^{\infty}$ with the property that $x_{n_k} \to x' \in X$ as $k \to \infty$. The lower semicontinuity of f now implies that

$$f(x') \le \lim_{k\to\infty}\inf f(x_{n_k}) = \lim_{j\to\infty} f(x_j) = A \tag{5.29}$$

but

$$f(x') \ge A \tag{5.30}$$

Therefore

$$f(x') = A. \tag{5.31}$$

□

As Hilbert spaces are a subset of topological spaces this theorem extends immediately to functionals. A lower semicontinuous functional therefore has a minimum.

Another definition which will be of use later is of convexity:

Definition 2. *A function f on a vector space V is convex if and only if*

$$f(\alpha x + (1-\alpha)y) \le \alpha f(x) + (1-\alpha) f(y) \tag{5.32}$$

for all x and y in V, and for any $\alpha \in [0,1]$.

As Hilbert spaces are vector spaces as well this definition automatically extends to functionals.

5.5 Fourier transforms

The **dual** is a much-used mathematical term; what we mean in this context is a set derived from an original set of interest by some (reversible) rule. Most often by the dual to a vector space V we mean the set of linear functions which map V to the real numbers. A "point" in the dual space corresponds not to a point in the original, but to a function on the original space. The analytic advantage to a dual space is that often problems difficult in one space are simple in the dual; we will see that the partition functions we want to evaluate are much simpler in a transformed space than in the original.

This is introduced as background to Fourier transforms. In the Fourier transform we examine a vorticity field $\omega(\mathbf{x})$ not as the site values, but as the spectrum of wave functions which added together approximate the original function. The transformed function $\hat{\omega}(\mathbf{k})$ is a function of the **wave-number** $\mathbf{k}$, which generalizes the frequency of a wave in ways which make it more convenient for multidimensional wave functions. The value of $\hat{\omega}(\mathbf{k})$ is the amplitude, the "strength" of that component of the wave.

The Fourier transform is the extension of a Fourier series to a system which is not periodic (or which has a period infinitely long), as opposed to one with finite period in time or in space. For a single-variable function $f(x)$ its Fourier transform $\hat{f}(k)$ is

$$F[f] = \hat{f}(k) = \int_{-\infty}^{\infty} f(x) e^{2\pi i k x} dx \tag{5.33}$$

and its inverse is

$$F^{-1}[f] = f(t) = \int_{-\infty}^{\infty} \hat{f}(k) e^{2\pi i k x} dk \tag{5.34}$$

Notice that we are using the transform with the exponential raised to $2\pi i k x$. It is common to use an alternate form in which the exponential is raised to ikx. That alternative has the effect of requiring the integrals to be scaled by constants. One of those constants scales the forward and another scales the inverse transforms and, together, these constants must have a product of $\frac{1}{2\pi}$. These constants are not interesting in themselves, but are a convenient source of scaling errors when one uses multiple books as references. By using an exponential factor of $2\pi i k x$ we avoid the scaling altogether.

This transform is linear; and any function $f(x)$ has Fourier transforms (and inverse transforms) provided there are at most finitely many discontinuities, that $\int_{-\infty}^{\infty} |f(x)| dx$ exists, and that the function satisfies the **Lipschitz**[10] **condition**

$$|f(x+h) - f(x)| \leq B |h|^{\beta} \tag{5.35}$$

for all x, for all h smaller than some margin ϵ, and for some positive B and β. (An alternate test one can use is whether $f(x)$ is of **bounded variation**, that is, if for all finite a and b there is an upper bound

$$\sum_{j=1}^{n} |f(x_j) - f(x_{j-1})| < M \tag{5.36}$$

for some constant M and for all collections of n points $a = x_0 < x_1 < x_2 < \cdots < x_n = b$.)

[10] Rudolf Otto Sigismund Lipschitz, 1832 - 1903, is best known for that condition, which in another guise guarantees the uniqueness of a solution to the differential equation $y' = f(t, y)$. He studied quaternions and rediscovered Clifford algebras, introducing in the process the spin groups; and his study of mechanics in differential geometry lead to special relativity. [288]

The Fourier transform of functions is linear; given functions $f(x)$ and $g(x)$ and a scalar α then

$$F\left[af(x) + g(x)\right] = a\hat{f}(x) + \hat{g}(x) \tag{5.37}$$

Given the **convolution** of two functions

$$(f * g)(t) = \int_{-\infty}^{\infty} f(t)g(x-t)dx \tag{5.38}$$

we can evaluate the Fourier transform and inverse transform of functions multiplied together:

$$F\left[f(x) * g(x)\right] = \hat{f}\hat{g} \tag{5.39}$$
$$f\left[f(x)g(x)\right] = \hat{f} * \hat{g} \tag{5.40}$$
$$f(x) * g(x) = F^{-1}\left[\hat{f}\hat{g}\right] \tag{5.41}$$
$$f(x)g(x) = F^{-1}\left[\hat{f} * \hat{g}\right] \tag{5.42}$$

Of particular analytic convenience is that the Fourier transform of the derivative of a function $f'(x)$ will be – provided $\lim_{x\to\pm\infty} f(x) = 0$, which is generally expected in functions modelling physical properties – easily found through integration by parts:

$$F\left[f'(x)\right] = \int_{-\infty}^{\infty} \frac{df}{dx} e^{2\pi ikx} dx \tag{5.43}$$
$$= e^{2\pi ikx} f(x)|_{x=-\infty}^{\infty} - 2\pi ik \int_{-\infty}^{\infty} f(x)e^{2\pi ikx} dx \tag{5.44}$$
$$= 0 - 2\pi ik \int_{-\infty}^{\infty} f(x)e^{2\pi ikx} dx \tag{5.45}$$
$$= -2\pi ik\hat{f} \tag{5.46}$$

This – assuming the derivatives of $f(x)$ vanish similarly – generalizes to

$$F\left[f^{(n)}(x)\right] = (-2\pi ik)^n \hat{f} \tag{5.47}$$

This points to the common use of Fourier series and transforms generally in solving differential equations, as they turn linear differential equations into polynomial equations.

In two dimensions the wave-number becomes a vector $\mathbf{k} = (k_1, k_2)$ with each component corresponding to the frequency in one direction. The Fourier transform of a function $f(x, y)$ becomes

$$F[f] = \hat{f}(k_1, k_2) = \int_{-\infty}^{\infty}\int_{-\infty}^{\infty} f(x, y)e^{2\pi i(k_1 x + k_2 y)} dx dy \tag{5.48}$$
$$F[f] = \hat{f}(\mathbf{k}) = \int_{-\infty}^{\infty}\int_{-\infty}^{\infty} f(\mathbf{x})e^{2\pi i\mathbf{k}\cdot\mathbf{x}} d\mathbf{x} \tag{5.49}$$

for vector $\mathbf{x}$. The inverse is

$$F^{-1}[\hat{f}] = f(x, y) = \int_{-\infty}^{\infty} \int_{-\infty}^{\infty} \hat{f}(k_1, k_2) e^{2\pi \mathrm{i}(k_1 x + k_2 y)} dk_1 dk_2 \quad (5.50)$$

$$F^{-1}[\hat{f}] = f(\mathbf{x}) = \int_{-\infty}^{\infty} \int_{-\infty}^{\infty} \hat{f}(\mathbf{k}) e^{2\pi \mathrm{i}\mathbf{k}\cdot\mathbf{x}} d\mathbf{k} \quad (5.51)$$

which form also gives us the generalized Fourier transform for an arbitrarily large number of dimensions.

Useful as Fourier transforms are for analytic study we will also want to examine such transforms on a mesh of finitely many points. This produces the **discrete Fourier transform**. Given a uniform mesh of points in a single variable x_0, x_1, x_2, $\cdots$, x_{n-1}, and letting the constant $f_j = f(x_j)$ for each j we define the discrete Fourier transform to be

$$\hat{f}_k = \sum_{j=1}^{N-1} f_k e^{-2\pi \mathrm{i} \frac{jk}{N}} \quad (5.52)$$

and the inverse

$$f_k = \frac{1}{N} \sum_{j=1}^{N-1} \hat{f}_k e^{2\pi \mathrm{i} \frac{jk}{N}} \quad (5.53)$$

The extension to a grid of regularly spaced points in several dimensions, for a point $\mathbf{x}$ and a wave vector of as many components $\mathbf{k}$, has the same form:

$$\hat{f}_{\mathbf{k}} = \sum_{\mathbf{j}=1}^{N-1} f_{\mathbf{j}} e^{-2\pi \mathrm{i} \frac{\mathbf{j}\cdot\mathbf{k}}{N}} \quad (5.54)$$

$$f_{\mathbf{k}} = \frac{1}{N} \sum_{\mathbf{k}=1}^{N-1} \hat{f}_{\mathbf{k}} e^{2\pi \mathrm{i} \frac{\mathbf{j}\cdot\mathbf{k}}{N}} \quad (5.55)$$

As with the continuous Fourier transform the discretized version presents a function which values correspond to the amplitude of a wave function with the matching frequency. On any particular mesh there is a maximum wave-number we can use; it corresponds to the wavelength which is twice the shortest distance between any two points. Higher frequencies exist, but correspond to higher harmonics of the frequencies already represented, and there is no way to solve uniquely for these components. The least troublesome treatment of them is therefore to ignore the higher wave-numbers; this corresponds to treating only the wave-numbers k less than some k_{max}, or to wave vectors $\mathbf{k}$ of norm less than some k_{max}.

With regular meshes we are also able to take advantage of the **fast Fourier transform,** a method of calculating the discrete Fourier coefficients at a considerable saving of computational cost. This algorithm is too involved to discuss here, particularly as our main interest will be in irregular, Monte Carlo-generated meshes, but it is easy to find described in texts about numerical computations.

Finding Fourier transforms for irregular meshes can be done, though we lose computationally efficient formulas available to regular meshes. We will not need to do much of this. Our particular interests are in whether the statistical equilibrium spectrum is strongest at the longest or the shortest wavelengths; detailed information is not particularly useful.

5.6 Spherical Harmonics

Our repeated use of the Laplace-Beltrami operator – or the inverse – in discretizing the continuous vorticity field to a discretized version recommends that choice to us. The eigenfunctions of the Laplace-Beltrami operator will depend on the domain, as the fundamental solutions do.

On the bounded plane – we will take the unit square $[0, 1] \times [0, 1]$ – the eigenfunctions of the Laplace-Beltrami operator are

$$\phi_{j,k}(\mathbf{x}) = C_{j,k} \sin(j\pi x) \sin(k\pi y) \tag{5.56}$$

where the Cartesian coordinates of $\mathbf{x}$ are (x, y) and j and k are positive integers. $C_{j,k}$ is a normalization constant. That we have two indices is slightly notationally different from our original base functions ϕ_j but does not substantially affect the summation or the argument. The extension of this summation to rectangles (with sides parallel to the x and y axes) is straightforward. The eigenvalues for each $\phi_{j,k}$ are equal to $j^2 k^2$.

What the eigenfunctions are on the unbounded plane is a deeper question and one we put off until after considering the unit sphere. On the unit sphere the eigenfunctions are the set of spherical harmonics:

$$\phi_{j,k} = Y_j^k(\mathbf{x}) = \sqrt{\left(\frac{2j+1}{4\pi}\right)\left(\frac{(j-k)!}{(j+k)!}\right)} P_j^k(z) e^{ik\theta} \tag{5.57}$$

where j is a positive integer and k is an integer between $-j$ and j. The functions P_j^k are the **associated Legendre[11] polynomials**, defined as

$$P_j^k(z) = \frac{(-1)^k}{2^j \cdot j!} (1 - z^2)^{\frac{k}{2}} \frac{d^{j+k}}{dz^{j+k}} \left(z^2 - 1\right)^j \tag{5.58}$$

[11] Adrien-Marie Legendre, 1752 - 1833, was one of the commitee which designed the Metric System. He was the first person to prove that π^2 is irrational. The Legendre functions he introduced in order to study the attraction on a point caused by an ellipsoid. [288]

(possibly with an additional scaling factor of $(-1)^k$, which is known as the **Condon**[12]**-Shortley**[13] **phase,** included or omitted at the preference of the author).

Though the symbols look intimidating the harmonics are not. The first several examples are

$$Y_0^0(z,\theta) = \frac{1}{2}\frac{1}{\sqrt{\pi}} \tag{5.59}$$

$$Y_1^{-1}(z,\theta) = \frac{1}{2}\sqrt{\frac{3}{2\pi}}\sqrt{1-z^2}e^{-i\theta} \tag{5.60}$$

$$Y_1^0(z,\theta) = \frac{1}{2}\sqrt{\frac{3}{\pi}}z \tag{5.61}$$

$$Y_1^1(z,\theta) = -\frac{1}{2}\sqrt{\frac{15}{2\pi}}\sqrt{1-z^2}e^{i\theta} \tag{5.62}$$

$$Y_2^{-2}(z,\theta) = \frac{1}{4}\sqrt{\frac{15}{2\pi}}\left(1-z^2\right)e^{-2i\theta} \tag{5.63}$$

$$Y_2^{-1}(z,\theta) = \frac{1}{2}\sqrt{\frac{15}{2\pi}}\sqrt{1-z^2}ze^{-i\theta} \tag{5.64}$$

$$Y_2^0(z,\theta) = \frac{1}{4}\sqrt{\frac{5}{\pi}}\left(3z^2-1\right) \tag{5.65}$$

$$Y_2^1(z,\theta) = -\frac{1}{2}\sqrt{\frac{15}{2\pi}}\sqrt{1-z^2}ze^{i\theta} \tag{5.66}$$

$$Y_2^2(z,\theta) = \frac{1}{4}\sqrt{\frac{15}{2\pi}}\left(1-z^2\right)e^{2i\theta} \tag{5.67}$$

While we write these spherical harmonics using complex numbers, the functions we approximate have only real components. We can treat this either by accepting complex-valued coefficients or by considering only the real parts of our functions Y_j^k.

A spherical harmonic is termed **zonal** if it fluctuates only in the latitudinal direction. The **tesseral** harmonics are those which fluctuate a different number of times in the latitudinal and longitudinal directions. The **sectorial** harmonics have the same number of fluctuations in latitudinal and in longitudinal directions.

[12] Edward Uhler Condon, 1902 - 1974, was an early master of quantum mechanics, examining alpha-particle tunnelling and proton-proton scattering and writing the era's definitive text on atomic spectra. The House Committee on Un-American Activities once asked if his leadership in the scientific revolution of quantum mechanics indicated a willingness to lead social revolution. [382] [435]

[13] George Hiram Shortley, born 1910. [435]

6

Discrete Models in Fluids

6.1 Introduction

With some basis now in statistical mechanics, we turn to fluid mechanics. We want a representation of a fluid's flow compatible with our Monte Carlo tools. We will develop these along two lines of thought. Our first is to represent the vorticity field of a fluid flow as a collection of point particles. This is the **vortex gas model**, a dynamical system we can treat just as we do any ordinary problem of mechanics. In the next chapter we will create the lattice gas model, a system based on approximating the vorticity field with a function continuous almost everywhere.

The vortex gas model introduces a particle discretization of the fluid, which produces an N-body Hamiltonian problem. Representing the fluid motion as a field of vorticity evokes gravitational and electrical fields, which we are accustomed to treating as collections of discrete vortex particles. So we represent the vorticity field, and thus the fluid velocity, as a set of point particles, each with a time-invariant "charge" called the vorticity.

Our discretization produces a Hamiltonian problem, so we may apply deterministic methods such as a Runge[1]-Kutta[2] symplectic integrator and derive molecular dynamics [324] [325]. It is different from standard dynamics problems: the velocities of these particles are not part of the Hamiltonian, and are not of any particular interest. Instead of position and momentum being conjugate variables we find the different directions of space are conjugate, with momentum never appearing. Nevertheless the mechanisms of studying Hamiltonian problems apply.

---

[1] Carle David Tolmé Runge, 1856 - 1927, was close friends with Max Planck, and outside his pure mathematics research sought to explain the spectral lines of elements besides hydrogen. [288]

[2] Martin Wilhelm Kutta, 1867 - 1944, is also renowned for the Zhukovski-Kutta airfoil and the Kutta-Joukowski theorem which describes the lift on an airfoil. [288]

6.2 Euler's Equations for Fluid Flows

We have already outlined the origins of our equation describing the fluid flow, by considering the fluid velocity at any point $\mathbf{u}$:

$$\frac{\partial}{\partial t}(\rho \mathbf{u}) = -\rho(\mathbf{u} \cdot \nabla)\mathbf{u} \tag{6.1}$$

Though this equation is its most common representation, we find it more convenient to rewrite Euler's equation as an interaction of vorticity and of stream functions.

Now we will rewrite it as an interaction of vorticity and of stream functions. The vorticity form is well-suited to statistical mechanics treatments. From the observation that the curl of a vector field corresponds to the rotation of that field – the speed with which a small disc at that spot would rotate – we let the vorticity $\boldsymbol{\omega}$ be represented as

$$\boldsymbol{\omega} = \nabla \times \mathbf{u} \tag{6.2}$$

Here we must make a decision that further limits the generality of our problem. We will consider only two-dimensional fluid flows for the moment. Three-dimensional fluid flows are quite interesting but fall outside the scope of our book. A problem like the ocean or the atmosphere of a planet we consider to be two-dimensional; to first approximation the altitude is negligible compared to the latitudinal and longitudinal widths. Later we will make things slightly more sophisticated without becoming too much more complex by creating a 2.5-dimensional problem. The "half dimension" reflects that we will let a property (like vorticity) depend on altitude at a single point, but treat longitudinal and latitudinal variations in that property as depending on the mean along an entire column of altitude at a point, rather than using the full complexity of three-dimensional dynamics.

For a two-dimensional problem, we define a stream function ψ to be the potential for the fluid velocity, the gradient with a transposition of element positions:

$$\mathbf{u} = \nabla^t \psi \tag{6.3}$$

This useful formulation relates the stream function to the vorticity through a Poisson[3] equation, $\nabla^2 \psi = -\omega$.

ψ can be found by making use of a Green's function for the Laplacian operator. This useful form relates the stream function to the vorticity through a Poisson equation,

[3] Siméon Denis Poisson, 1781 - 1840, studied potential theory, elasticity, electricity, and differential equations. He is quoted as saying, "Life is good for only two things: to study mathematics and to teach it." He introduced the expression "Law of Large Numbers". [288]

$$\nabla^2 \psi = \omega \tag{6.4}$$

ψ can be solved for by a Green's function for the Laplace-Beltrami operator: given $G(\mathbf{x}, \mathbf{y})$ then for any given ω

$$\psi(\mathbf{x}) = \int G(\mathbf{x}, \mathbf{y}) \omega(\mathbf{y}) d\mathbf{y} \tag{6.5}$$

for an integration over the entire domain.Though a different function needs to be used for each domain, there are many interesting domains for which the Green's functions are known and so for which the methods outlined in this book can study [324].

Equation 1.10 can now be written

$$\frac{D}{Dt}\omega = 0 \tag{6.6}$$

In chapter 1 we mentioned three physical properties used to construct our model, and then built the equation describing inviscid fluid flow with only the conservation of mass and the conservation of momentum; our recasting of this equation used no new physics properties, merely our understanding of vorticity and the curl of a velocity being related.

We complete the derivation of our model by introducing the third property, kinetic energy. If the fluid's motion is described by $\mathbf{u}(\mathbf{x}, t)$ and its density is constant, then

$$E = \frac{1}{2}\rho \int_{\Omega} |\mathbf{u}(\mathbf{x}, t)|^2 \, d\mathbf{x} \tag{6.7}$$

$$= \frac{1}{2}\rho \int_{\Omega} |\nabla \phi|^2 \, d\mathbf{x} \tag{6.8}$$

which can be integrated by parts to provide

$$E = -\frac{1}{2}\rho \int_{\Omega} \phi \Delta \phi d\mathbf{x} + \frac{1}{2}\rho \int_{\Omega} \nabla \cdot (\phi \nabla \phi) \, d\mathbf{x} \tag{6.9}$$

The density of the fluid is constant, and we may assume without loss of generality $\rho = 1$. We do this, and leave ρ out of the rest of this derivation.

Next we observe $\omega = -\Delta\phi$. By Stokes's theorem the second integral in the above step equals an integral over the boundary $\partial\Omega$ instead. Letting $d\mathbf{S}$ represent the differential over the boundary then

$$E = \frac{1}{2} \int_{\Omega} \phi \omega d\mathbf{x} + \frac{1}{2} \int_{\partial\Omega} (\phi \boldsymbol{\nabla} \boldsymbol{\phi}) \cdot d\mathbf{S} \tag{6.10}$$

The rightmost integral is zero: $\phi \boldsymbol{\nabla} \boldsymbol{\phi}$ vanishes on the boundary of the domain Ω.

As $\omega = \Delta\phi$, then $\phi = \Delta^{-1}\omega$, with Δ^{-1} the inverse Laplace-Beltrami operator. This operator depends on the domain Ω – and from it we will get the different interaction potentials. Replacing ϕ with this inverse operator we have

$$E = \frac{1}{2}\int_{\Omega} \left(\Delta^{-1}\omega\right)\omega d\mathbf{x} \tag{6.11}$$

$$= -\frac{1}{2}\int_{\Omega} \omega\Delta^{-1}\omega d\mathbf{x} \tag{6.12}$$

Now we discretize the vorticity. We use the **Dirac delta function** $\delta(\mathbf{z})$, which is zero for every point not the origin, but is sufficiently large at the origin that the integral over any domain containing the origin is one.

Let there be N discrete particles, each at the position $\mathbf{z}_j$ with vorticity s_j. We approximate $\omega(\mathbf{z}, t)$ with the summation

$$\omega(\mathbf{z}, t) = \sum_{j=1}^{N} s_j \delta\left(\mathbf{z} - \mathbf{z}_j\right) \tag{6.13}$$

Now substitute equation 6.13 into equation 6.12. The kinetic energy is

$$E = -\frac{1}{2}\int_{\Omega} \left(\sum_{j=1}^{N} s_j \delta\left(\mathbf{z} - \mathbf{z}_j\right)\right) \Delta^{-1} \left(\sum_{k=1}^{N} s_k \delta\left(\mathbf{z} - \mathbf{z}_k\right)\right) d\mathbf{z} \tag{6.14}$$

By the definition of the Dirac delta function this simplifies to

$$E = -\frac{1}{2}\sum_{j=1}^{N} s_j \Delta^{-1} \left(\sum_{k=1}^{N} s_k \delta\left(\mathbf{z}_j - \mathbf{z}_k\right)\right) \tag{6.15}$$

To continue we need the solution to the inverse Laplace-Beltrami operator on a Dirac delta function, called the fundamental solutionor the Green's function on the domain Ω. We label it $G(\mathbf{z}_k, \mathbf{z}_j)$; the kinetic energy is

$$E = -\frac{1}{2}\sum_{j=1}^{N} s_j \sum_{k=1}^{N} s_k G\left(\mathbf{z}_k, \mathbf{z}_j\right) \tag{6.16}$$

$$= -\frac{1}{2}\sum_{j=1}^{N}\sum_{k=1}^{N} s_j s_k G\left(\mathbf{z}_k, \mathbf{z}_j\right) \tag{6.17}$$

(with the note $G(\mathbf{z}, \mathbf{z}) = 0$, so a particle does not interact with itself).

Finding a Green's function is often challenging. But for many common domains it is already known. On the infinite, unbounded plane, the function is

$$G(\mathbf{z}_k, \mathbf{z}_j) = \frac{1}{2\pi} \log\left|\mathbf{z}_k - \mathbf{z}_j\right| \tag{6.18}$$

where $|\mathbf{z}_k - \mathbf{z}_j|$ is the Euclidean distance between the two points $\mathbf{z}_k$ and $\mathbf{z}_j$. On the sphere of radius R the operator becomes [324]

$$G(\mathbf{z}_k, \mathbf{z}_j) = \frac{1}{4\pi}\left(\log\left(R^2 - \mathbf{z}_k \cdot \mathbf{z}_j\right) - \log(R^2)\right) \tag{6.19}$$

and note again $R^2 - \mathbf{z}_k \cdot \mathbf{z}_j$ is the distance between points j and k.

Substituting the Green's function of equation 6.18 into the kinetic energy of the fluid from equation 6.17 we get this energy of the N vorticity particles:

$$E = -\frac{1}{4\pi}\sum_{j=1}^{N}\sum_{k\neq j}^{N} s_j s_k \log|\mathbf{z}_k - \mathbf{z}_j| \tag{6.20}$$

(with the second summation indicates k running from 1 to N except for j). On the sphere of radius R, the energy becomes

$$E = -\frac{1}{8\pi R^2}\sum_{j=1}^{N}\sum_{k\neq j}^{N} s_j s_k \left(\log\left(R^2 - \mathbf{z}_k \cdot \mathbf{z}_j\right) - \log(R^2)\right) \tag{6.21}$$

The constant term in equation 6.21 has no effect on the dynamics, and is often overlooked. This reflects a problem of notation which may confuse one reading papers on vortex dynamics: constant terms or multiples may be included or omitted at the convenience of the author. Conventions must be checked before results from different sources are compared.

The result of these transformations is a Hamiltonian function. The tools this fact offers us – and the evidence this is indeed a Hamiltonian – require some exposition.

6.3 N-body Hamiltonians

Hamiltonian functions are convenient ways to study many dynamics problems. They derive from systems with a potential and a kinetic energy dependent on the positions $q^1, q^2, q^3, \cdots, q^N$ and momenta $p_1, p_2, p_3, \cdots, p_N$ of N particles, and (possibly) the time t.

The Hamiltonian $H(t, q^1, q^2, q^3, \cdots, q^N, p_1, p_2, p_3, \cdots, p_N)$ is the total energy of the system, and the evolution of positions and momenta is governed by these elegant near-symmetric equations [146] [295] [397]:

$$\frac{dq^j}{dt} = \frac{\partial H}{\partial p_j} \tag{6.22}$$

$$\frac{dp_j}{dt} = -\frac{\partial H}{\partial q^j} \tag{6.23}$$

These are Hamilton's equations. We find it useful to examine the phase space, the set of all possible values of position and momentum which the particles in the system may have. The time-evolution of the system traces out of a curve – the **orbit** or **trajectory** – in phase space. If the Hamiltonian is independent of time, then the energy is constant and the trajectory is a level curve – the set of points satisfying $H(q^1, q^2, q^3, \cdots, q^N, p_1, p_2, p_3, \cdots, p_N) = C$. [146] [295]

We view motion now not as a set of particles responding to forces, but as the parametrized tracing out of a curve, with time the parameter. We can pick the positions and momenta independently of one another; it simply selects a new trajectory. Satisfying an initial condition is just verifying our initial coordinates are on the level curve we choose. We can even be casual about the dimensions of space; in the Hamiltonian we can write the x, y, and z coordinates of a particle as q^1, q^2, and q^3, and the corresponding components of momentum as p_1, p_2, and p_3. The values of positions do determine the evolution of the momenta, and vice-versa: each q^j is **conjugate** to the corresponding p_j [146] [295].

Though the Hamiltonian form offers a correct set of differential equations describing a problem its great utility is encouraging the habit of considering entire trajectory of a problem, or in comparing different trajectories. (This is the philosophy describing, for example, a satellite's motion based on whether it has enough energy for an elliptic or a circular orbit, or whether it has escape velocity.) It also examines the geometry of a system – the symmetries of its particles and of the interaction laws – which proves to be an enormously fruitful approach.

We turn to vortices of uniform strength on the plane, as described by equation 6.20. We rewrite vector $\mathbf{z}_j$ as the complex number $p_j + \mathrm{i}q^j$. Each p_j and q^j are a **conjugate pair** (to be explained shortly); so, for each of the N particles

$$\frac{dq^j}{dt} = \frac{\partial H}{\partial q^j} \tag{6.24}$$

$$\frac{dp_j}{dt} = -\frac{\partial H}{\partial p_j} \tag{6.25}$$

which can also be written more compactly as

$$\frac{dz_j}{dt} = -\mathrm{i}\frac{\partial H}{\partial \bar{z}_j} \tag{6.26}$$

where the derivative with respect to the complex variable $\bar{z}$ means $\left(\frac{\partial}{\partial q} - \mathrm{i}\frac{\partial}{\partial p}\right)$. The norm of the distance $\mathbf{z}_j - \mathbf{z}_k$ is the Euclidean distance between these points, $\sqrt{(p_j - p_k)^2 + (q^j - q^k)^2}$. So the evolution of position in time is

$$\frac{dp_j}{dt} = -\sum_{k \neq j}^{N} \frac{1}{|z_j - z_k|}(q^j - q^k) \tag{6.27}$$

$$\frac{dq^j}{dt} = -\sum_{k \neq j}^{N} \frac{1}{|z_j - z_k|}(p_j - p_k) \tag{6.28}$$

At this point it appears our intuition of what our symbols mean has diverged from what we are modelling. The understanding of q as position and p as momentum clashes with a position of $q + ip$. It seems to call a dimension of space a momentum, and choosing which dimension is which seems to tie the dynamics to the basis vectors one uses to place the q and p axes. How can we reconcile our symbols to the problem we mean to model?

The answer is that conjugate variables are part of the structure of the phase space of a problem. In traditional mechanical problems this structure makes the conjugate variables correspond to positions and momenta, but that is our interpretation. It is not an inherent property of the variables [397]. Were we to replace each q with P and replace each p with $-Q$ we would still have a set of Hamiltonian equations, and nothing in the equations would indicate we were "incorrectly" reading what is "really" position as momentum and what is "really" momentum as position [146].

There is another minor divergence between what we wrote and what we mean by it. The N-body Hamiltonian is a representation of the kinetic energy of a fluid (equation 6.8 et seq). It must always be a positive number (or, if the fluid does not move, zero). But an N-body system can easily have negative energy. Why does negative energy occur, and why does it not reflect a problem in our derivation?

Equations 6.27 and 6.28 apply to vortices of uniform strength. If vortex j is at the coordinates (x_j, y_j) and has strength s_j then the conjugate pair are [46]

$$q^j = \sqrt{|s_j|} sign(s_j) x_j \tag{6.29}$$

$$p_j = \sqrt{|s_j|} sign(s_j) y_j \tag{6.30}$$

with the Hamiltonian [324]

$$H(\mathbf{q}, \mathbf{p}) = -\frac{1}{8\pi} \sum_{j=1}^{N} \sum_{k \neq j}^{N} s_j s_k \log\left(q^k - p_j\right) \tag{6.31}$$

which is like scaling the variables in an ordinary mechanics problem so the masses are all 1. It is common to make strength "part" of the variables.

On the sphere of radius R, again writing positions as $\mathbf{x}_j$ with potential s_j the potential energy is

$$H = -\frac{1}{8\pi R^2} \sum_{j=1}^{N} \sum_{k=1}^{N} s_j s_k \log(R^2 - \mathbf{x}_j \cdot \mathbf{x}_k) \tag{6.32}$$

The derivative of the position $\mathbf{x}_j$ can be written compactly as

$$\frac{d}{dt}\mathbf{x}_j = -\frac{1}{8\pi R}\sum_{k\neq j} s_j \frac{\mathbf{x}_j \times \mathbf{x}_k}{R^2 - \mathbf{x}_j \cdot \mathbf{x}_k} \tag{6.33}$$

Writing positions in terms of the longitude ϕ and co-latitude θ the motion of particle j is described by

$$\frac{d\phi_j}{dt} = -\frac{1}{8\pi R}\sum_{k\neq j}^{N} s_k \frac{\sin(\theta_k)\sin(\phi_j - \phi_k)}{1 - \cos(\gamma_{j,k})} \tag{6.34}$$

$$\sin(\theta_j)\frac{d\phi_j}{dt} = -\frac{1}{8\pi R}\sum_{k\neq j}^{N} s_k \times \frac{\sin(\theta_j)\cos(\theta_k) - \cos(\theta_j)\sin(\theta_k)\cos(\phi_j - \phi_k)}{1 - \cos(\gamma_{j,k})} \tag{6.35}$$

where $\gamma_{j,k} = \cos(\theta_j)\cos(\theta_k) + \sin(\theta_j)\sin(\theta_k)\cos(\phi_j - \phi_k)$ – the angular separation between points j and k.

We may change to a symplectic pair of coordinates [46] [295] [324] defined by

$$p_j = \sqrt{|s_j|}\cos(\theta_j) \tag{6.36}$$

$$q^j = \sqrt{|s_j|}\phi_j \tag{6.37}$$

which recasts the problem into a set of Hamiltonian equations, where

$$H = -\frac{1}{8\pi R^2}\sum_{j=1}^{N}\sum_{k\neq j}^{N} s_j s_k \log(d_{j,k}^2) \tag{6.38}$$

with $d_{j,k}$ the distance between $\mathbf{x}_j$ and $\mathbf{x}_k$.

6.4 Symplectic Variables

We ask how to recognize the conjugate variables in a problem, and how to test whether a function is Hamiltonian. This requires background in linear algebras and in vector functions. We will have to find first a **symplectic manifold**, the pairing of conjugate coordinates, and then find an energy function which is Hamiltonian.

The components of a symplectic manifold are a vector space – the coordinates describing the system; and a **differential two-form** – describing what pairs of variables are conjugate and may be used for Hamiltonian functions. Whether a function is Hamiltonian depends on whether its differential equals a particular property of those symplectic coordinates applied to the function.

Let Z be the phase space of the problem, and suppose it can be written as the direct product $W \times W^*$, where W is some vector space and W^* is its dual. These are the part of phase space which represents the variables p and the part which represents q. The canonical symplectic form ω is

$$\omega((w_1, \alpha_1), (w_2, \alpha_2)) = \alpha_2(w_1) - \alpha_1(w_2) \tag{6.39}$$

where by $\alpha_2(w_1)$ we mean the inner product between a vector in W and one in W^*. Typically this is the dot product, but if the phase space is something like the complex numbers this product may be a different form [295].

Consider elements of $\mathbf{C}^n$, vectors of n complex-valued numbers. The inner product between two complex n-tuples z and w is

$$\langle z, w \rangle = \sum_{j=1}^{n} z_j \bar{w}_j = \sum_{j=1}^{n} (x_j u_j + y_j v_j) + \sum_{j=1}^{n} (u_j y_j - v_j x_j) \tag{6.40}$$

taking $z_j = x_j + \mathrm{i}y_j$ and $w_j = u_j + \mathrm{i}v_j$. The real part of the above is the real inner product of two n-tuples z and w; negative the imaginary part is the symplectic form – if we take $\mathbf{C}^n$ to equal $\mathbf{R}^n \times \mathbf{R}^n$, and let the x_j and u_j be in the first $\mathbf{R}^n$ and y_j and v_j the second. Notice we can view this problem either in terms of n complex-valued "positions" z_j or in terms of $2n$ real-valued coordinates p_j and q^j.

An alternate way to find ω is the sum of wedge products of differentials $dp_j \wedge dq^j$ – the differential volume for the phase space defined by coordinates p_j and q^j. It is therefore common to find in symplectic forms the **Jacobian**[4] representing a continuous transformation from $\mathbf{R} \times \mathbf{R}$ to the space (p_j, q^j). If Z is the cylinder, $S^1 \times \mathbf{R}$[5], the symplectic form is $\Omega = d\theta \wedge dp$. On the two-sphere[6], S^2, the two-form (using the coordinates ϕ for longitude and θ for co-latitude – angle from the North pole) is $\cos(\theta) d\theta \wedge d\phi$.

The **symplectic manifold** is the phase space with its symplectic form, the pairing (Z, ω). The choice of coordinates is implied by the phase space; thus we have many functions which are Hamiltonian on the same vector space.

Given a set of symplectic coordinates, how do we know whether a function is Hamiltonian? To test whether a function $H(z)$ is we need first to define the Hamiltonian vector field X_H [295]:

$$X_H(\mathbf{z}) \equiv \left(\frac{\partial H}{\partial p_j}, -\frac{\partial H}{\partial q^j} \right) |_{\mathbf{z}} \tag{6.41}$$

$$= J \nabla H|_{\mathbf{z}} \tag{6.42}$$

where ∇H is the gradient of H and

[4] Carl Gustav Jacob Jacobi, 1804 - 1851, was expert in many fields, including determinants, differential equations, and number theory (he had results on cubic residues which impressed Gauss). He introduced the seminar method for teaching the latest mathematics to students. [288]

[5] The space S^1 is the unit circle – the "sphere" in two dimensions.

[6] The space S^2 is the unit sphere in three dimensions, using two coordinates.

$$J = \begin{pmatrix} 0 & I \\ -I & 0 \end{pmatrix} \tag{6.43}$$

with I the n-by-n identity matrix, and 0 the n-by-n zero matrix.

Notice there is a resemblance to the differential dH, the gradient one-form of the Hamiltonian [295]

$$dH = \frac{\partial H}{\partial p_1}dp_1 + \frac{\partial H}{\partial q^1}dq^1 + \frac{\partial H}{\partial p_2}dp_2 + \frac{\partial H}{\partial q^2}dq^2 + \cdots + \frac{\partial H}{\partial p_n}dp_n + \frac{\partial H}{\partial q^n}dq^n \tag{6.44}$$

A function H is Hamiltonian if, for all $\mathbf{v}$ and $\mathbf{z}$ in the phase space Z,

$$dH(\mathbf{z}) \cdot \mathbf{v} = \omega(X_H(\mathbf{z}), \mathbf{v}) \tag{6.45}$$

[1] [295] [397]. The function is Hamiltonian if the vector field describing its evolution in time equals the gradient with respect to the symplectic pairs of variables.

We may let the Hamiltonian depend on time. We do this by assigning a new "position" coordinate q^0 to equal t. Its conjugate p_0 is the energy [1]. In this way we do not need to treat time-dependent and time-independent systems separately; the analytic structure we have is abstract enough to address both cases. Notice if the Hamiltonian does not depend on time, then $\frac{\partial H}{\partial t} = \frac{dE}{dt} = 0$, or, energy is conserved.

Once we have these canonical coordinates we can construct the Poisson bracket for any two functions F and G; for a finite number of conjugate variables p_j and q^j, with i from 1 to n, this bracket is

$$\{F, G\} = \sum_{j=1}^{n} \left(\frac{\partial F}{\partial q^j}\frac{\partial G}{\partial p_j} - \frac{\partial F}{\partial p_j}\frac{\partial G}{\partial q^j} \right) \tag{6.46}$$

The Poisson bracket is anti-commutative: $\{F, G\} = -\{G, F\}$. If F and G commute then their Poisson bracket is zero [1] [146] [295].

If the second function is a Hamiltonian H and the first is any function on the phase space Z, then there is an interesting physical interpretation of the Poisson bracket of F and H:

$$\frac{d}{dt}F = \{F, H\} \tag{6.47}$$

The Poisson bracket of quantity F with the Hamiltonian is the change of F as the system evolves [146] [295]. If F commutes with H it is a quantity conserved by the system.

There are functions C which commute with every function F; if $\{C, F\} = 0$ for all F then C is a **Casimir**[7] **invariant**. These Casimir invariants provide

[7] Hendrik Bugt Casimir, 1909 - 2000, is best known for predicting the Casimir Effect, a consequence of virtual particles which drives together two parallel uncharged planes. [340]

a method, the **Energy-Casimir method**, for studying the stability of dynamical systems [295] [324].

The Poisson brackets also provide a compact way of writing Hamilton's equations [146]:

$$\frac{dq^j}{dt} = \{q^j, H\} = \frac{\partial H}{\partial p_j} \tag{6.48}$$

$$\frac{dp_j}{dt} = \{p_j, H\} = -\frac{\partial H}{\partial q^j} \tag{6.49}$$

Finally the Poisson brackets provide a way to determine whether a particular Hamiltonian may be solved simply by integrating with respect to time. Suppose we have several quantities A, B, and C, all conserved so $\{A, H\} = 0$ and so on. If these quantities are also conserved with respect to each other, so $\{A, B\} = 0$, $\{B, C\} = 0$, and $\{C, A\} = 0$, these conserved quantities are in **involution** with one another.

Liouville's[8] theorem tells us every constant of motion which is involutive allows us to remove one degree of freedom, and so cut a pair p_j and q^j. If we have N involutive conserved quantities we can solve up to an N-body problem by quadrature – that is, by algebraic operations, inversions of functions, and integrals of a function of a single variable.

6.5 Coordinates and Stereographic Projection

We mentioned in section 6.3 we can take a Hamiltonian with variables p_j and q^j and swap them – set for each j, $P_j = q^j$ and $Q^j = -p_j$. After this we have another Hamiltonian system with new variables P_j and Q^j. This example shows there are some changes of variables which preserve the symplectic structure and the Hamiltonian properties of a function. When does a transformation do this?

A function $F(p, q)$ which transforms the original coordinates to (P, Q) is a **canonical transformation** if it preserves the symplectic structure. It must preserve volume in phase space and must be locally **diffeomorphic** – F must have at least a local inverse and both F and its inverse are differentiable. The requirement to preserve volume means the Jacobian of the transformation has determinant equal to 1. [146] [295] The Hamiltonian in the new coordinate system is occasionally referred to as the **Kamiltonian**, as $K(t, P, Q)$ is common shorthand for the transformed Hamiltonian.

A canonical transformation also must preserve the handedness of the coordinate system, which is why a minus sign appears in swapping p_j and q^j.

[8] Joseph Liouville, 1809 - 1882, was the first person to prove the existence of a transcendental number, the number $0.110001000000000000000001 \cdots$, which has a 1 in the $n!$-th decimal place and zero otherwise. He brought the papers of Evariste Galois, 1811-1832, out of obscurity. [69] [288]

If we overlook the handedness, the transformed Hamiltonian we would derive describes motion backwards in time. While not our original problem it is still a Hamiltonian system. When energy is conserved there is no obvious difference between time running forwards and backwards.

This creates an apparent contradiction, as there are problems – such as the mixing of two initially separated gases – in which each particle obeys Hamiltonian dynamics, but in which we can tell from observing the system for a time whether it is moving forwards or backwards. The resolution of this apparent contradiction took decades to work out in its full implications, and will be addressed in another chapter on statistical mechanics [64].

Stereographic projection is a mapping from points (ϕ, θ) on the unit sphere onto the complex plane. Here ϕ represents the longitude and θ the co-latitude, the angular separation of the point from the North pole. When we take the Hamiltonian for the interaction of vortices on the unit sphere and rewrite it in terms of the new stereographic projections we have

$$r_j = \tan\left(\frac{\theta_j}{2}\right) \tag{6.50}$$

$$\phi_j = \phi_j \tag{6.51}$$

$$x_i = r_j \cos(\phi_j) \tag{6.52}$$

$$y_i = r_j \sin(\phi_j) \tag{6.53}$$

The Hamiltonian in terms of r_j and ϕ_j (and allowing once again vortices to have different strengths s_j) is

$$H_p = \frac{1}{8\pi R^2} \sum_{j=1}^{N} \sum_{k \neq j}^{N} s_j s_k \log\left(\frac{r_j^2 + r_k^2 - 2 r_j r_k \cos(\phi_k - \phi_j)}{(1 + r_j)^2 (1 + r_k)^2}\right) \tag{6.54}$$

which provides equations of motion

$$s_j \frac{d}{dt} (r_j)^2 = \left(1 - r_j^2\right)^2 \frac{\partial}{\partial \phi_j} H_p \tag{6.55}$$

$$s_j \frac{d}{dt} \phi_j = -\left(1 - r_j^2\right)^2 \frac{\partial}{\partial r_j^2} H_p \tag{6.56}$$

Since points on the plane may have their positions described by coordinates x_j and y^j, we may then write the evolution of x and y by

$$\frac{d}{dt} x_j = -\frac{(1 + r_j^2)^2}{2} \frac{\partial H_p}{\partial y^j} \tag{6.57}$$

$$\frac{d}{dt} y^j = \frac{(1 + r_j^2)^2}{2} \frac{\partial H_p}{\partial x_j} \tag{6.58}$$

or, using $z_j = x_j + \mathrm{i} y^j$, in the form

$$\frac{d}{dt}\bar{z}_j = -\mathrm{i}\frac{(1+|z_j|^2)^2}{2}\frac{\partial H}{\partial z_j} \tag{6.59}$$

Using the quantity $\sigma = \sum_{j=1}^{n} s_j$ we can write as the dynamics on the surface of the sphere

$$\frac{d}{dt}\bar{z}_j = -\mathrm{i}\frac{(1+|z_j|^2)^2}{8\pi R^2}\left[\sum_{k\neq j}^{N}\frac{s_j}{z_k - z_j} - \frac{\sigma\bar{z}_j}{1+|z_j|^2}\right] \tag{6.60}$$

As the radius of the sphere increases these should become equivalent to motion on the plane. This can be done by substituting $w = 2Rz$ [295] [324]; with this the equation of motion is rewritten

$$\frac{d}{dt}\bar{w}_j = -\mathrm{i}\frac{(4+\frac{|w_j|^2}{R^2})^2}{32\pi}\left[\sum_{k=1}^{N}\frac{s_k}{w_j - w_k} - \frac{\sigma\bar{w}_j}{R^2(r+\frac{|w_j|^2}{R^2})}\right] \tag{6.61}$$

In the limit as $R \to \infty$, this becomes – to the first order – the equations

$$\frac{d}{dt}\bar{w}_j = -\frac{\mathrm{i}}{2\pi}\sum_{k\neq j}^{N}\frac{s_k}{w_j - w_k} \tag{6.62}$$

showing this chain of transformations has not altered the underlying physics. The problems of vortex motion on the plane and the sphere are not completely different.

6.6 Dynamics on the Plane

With this much abstract discussion it is easy to lose the actual physics. The deterministic dynamics can be quite well numerically integrated, and we can develop an intuition of its workings. So we examine some behaviors of N vortices on the plane and on the sphere.

Quantities in involution to one another are useful in making problems exactly solvable. Ordinarily vortex gas problems, on the plane and the sphere, have three such quantities. Thus we can analytically solve problems with three vortices. On the plane we have three such quantities [46]: the "linear momentum" in both the x and the y directions,

$$P_x = \sum_{j=1}^{N} s_j x_j \tag{6.63}$$

$$P_y = \sum_{j=1}^{N} s_j y_j \tag{6.64}$$

and the "angular momentum",

$$L = \sum_{j=1}^{N} s_j (x_j + y_k)^2 \tag{6.65}$$

In special cases we have another involutive quantity. P_x and P_y are almost in involution with the Hamiltonian [46] and L.

$$\{P_x, P_y\} = \sum_{j=1}^{n} s_j \tag{6.66}$$

$$\{P_x, H\} = -2P_y \tag{6.67}$$

$$\{P_y, H\} = 2P_x \tag{6.68}$$

would be involutive if P_x, P_y, and $\sum_{j=1}^{n} s_j$ were all zero. In this case where total vorticity is zero and the "center of vorticity" is at the origin, we can solve by quadrature problems with four independent vortices.

That we do not have more quantities does not mean problems with more than four vortices are insoluble. These are deterministic, Hamiltonian systems and may be solved numerically or by the same analytical techniques open to any N-body problem.

We look first at the plane. Given N vortices, coordinates $z_j = x_j + \mathrm{i}y_j$, and vorticities s_j, the Hamiltonian is [46]

$$H = -\frac{1}{4\pi} \sum_{j=1}^{N} \sum_{k=1}^{N} s_j s_k \log(|z_j - z_k|) \tag{6.69}$$

and the particles follow by the equations

$$s_j \frac{dx_j}{dt} = \frac{\partial H}{\partial y_j} = -\frac{1}{4\pi} \sum_{k \neq j}^{N} s_j s_k \frac{2(y_j - y_k)}{(x_j - x_k)^2 + (y_j - y_k)^2} \tag{6.70}$$

$$s_j \frac{dy_j}{dt} = -\frac{\partial H}{\partial x_j} = \frac{1}{4\pi} \sum_{k \neq j}^{N} s_j s_k \frac{2(x_j - x_k)}{(x_j - x_k)^2 + (y_j - y_k)^2} \tag{6.71}$$

Notice we do not have an actual force here. But we can build an idea of how the particles interact by thinking of the direction particle k pushes particle j in: particle j is sent perpendicular to the line segment drawn between points j and kk, with a strength decreasing as the reciprocal of the distance between them. As viewed from above particle k, the motion of particle j is counter-clockwise if s_k is positive, and clockwise if s_j is negative. The net motion of particle j is the sum of the motions imposed by all the other particles.

Example 1. Two Equal Vortices

The simplest problem to consider is on the plane, examining the motion of two equal vortices a distance d apart. Take the positions to be (x_1, y_1) and

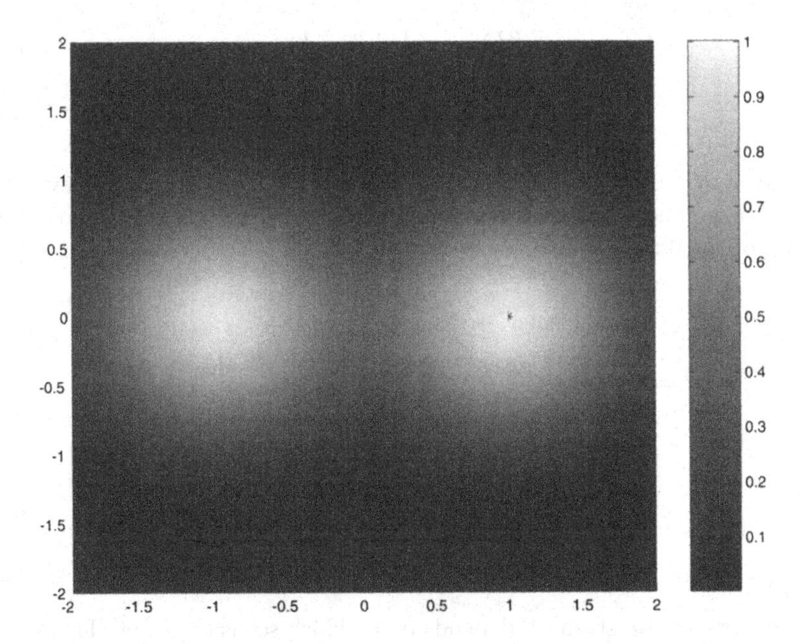

Fig. 6.1. These two vortices of equal strength will rotate around the origin at constant speed.

(x_2, y_2). Letting, without the loss of generality, both vortices have strength 1, the Hamiltonian is

$$H = -\frac{1}{4\pi}\log\left((x_2 - x_1)^2 + (y_2 - y_1)^2\right) \tag{6.72}$$

and the equations of motion are:

$$\frac{dx_1}{dt} = \frac{\partial H}{\partial y_1} = -\frac{1}{4\pi}\frac{-2(y_2 - y_1)}{(x_2 - x_1)^2 + (y_2 - y_1)^2} \tag{6.73}$$

$$\frac{dy_1}{dt} = -\frac{\partial H}{\partial x_1} = \frac{1}{4\pi}\frac{-2(x_2 - x_1)}{(x_2 - x_1)^2 + (y_2 - y_1)^2} \tag{6.74}$$

$$\frac{dx_2}{dt} = \frac{\partial H}{\partial y_2} = -\frac{1}{4\pi}\frac{2(y_2 - y_1)}{(x_2 - x_1)^2 + (y_2 - y_1)^2} \tag{6.75}$$

$$\frac{dy_1}{dt} = -\frac{\partial H}{\partial x_2} = \frac{1}{4\pi}\frac{2(x_2 - x_1)}{(x_2 - x_1)^2 + (y_2 - y_1)^2} \tag{6.76}$$

As d equals $\sqrt{(x_2 - x_1)^2 + (y_2 - y_1)^2}$ we can simplify these to

$$\frac{dx_1}{dt} = \frac{1}{2\pi}\frac{1}{d}\frac{y_2 - y_1}{d} \tag{6.77}$$

$$\frac{dy_1}{dt} = -\frac{1}{2\pi}\frac{1}{d}\frac{x_2 - x_1}{d} \tag{6.78}$$

$$\frac{dx_2}{dt} = -\frac{1}{2\pi}\frac{1}{d}\frac{y_2 - y_1}{d} \tag{6.79}$$

$$\frac{dy_2}{dt} = \frac{1}{2\pi}\frac{1}{d}\frac{x_2 - x_1}{d} \tag{6.80}$$

This is symmetrical movement. We can study it by looking at the evolution of d in time. The derivatives needed are simpler to calculate if we look at the square of the distance instead, so we will take its time derivative:

$$\frac{d}{dt}d^2 = \frac{d}{dt}((x_2 - x_1)^2 + (y_2 - y_1)^2) \tag{6.81}$$

$$= 2(x_2 - x_1)(\frac{dx_2}{dt} - \frac{dx_1}{dt}) + 2(y_2 - y_1)(\frac{dy_2}{dt} - \frac{dy_1}{dt}) \tag{6.82}$$

$$= 2(x_2 - x_1)(-\frac{1}{\pi})(\frac{y_2 - y_1}{d^2}) + 2(y_2 - y_1)(\frac{1}{\pi})(\frac{x_2 - x_1}{d^2}) \tag{6.83}$$

$$= \frac{2}{\pi}\left(-\frac{(x_2 - x_1)(y_2 - y_1)}{d^2} + \frac{(x_2 - x_1)(y_2 - y_1)}{d^2}\right) \tag{6.84}$$

$$= 0 \tag{6.85}$$

The interaction strength depends on d, which see is constant. The motion of both is perpendicular to the line segment between them. So the vortices rotate at constant speed around the midpoint.

Let θ be the angle the line segment between the points makes with the horizontal axis; $\cos\theta = \frac{1}{d}(x_2 - x_1)$. Then

$$\frac{d}{dt}(x_2 - x_1) = -\frac{1}{\pi}\frac{1}{d}\frac{y_2 - y_1}{d} \tag{6.86}$$

$$\frac{d}{dt}(y_2 - y_1) = \frac{1}{\pi}\frac{1}{d}\frac{x_2 - x_1}{d} \tag{6.87}$$

$$\frac{d}{dt}\cos(\theta) = -\frac{1}{\pi d^2}\sin(\theta) \tag{6.88}$$

$$\frac{d}{dt}\sin(\theta) = \frac{1}{\pi d^2}\cos(\theta) \tag{6.89}$$

From the last two equations θ must be linear in time; one can verify this by setting $\theta = \frac{1}{\pi d^2}t + C$, with C a constant. One can also evaluate the derivative of θ directly, and find it constant. □

Example 2. Two Unequal Vortices

Consider vortices of strength s_1 and s_2 a distance d apart. Again let positions be (x_1, y_1) and (x_2, y_2). The motions of both particles are perpendicular to the line connecting them, so the distance d is constant. The vortices (with one exceptional case) rotate at a uniform angular velocity; the algebra is substantially as above.

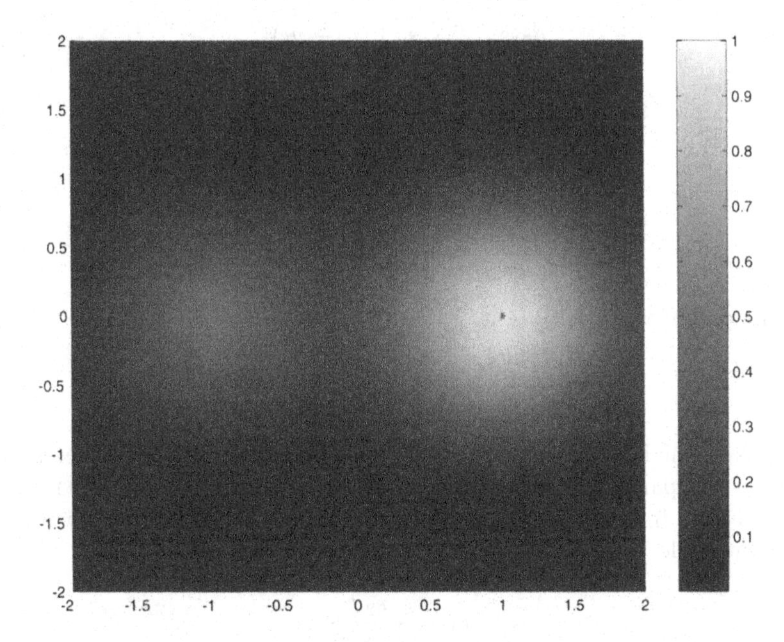

Fig. 6.2. The left vortex is a quarter the strength of the right. They will rotate around their barycenter.

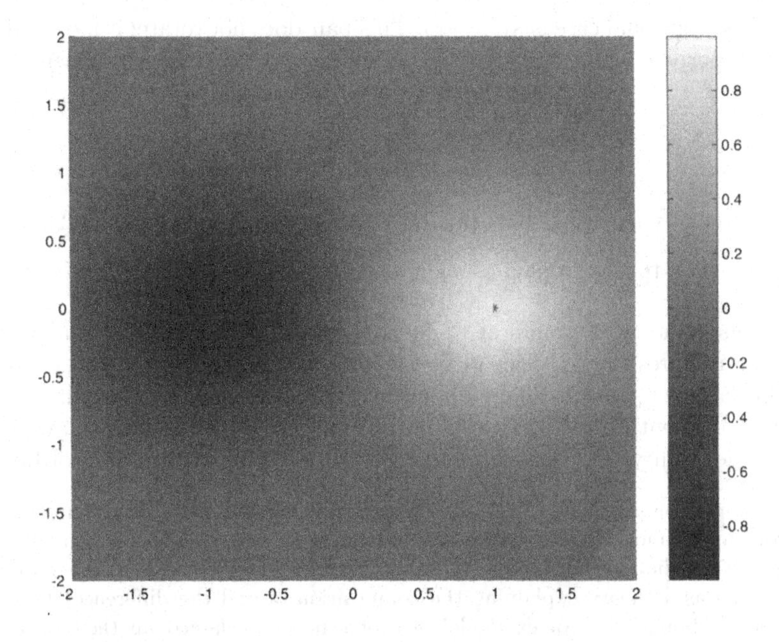

Fig. 6.3. These vortices have equal but opposite strength. They will move in a straight line perpendicular to the axis connecting them.

$$\frac{ds_1 x_1}{dt} = \frac{s_1 s_2}{2\pi} \frac{1}{d} \frac{(y_2 - y_1)}{d} \tag{6.90}$$

$$\frac{ds_1 y_1}{dt} = -\frac{s_1 s_2}{2\pi} \frac{1}{d} \frac{(x_2 - x_1)}{d} \tag{6.91}$$

$$\frac{ds_2 x_2}{dt} = -\frac{s_1 s_2}{2\pi} \frac{1}{d} \frac{(y_2 - y_1)}{d} \tag{6.92}$$

$$\frac{ds_2 y_2}{dt} = \frac{s_1 s_2}{2\pi} \frac{1}{d} \frac{(x_2 - x_1)}{d} \tag{6.93}$$

The center of vorticity is at

$$x_c = \frac{s_1 x_1 + s_2 x_2}{s_1 + s_2} \qquad y_c = \frac{s_1 y_1 + s_2 y_2}{s_1 + s_2} \tag{6.94}$$

which is on the line connecting the two particles, thought it is not necessarily between the particles. Its position is, with one exception, is fixed in time, which follows from its definition. Measured from this center the two vortices trace out circles (except in one case) of radii

$$r_1 = \frac{s_2 d}{s_1 + s_2} \tag{6.95}$$

$$r_2 = \frac{s_1 d}{s_1 + s_2} \tag{6.96}$$

The exceptional case is $s_1 = -s_2$. This pair does not rotate; it moves along the line perpendicular the segment connecting the two, at a uniform linear velocity of

$$\frac{s_2}{2\pi d} \tag{6.97}$$

The center of vorticity bisects the line segment between the points [226]. □

Example 3. A Regular Polygon

The symmetry of points of uniform strength at the vertices of a regular polygon requires the particles move together at a uniform rate. This example has been known since William Thomson[9] to be a relative equilibrium.

We begin with three vortices uniformly spaced. Place the vortices, without loss of generality, at the coordinates $(1, 0)$ for vortex A, at the coordinates

[9] William Thomson, 1824 - 1907, the Lord Kelvin. Justly famous for his far-reaching work in thermodynamics, and mildly infamous for overconfident statements like an 1895 prediction that heavier-than-air flight would be impossible, he researched vortices as a theory explaining the composition of and the differences between atoms. Though the vortex model of atoms never explained all their observed properties (no model could before quantum mechanics), this research influenced fluid dynamics and inspired knot theory. He also developed much of the theory and some key instrumentation for the transatlantic telegraph cable. [25] [288]

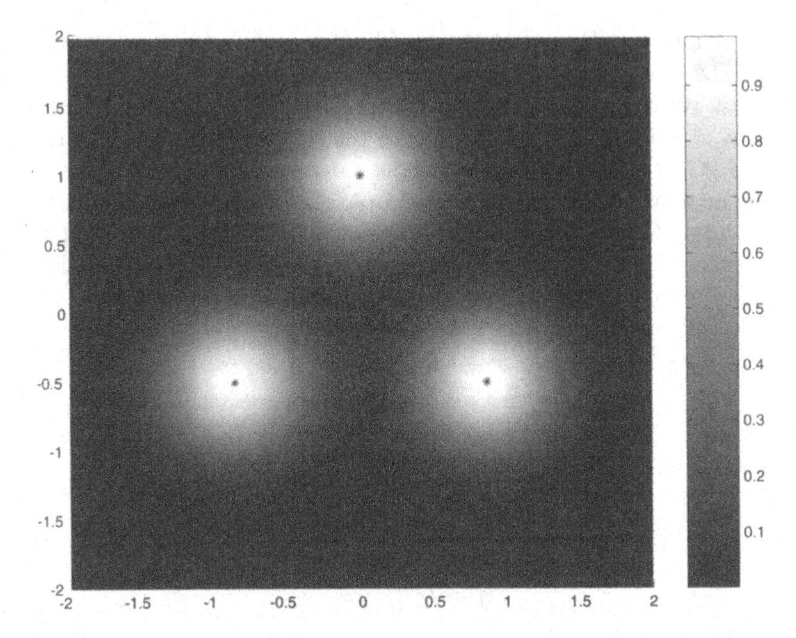

Fig. 6.4. A regular triangle is a dynamically stable arrangement of vortices, as are other regular polygons with few vertices.

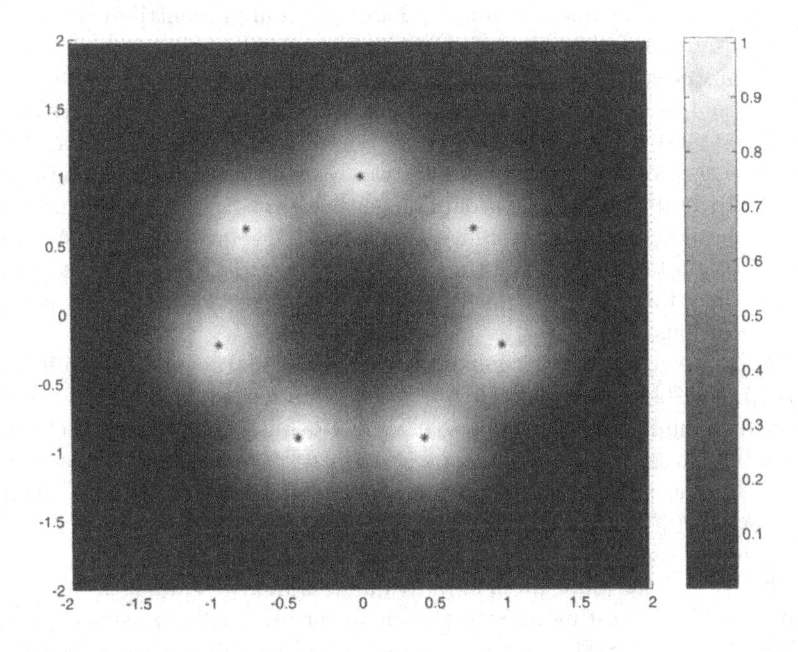

Fig. 6.5. A heptagon is neutrally stable. Regular polygons with more points will be unstable.

$(-\frac{1}{2}, \frac{\sqrt{3}}{2})$ for vortex B, and at the coordinates $(-\frac{1}{2}, -\frac{\sqrt{3}}{2})$ for vortex C. The distance BA is equal to the distance CA, both equal to $2\sin\left(\frac{1}{2}(\frac{2\pi}{3})\right)$. Since the motion on A from B is perpendicular to the line segment BA, and the motion on A from C is perpendicular to the line segment CA. The resulting motion of A is tangent to the unit circle.

If the angular separation, measured from the origin, between A and B and between A and C is the angle ϕ, then the net motion of A is $2f\cos\left(\frac{1}{2}(\pi - \phi)\right)$, with f is the magnitude of the interaction between B or C and A. This falls off as $\frac{1}{d}$, reciprocal of the distance between B and A; d equals $2\sin\left(\frac{1}{2}\phi\right)$. Thus the motion of vortex A will be

$$\frac{dx_A}{dt} = 0 \tag{6.98}$$

$$\frac{dy_A}{dt} = -\frac{1}{4\pi}\frac{1}{2\sin(\frac{\phi}{2})}2\cos(\frac{1}{2}(\pi - \phi)) \tag{6.99}$$

$$= -\frac{1}{4\pi}\frac{\cos(\frac{1}{2}(\pi - \phi))}{\sin(\frac{\phi}{2})} \tag{6.100}$$

$$= -\frac{1}{4\pi} \tag{6.101}$$

We may immediately expand these results to the rotation rate of any regular polygon of N vortices. Every pair of vortices, one at angle ϕ clockwise from A, and one at angle ϕ counterclockwise from A, contributes the same net impulse to the movement of A. The net rate of rotation is $\frac{1}{4\pi}(N-1)$. □

Thompson went further, finding a polygon will be stable – minor deviations from the polygon will not break up the shape – for up to six vortices on the plane. For more than seven vortices the shape is unstable, and small errors will cause the polygon to shatter. For exactly seven vortices the configuration is neutrally stable, some perturbations neither growing nor declining.

One method for testing stability comes from T H Havelock [378]. We translate the origin to the center of the polygon. Then write the motion equations in polar coordinates, and consider displacements (r_n, θ_n) from the polygon vertex positions of each vortex.

The only eigenfunctions compatible with rotational symmetry require $r_n = \alpha(t)\exp\left(\frac{2\pi ikm}{N}\right)$ and $\theta_n = \beta(t)\exp\left(\frac{2\pi ikm}{N}\right)$ for integers k from 0 to $N-1$. The k is a wave-number, the "wavelength" of the perturbation, which is clearer if one plots r_n or θ_n versus n. Examining $\alpha(t)$ and $\beta(t)$ finds the displacements decrease – the system is (linearly) stable – if for a given N, the quantity $k(N-k) - 2(N-1)$ is negative for each k. It grows if at least one k makes that quantity positive.

This is not the only method to examine stability. Adding a "periodic" displacement and testing its growth is however often the easiest one to algebraically evaluate. Stability is an entire field to itself and reaches far beyond the scope of this book. A polygonal ring almost inevitably shatters when one numerically integrates to explore its motion. The compromise between the real numbers and the floating point numbers on the computer serve as constant perturbations, so even quite good numerical integrators tend to see larger polygons break up.

There is also another viewpoint, in which the motion of a set of vortices is characterized by the sum of the products of vortex strength; for three points this would be simply $S = s_1 s_2 + s_2 s_3 + s_3 s_1$. [423] This approach becomes particularly advantageous when combined with a local coordinate system scheme, measuring the positions of vortices from a center of vorticity. Were we more interested in exactly solving the dynamics of these systems, rather than picking specific points for physical intuition or for an understanding of what we should see in statistical mechanics studies, we would spend more time considering these sorts of characterization and how the evolution of a system is seen in phase space. □

Example 4. The Infinite Line

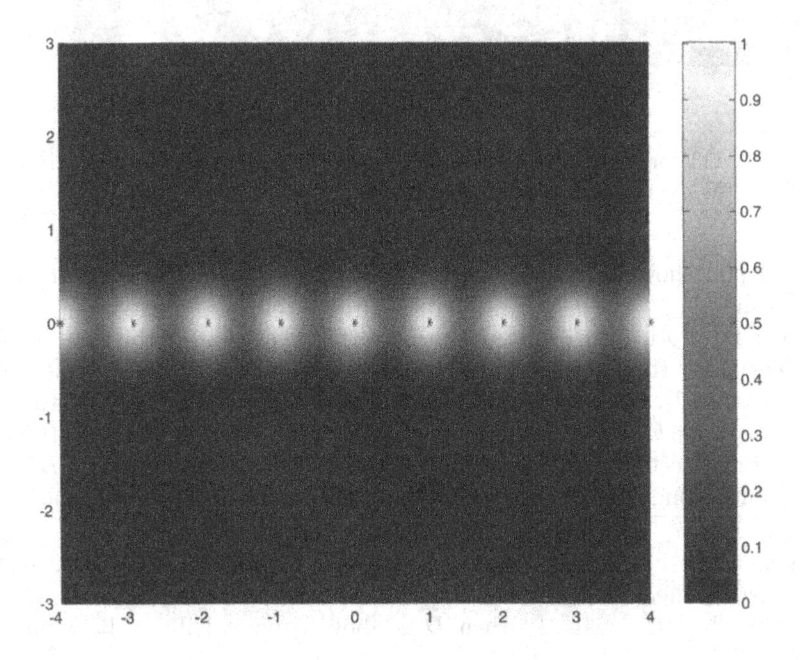

Fig. 6.6. An infinite line of vortices with uniform strength is an equilibrium, but it is not stable. A finite line is a configuration that rotates around its center.

An interesting phenomenon occurs with an infinitely long string of uniformly spaced vortices of uniform strength on a single line: we have a fixed equilibrium. From any one vortex A the two immediately adjacent it are the same strength and on opposite sides; the net motion of A is zero. The next

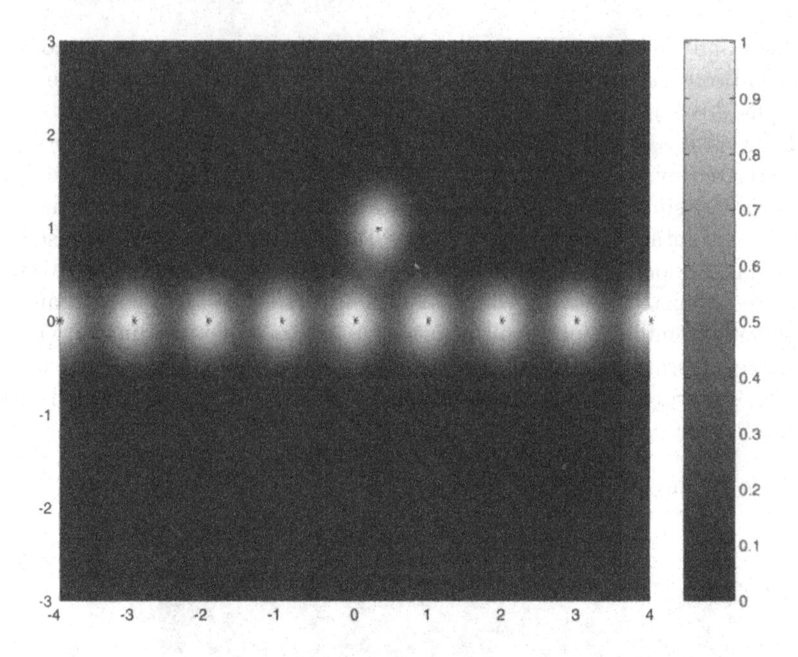

Fig. 6.7. This perturbation of the infinite line will rapidly break up its structure. However, the dynamics of the perturbing vortex suggest another system.

nearest pair show the same geometry. This continues along every pair, so no vortex moves.

Also interesting is the motion of test vortex D away from this line. Let the vortices along the line be separated by the distance a. There is some point B nearest to D. Let c be the perpendicular distance between D and the line of vortices; let b be the distance from where the perpendicular line dropped from D intersects that line to B. For each integer j, there is a point B_j that is j points along from B (also called B_0). The distance between D and B_j is $d_j = \sqrt{c^2 + (a - b + ja)^2}$.

For two positions of D symmetries make the motion of D easy to calculate. The first is when b equals zero, when D is "above" one of the particles; the other is when B equals $\frac{a}{2}$, when D is above a point halfway between two particles.

As with the regular polygon the motion of any pair on D is $2f\cos(\phi)$, with f the strength of interaction, and ϕ the angle with respect to the infinite line between either particle and D.

With $b = 0$ let the point underneath D be $j = 0$. The distance from D to the pair B_j and B_{-j} is $d_j = \sqrt{c^2 + (ja)^2}$; the cosine of the angle ϕ_j is $\frac{c}{d_j}$. The result is an infinite sum:

$$\frac{d}{dt}x_D = \frac{1}{4\pi} \sum_{j=-\infty}^{\infty} 2 \frac{1}{\sqrt{c^2 + (ja)^2}} \frac{c}{\sqrt{c^2 + (ja)^2}} \tag{6.102}$$

$$= \frac{2c}{4\pi} \sum_{j=-\infty}^{\infty} \frac{1}{c^2 + (ja)^2} \tag{6.103}$$

$$= \frac{1}{2a} \coth\left(\frac{c\pi}{a}\right) \tag{6.104}$$

which uses the hyperbolic cotangent[10].

With $b = \frac{a}{2}$ and continuing as above (and picking B_0 to be either particle nearest D) we have another infinite sum:

$$\frac{d}{dt}x_D = \frac{1}{4\pi} \sum_{j=-\infty}^{\infty} 2 \frac{1}{\sqrt{c^2 + ((j - \frac{1}{2})a)^2}} \frac{c}{\sqrt{c^2 + (ja)^2}} \tag{6.105}$$

$$= \frac{2c}{4\pi} \sum_{j=-\infty}^{\infty} \frac{1}{c^2 + ((j - \frac{1}{2})a)^2} \tag{6.106}$$

$$= \frac{1}{2a} \tanh\left(\frac{c\pi}{a}\right) \tag{6.107}$$

using the hyperbolic tangent[11].

This is not an equilibrium; the influence of D will break up the line. But this does lead to the next example. □

An interesting phenomenon which will affect later work appears here. The logarithmic potential between two points is long-range: there is no vortex which can be so far away its contribution to the potential becomes ignorable (though the "force" it exerts gets close to zero).

Example 5. The Von Kármán Trail

The von Kármán[12] trail is a pair of infinitely long strings of evenly spaced vortices. One string's vortices are all of strength s; the other's are all $-s$. The lines are a distance c apart. Each vortex on a single line is a distance a from its neighbors.

When the lines are evenly staggered this is an equilibrium [399]. This configuration has long fascinated physicists and mathematicians; these paired string of positive and negative vortices are produced by a smooth fluid flowing

---

[10] $\coth(z) = \frac{\exp(z)+\exp(-z)}{\exp(z)-\exp(-z)}$, quotient of the hyperbolic cosine and hyperbolic sine.

[11] $\tanh(z) = \frac{\exp(z)-\exp(-z)}{\exp(z)+\exp(-z)}$, quotient of the hyperbolic sine and hyperbolic cosine.

[12] Theodore von Kármán, 1881 - 1963, is one of the founders of the study of aerodynamics and supersonic aerodynamics. He demonstrated in 1940 that a stable, long-burning solid rocket was buildable; and he was a founder of both the Jet Propulsion Laboratory and the Aerojet rocket corporation, which produced the Service Propulsion System engine on the Apollo Service Module and the Orbital Maneuvering Subsystem rockets on the Space Shuttle. [288]

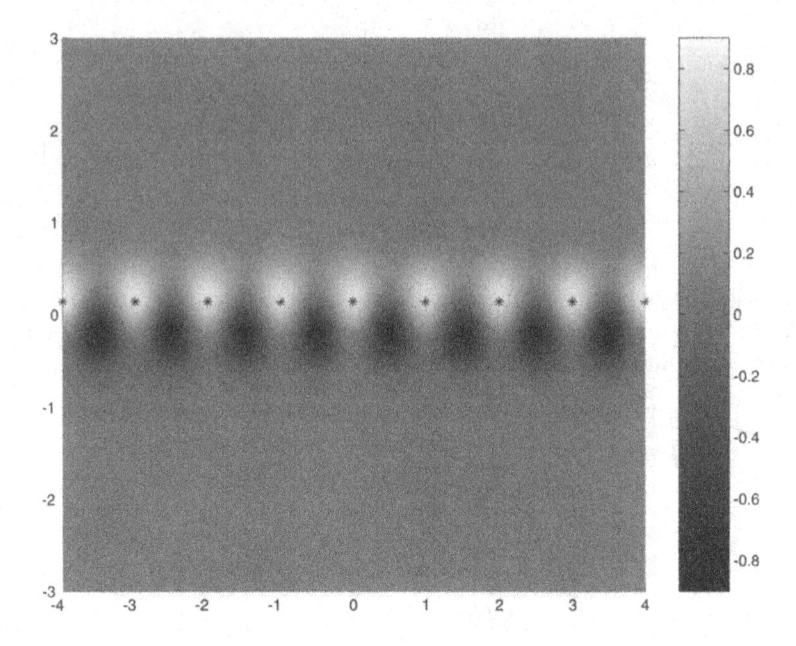

Fig. 6.8. A von Kármán trail is a particularly fascinating arrangement of trails of vortices. It is a common feature of fluids moving past an obstruction.

past a cylindrical obstruction. These von Kármán trails are observed in the atmosphere; many orbital photographs show examples. (Von Kármán was not the first to observe them, of course; this staggered double row can be observed by anyone who has drawn a hand through water. An example of it is depicted, apparently, in an Assyrian relief from the ninth century BC [378], citing [284].)

Finding the speed of the vortices requires no new work. Each vortex has a net zero motion from its own line, and receives only a motion from the opposite line. If the lines are "staggered", so each vortex is halfway between the two nearest opposing ones, then both lines move at speed $\frac{s}{2a} \tanh\left(\frac{c\pi}{a}\right)$.

□

The von Kármán trail is unstable for most c and a. The only arrangement even neutrally stable is when $\frac{c}{a}$ equals $\frac{1}{\pi} \sinh^{-1}(1)$, approximately 0.281. [274] [399]

We may wonder if fixed equilibria exist for finitely many vortices. They do: V V Kozlov proved [226] there exist non-rotating equilibria for every integer N. Kozlov also proved the existence of stable, non-rotating equilibria for every perfect square N of points, and found stable equilibria for up to 50 points.

6.7 Dynamics on the Sphere

On the surface of the sphere the derivative of $\mathbf{x}_j$ may be written as in equation 6.33. The motion exerted by k on j is tangent to the unit sphere at the point $\mathbf{x}_j$, is perpendicular to the vector between them, and has strength

$$\frac{\sin(d)}{\sqrt{2\left(1-\cos\left(d\right)\right)}} \tag{6.108}$$

where d is the distance (on the unit sphere, the angular separation) between them. In the symplectic coordinates

$$p_j = \sqrt{|s_j|}\cos\left(\theta_j\right) \tag{6.109}$$

$$q^j = \sqrt{|s_j|}\phi_j \tag{6.110}$$

the Hamiltonian is

$$H = -\frac{1}{8\pi R^2}\sum_{j=1}^{N}\sum_{k=1}^{N} s_j s_k \log(d_{j,k}^2) \tag{6.111}$$

with $d_{j,k}$ the distance between $\mathbf{x}_j$ and $\mathbf{x}_k$.

In understanding the dynamics we remember if the sphere is very large the space approximates a plane, so particle motion will resemble that above. This is most easily demonstrated by the regular polygon. Without loss of generality we set it on a single line of latitude. The equilateral triangle is a stable equilibrium on all lines of latitude, though it does not rotate only on the equator. Similarly the square, pentagon, hexagon, and so on are relative equilibria anywhere, rotating on every latitude except the equator.

Example 6. Two points

We begin with two vortices, both of strength 1, placed at (p_1, q_1) and (p_2, q_2). From equation 6.7 the motion of the first vortex is

$$\frac{dq_1}{dt} = \frac{\partial H}{\partial p_1}$$

$$= -\frac{1}{8\pi R^2}\frac{-p_2 + p_1\frac{\sqrt{1-p_2^2}}{\sqrt{1-p_1^2}}\cos(q_2-q_1)}{1-p_1p_2-\sqrt{1-p_1^2}\sqrt{1-p_2^2}\cos(q_2-q_1)} \tag{6.112}$$

$$\frac{dp_1}{dt} = -\frac{\partial H}{\partial q_1}$$

$$= -\frac{1}{8\pi R^2}\frac{\sqrt{1-p_1^2}\sqrt{1-p_2^2}\sin(q_2-q_1)}{1-p_1p_2-\sqrt{1-p_1^2}\sqrt{1-p_2^2}\cos(q_2-q_1)} \tag{6.113}$$

and the second is similar.

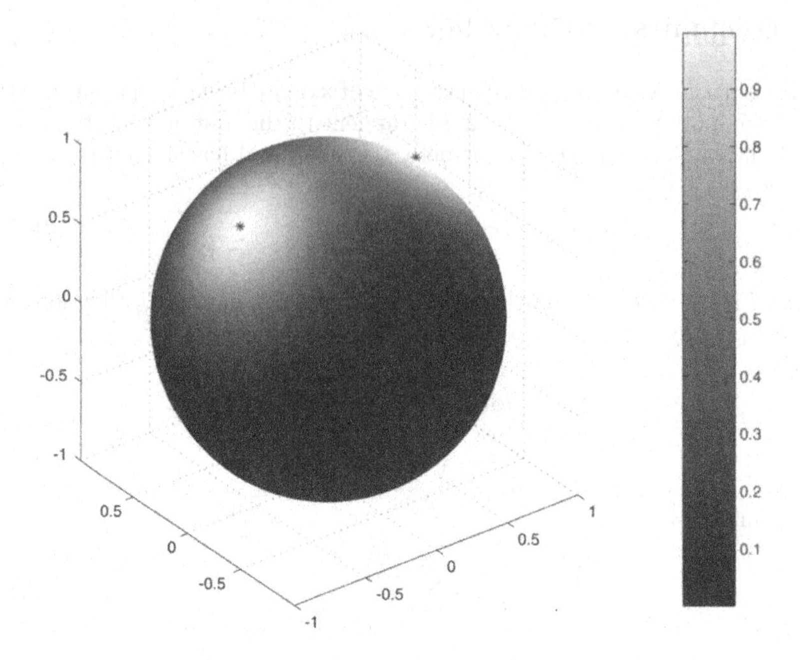

Fig. 6.9. Two points of uniform strength on the sphere will have dynamics very similar to those on the plane, though the speed at which they rotate will depend on their distance in a more complicated way than it does on the sphere. In particular, an antipodal pair no longer rotates.

Set both on the same line of latitude, $p_2 = p_1$, and space them evenly along that line, $q_2 = q_1 + \pi$. The equations of motion simplify to

$$\frac{dq_1}{dt} = \frac{1}{8\pi R^2} \frac{p_1}{1 - p_1^2} \tag{6.114}$$

$$\frac{dp_1}{dt} = 0 \tag{6.115}$$

a relative equilibrium. Both q_1 and q_2 increase linearly in time neither latitude changes. □

Naïvely taking the "large sphere" limit by letting only R grow infinitely large seems to give the wrong result of no rotational motion on the plane. In fact p_1, the cosine of the co-latitude, also depends on R, so our limit must take this in consideration. An equivalent "large sphere" limit holds R equal to one and lets p_1 approach either positive or negative one. This yields rotation around the midpoint of the two vortices at a uniform rate, the behavior on the plane.

If the vortices are on the equator we have a fixed equilibrium between the "antipodal" pairs.

Example 7. Triangle

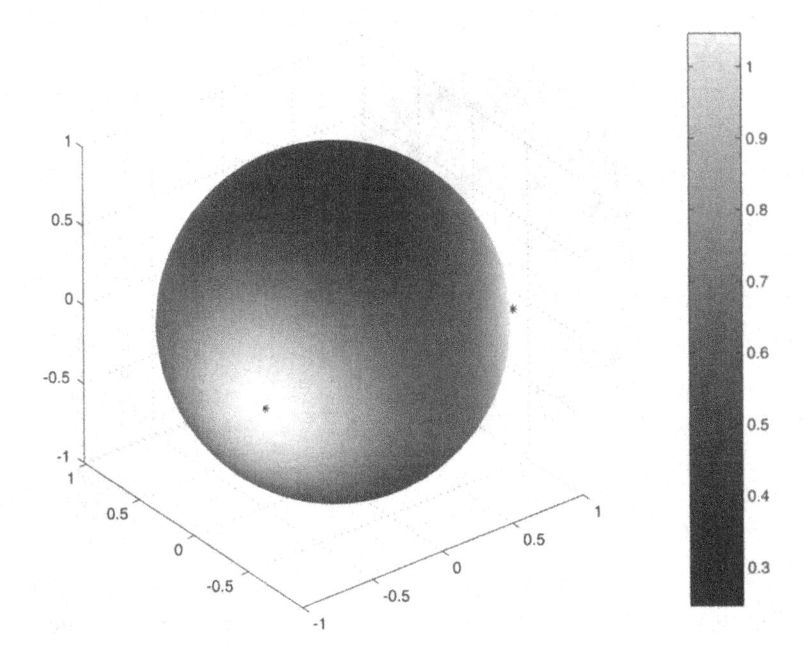

Fig. 6.10. A triangle is always dynamically stable. An equilateral triangle will rotate around its center with a uniform speed that depends on how large the triangle is. If the triangle is on a great circle such as the equator then the system is in equilibrium.

Next consider points of uniform strength at the vertices of an equilateral triangle, all at the same latitude. Label the vertices A, B, and C. We will consider this geometrically first and then examine the algebra.

The motion on A must be equal from both B and C, as their separations are equal. The direction of the motions are on the sphere and perpendicular to the line segments BA and CA. Their sum then is tangent to the little circle containing A, B, and C, so the motion is entirely longitudinal, with no change in latitude. By symmetry B and C move in the same (longitudinal) direction. With latitude unchanging the triangle rotates at uniform speed.

Now we use the Hamiltonian to calculate the motion of A:

$$H = -\frac{1}{8\pi R^2} \sum_{j=1}^{3} \sum_{k \neq j}^{3} \log(d_{j,k}) \tag{6.116}$$

$$d_{j,k} = 1 - p_j p_k - \sqrt{1 - p_j^2}\sqrt{1 - p_k^2} \cos(q_k - q_j) \tag{6.117}$$

We will use the simplifications that p_j are set to p_A, and the longitudes q_j are $\frac{2\pi}{3}$ greater than and less than q_A.

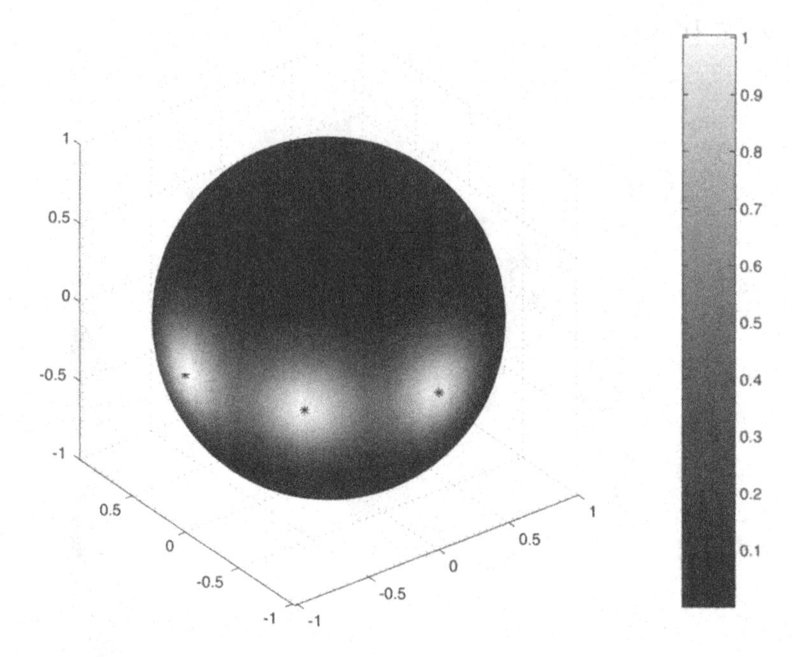

Fig. 6.11. Any three points along a great circle are also an equilibrium.

$$\frac{d}{dt}p_A = \frac{\partial H}{\partial q_A} = \frac{1}{8\pi R^2}\left(\frac{\sin\left(\frac{2\pi}{3}\right)}{1-\cos\left(\frac{2\pi}{3}\right)} - \frac{\sin\left(\frac{2\pi}{3}\right)}{1-\cos\left(\frac{2\pi}{3}\right)}\right) \tag{6.118}$$

$$= 0 \tag{6.119}$$

$$\frac{d}{dt}q_A = -\frac{\partial H}{\partial p_A} = -\frac{1}{8\pi R^2}\left(\frac{p_A}{1-p_A^2} - \frac{p_A}{1-p_A^2}\right) \tag{6.120}$$

$$= -\frac{1}{4\pi R^2}\left(\frac{p_A}{1-p_A^2}\right) \tag{6.121}$$

Kidambi and Newton [324] [325] continue the geometric view, examining the change in time of the (square of the) distances between points $d_{j,k}$. The evolution of their relative distances are

$$\frac{d}{dt}(d_{1,2}^2) = \frac{V}{\pi R}\left(\frac{1}{d_{2,3}^2} - \frac{1}{d_{3,1}^2}\right) \tag{6.122}$$

$$\frac{d}{dt}(d_{2,3}^2) = \frac{V}{\pi R}\left(\frac{1}{d_{3,1}^2} - \frac{1}{d_{1,2}^2}\right) \tag{6.123}$$

$$\frac{d}{dt}(d_{3,1}^2) = \frac{V}{\pi R}\left(\frac{1}{d_{1,2}^2} - \frac{1}{d_{2,3}^2}\right) \tag{6.124}$$

where V is the vector triple product

$$V = \mathbf{x}_1 \cdot (\mathbf{x}_2 \times \mathbf{x}_3) \tag{6.125}$$

(that is, the volume of the parallelepiped traced out by the vectors from the origin to each of the three vortices). Written in this form the triangle being a relative equilibrium is obvious. It also highlights a something new: if the three points lie on a great circle then they are in relative equilibrium. □

Though we will not prove it, the triangle is a stable configuration of vortices – as on the plane, a small displacement of the vortices will remain small, and the shape of the triangle will remain.

Example 8. Regular Polygons

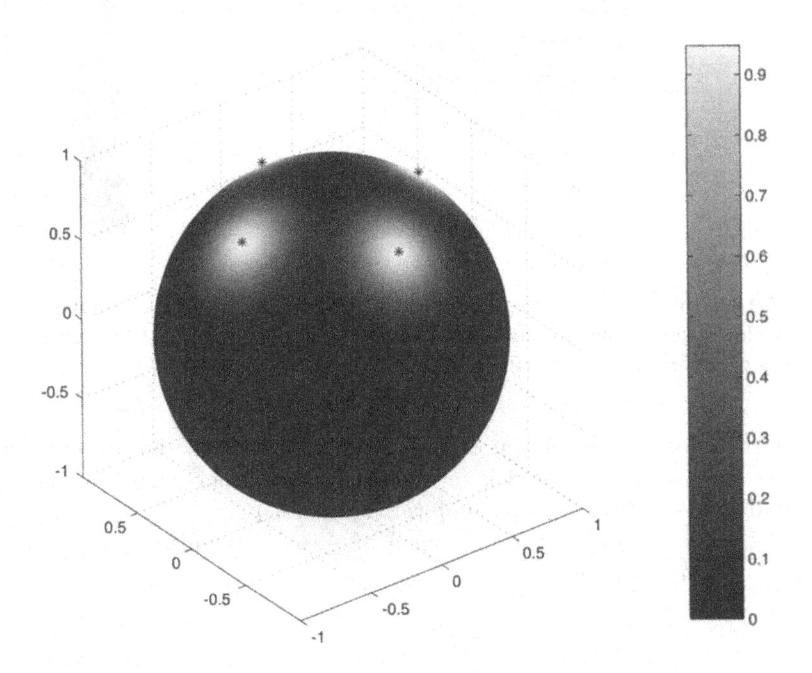

Fig. 6.12. Regular polygons are rotating equilibrium states, although very few of them are stable. In fact, squares, pentagons, and hexagons are only stable if they are sufficiently small – effectively, if the sphere is so big as to act like a plane.

Points of uniform strength at the vertices of a regular polygon we again treat by placing on a single latitude. From a point A, we pair off its neighboring vortices. The pair j vortices clockwise and counterclockwise of A are equal distances from A; their net motion imparted on A is tangent to the little circle containing the polygon.

If there are an even number of points then one point does not pair off with any others; it is at the same latitude and at longitude π greater than A. But

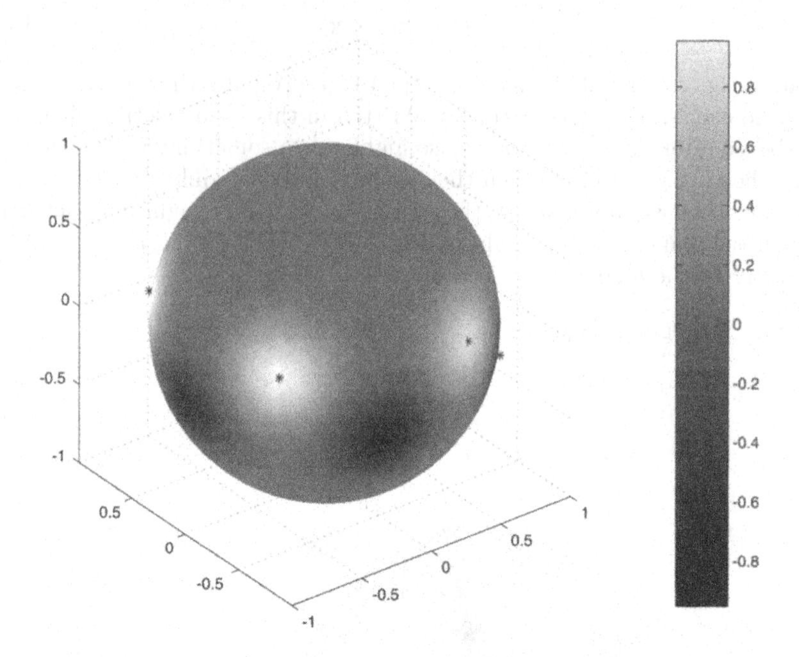

Fig. 6.13. Obviously patterns like a von Kármán trail may form on the sphere, even if they are not long-lasting phenomena. Supposing we can create rigid rings of vortices of uniform strength will open several interesting questions.

its contribution to the motion of A is still tangent to the latitudinal circle. And the geometry of point A is repeated for every other point. Just as with two and three vortices we have uniform rotation.

Now we can calculate the movement of A, and use this pairing of points around our chosen point.

$$H = -\frac{1}{8\pi R^2} \sum_{j=1}^{N} \sum_{k \neq j}^{N} \log(d_{j,k}) \tag{6.126}$$

$$d_{j,k} = 1 - p_j p_k - \sqrt{1 - p_j^2}\sqrt{1 - p_k^2} \cos(q_k - q_j) \tag{6.127}$$

There are either $2K$ or $2K + 1$ vortices in the ring other than A. If there are $2K$ then

$$\frac{d}{dt} p_A = \frac{\partial H}{\partial q_A} = \frac{1}{8\pi R^2} \sum_{J=1}^{K} \left(\frac{\sin\left(\frac{2\pi J}{N}\right)}{1 - \cos\left(\frac{2\pi J}{N}\right)} - \frac{\sin\left(\frac{2\pi J}{N}\right)}{1 - \cos\left(\frac{2\pi J}{N}\right)} \right) \tag{6.128}$$

$$= 0 \tag{6.129}$$

$$\frac{d}{dt} q_A = -\frac{\partial H}{\partial p_A} = -\frac{1}{8\pi R^2} \sum_{J=1}^{K} \left(\frac{p_A}{1 - p_A^2} - \frac{p_A}{1 - p_A^2} \right) \tag{6.130}$$

$$= -\frac{2K}{8\pi R^2}\left(\frac{p_A}{1-p_A^2}\right) \tag{6.131}$$

$$= -\frac{N-1}{8\pi R^2}\left(\frac{p_A}{1-p_A^2}\right) \tag{6.132}$$

If there are $2K+1$ vortices we add one term each to equations 6.128 and 6.130. The term added to equation 6.128 is $\frac{1}{8\pi R^2}\frac{\sin(\pi)}{1-\cos(\pi)}$, which is zero. The term added to equation 6.130 is $-\frac{1}{8\pi R^2}\frac{p_A}{1-p_A^2}$. The angular rotation is again $-\frac{N-1}{8\pi R^2}\left(\frac{p_A}{1-p_A^2}\right)$.

Kidambi and Newton [324] [325] again examine the evolution of the (square of the) pair distances $d_{j,k}$. The formulas of the above section generalize; for any two vortices j and k

$$\frac{d}{dt}(d_{j,k})^2 = \frac{1}{8\pi R}\sum_{l\neq j}^{N}\sum_{l\neq k}^{N} V_{j,k,l}\left(\frac{1}{d_{k,l}^2} - \frac{1}{d_{l,j}^2}\right) \tag{6.133}$$

where, as above,

$$V_{j,k,l} = \mathbf{x}_j \cdot (\mathbf{x}_k \times \mathbf{x}_l) \tag{6.134}$$

So we know a regular polygon with points of uniform strength is a relative equilibrium. □

Unlike Thompson's results on the plane, the square is only a stable equilibrium near the north and south poles – only if the square covers a small enough area. The square is stable at co-latitude θ only if $\cos^2(\theta) > \frac{1}{3}$. Closer to the equator perturbations break up the shape. For five points we must have $\cos^2(\theta) > \frac{1}{2}$, and for six points we must have $\cos^2(\theta) > \frac{4}{5}$. Greater than six vortices are never stable in a single ring.

We may wonder what happens if the number of vortices grows infinitely long. It is tempting to imagine the banded clouds of Jupiter and Saturn might be such vortex rings. To take such a "continuum limit" requires careful setting up of our limits, something we will explore in a later chapter.

Example 9. Tetrahedron

Our first three-dimensional example is created by placing vortices of uniform strength at the corners of a tetrahedron. The vortices cannot be put on a single line of latitude. (They can be placed on two lines, symmetric around the equator with points "staggered" longitudinally.) This is an equilibrium, which we will argue geometrically.

Consider point A. The other three are a uniform distance and are equally strong. The other three vortices are also placed at the vertices of an equilateral triangle. The motion from each of them is equal in strength and each $\frac{2\pi}{3}$ out of phase with the others, so the net motion is zero. By symmetry none of the three other points will move. □

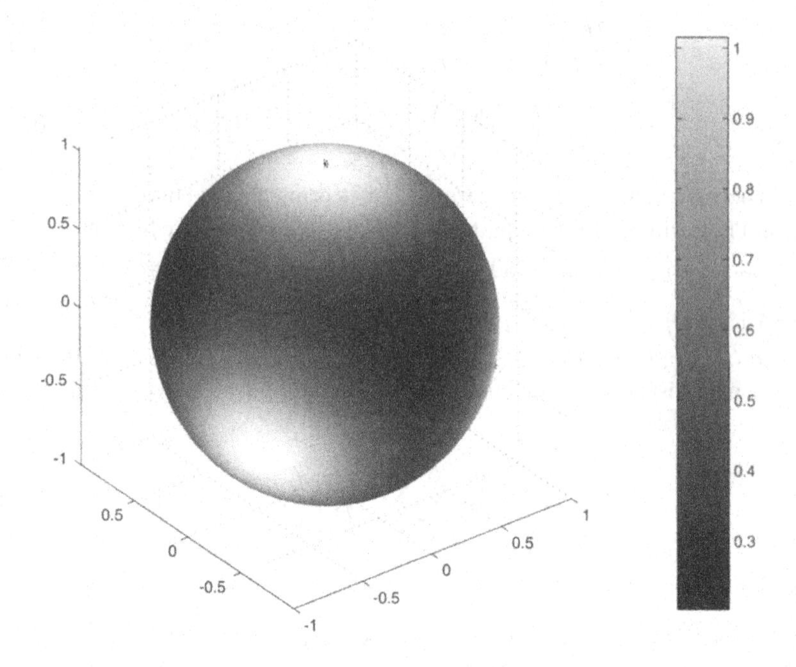

Fig. 6.14. The corners of a tetrahedron are an equilibrium, and are in positions which cannot be put on a single line of latitude.

Example 10. Double Pyramid

The double pyramid has five points. Three are an equilateral triangle on a great circle; without loss of generality we place them on the equator. The other two are on the axis perpendicular to the great circle; here, the north and south poles.

Consider either pole; take the south. The north pole is antipodal, so does not affect the south. The equatorial points are equally distant from the south pole, and each one's motion is $\frac{2\pi}{3}$ out of phase with its neighbors. The net motion is zero, and the polar points do not move.

Consider equatorial point A. Its nearest neighbors are the poles, which are equidistant and in opposite directions. Their net contribution to the movement of A is zero. The other equatorial vortices are the same distance from A and in opposite directions, so their motion adds to zero. So the equatorial vortex does not move. □

Example 11. Octahedron

The octahedron is the last polyhedron we address for now, though this sort of argument can be extended to more shapes. The octahedron has six points (the name reflects it having eight *faces*); without loss of generality we put them at the points with Cartesian coordinates $(1, 0, 0)$, $(0, 1, 0)$, $(0, 0, 1)$,

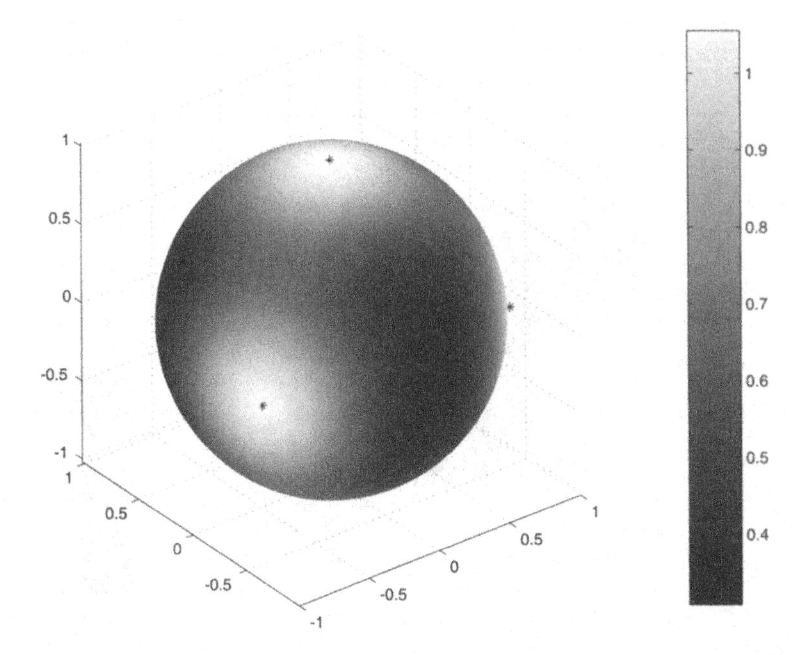

Fig. 6.15. The corners of a bipyramid also form an equilibrium, something easy to argue geometrically. There are five points in the system, an equilateral triangle on the equator and two polar vortices.

$(-1, 0, 0)$, $(0, -1, 0)$, and $(0, 0, -1)$. (We might also represent it as two equilateral triangles staggered around the equator.)

From the arbitrary point A we see the same pattern as above. An antipodal vortex contributes no motion to A. The remaining vortices are a ring of four, all the same distance from A and each a quarter-turn of the sphere from its nearest neighbors. The net motion is zero. And therefore the octahedron is an equilibrium. □

Similar arguments can be advanced about shapes such as the vertices of the cube and many additional familiar polyhedra.

Let us put aside listing particular cases and try to describe what more general conditions for equilibria. We will review the symmetries of the Hamiltonian: how can we change a set of vortex positions and strengths without changing the dynamics?

We have several conserved and several involutive quantities. Conserved besides energy are:

$$X = \sum_{j=1}^{N} s_j \sqrt{1 - p_j^2} \cos(q^j) \tag{6.135}$$

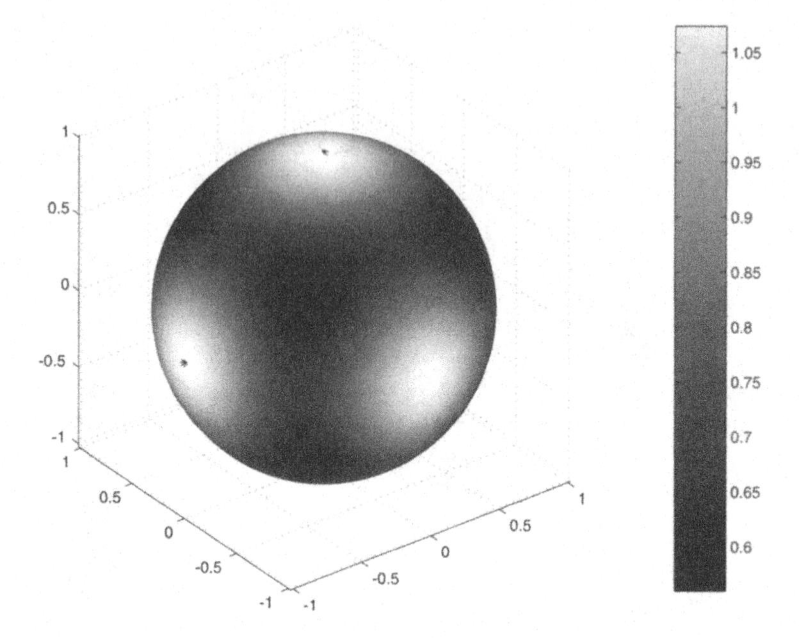

Fig. 6.16. The six vertices which form the corners of an octahedron are an equilibrium. Though the placement of points at the poles obscures it, the vertex placement may be considered as two staggered rings of three vertices each. Rings of uniformly spaced points and polar vortices make up a large number of interesting classical dynamics problems for vortex motion.

$$Y = \sum_{j=1}^{N} s_j \sqrt{1 - p_j^2} \sin(q^j) \tag{6.136}$$

$$Z = \sum_{j=1}^{N} s_j p_j \tag{6.137}$$

Only three, as on the plane, are involutive:

$$\{H, X^2 + Y^2\} = 0 \tag{6.138}$$

$$\{X^2 + Y^2, Z\} = 0 \tag{6.139}$$

$$\{Z, H\} = 0 \tag{6.140}$$

As on the plane,

$$\{X, Y\} = Z \tag{6.141}$$

$$\{Y, Z\} = X \tag{6.142}$$

$$\{Z, X\} = Y \tag{6.143}$$

so if X, Y, and Z are all zero then we have four conserved, involutive quantities. This condition is that the center of vorticity $\mathbf{M} = \sum_j s_j \mathbf{x}_j$ is the zero vector. Unlike the plane we do not need the sum of vortices to equal zero.

The search for equilibria runs the risk of becoming an exhausting list of candidates. Some are obvious, like a Von Kármán- like trail of vortices staggered above and below the equator, or the vertices of select polyhedra. Rather than reiterate papers such as Lim, Montaldi, and Roberts [270] which classify equilibria by their symmetries we note many of them can be found by considering what arrangements of points respect the finite subgroups of $O(3)$ – the group (in the abstract algebra sense) of orthogonal matrices with three rows and columns, which corresponds to the rotations and reflections one may perform on the sphere – and of $SO(3)$ – the group of orthogonal matrices with determinant one, the rotations on the sphere.

The subgroups of $SO(3)$ which are useful here begin with C_n, the group of order n created by rotating the sphere around the z-axis. This shape is vertices of a regular polygon on a single latitude. These may further include points added at either or both poles.

Next is D_n, the dihedral group with order $2n$, created by two rings with n points above and below the equator, and optional points at the poles. It also includes the equatorial regular polygon of n vortices.

We have three subgroups inspired by polyhedra. The symmetries of a regular tetrahedron are $\mathbf{T}$. The symmetries of the octahedron are $\mathbf{O}$. Those of the icosahedron are $\mathbf{I}$. These are Platonic solids, shapes which will inspire further work. The cube is representable with D_n. The dodecahedron we leave alone for the moment.

If the vertices of a polyhedron represent an equilibrium, then adding new points of the same strength at the center of each face is again an equilibrium. This sort of extension will lead us to mesh-generating algorithms later.

While we find many equilibria from these symmetries, we do not find all of them. Indeed, the equilibria with the lowest possible energies – which we know must be stable – we will see generally are not symmetric patterns.

6.8 Remarks

This chapter outlined the turning of a fluid dynamics problem into one of particle dynamics, the vortex gas problem. From the partial differential equations for inviscid fluids we derive a Hamiltonian, which makes clear conserved quantities such as energy, momentum, and angular momentum [324]. We find this by approximating the vorticity field as a set of Dirac delta functions.

This approach is useful for many differential equations. We will return to this strategy to produce models such as the heton model, a representation of how temperature can be transported by fluids of several layers. We can also add properties like the rotation of a planet, or could add the interference that islands and continents will have on an ocean or an atmospheric flow, each time extending the same technique by the addition of a boundary condition or of an extra "field" term. It is straightforward and it leaves us with pairwise particle interactions familiar to any mathematics or physics student.

It has a weakness in Monte Carlo as it only works well for problems in which all vortices have the same sign. Mixed-sign vortex gases tend numerically to clump together into a few spots of positive and negative vortices near one another. There is considerable work of interest in single-species vortex gases, and they will be a foundation of the chapter on mesh generation. To address mixed-sign gases, though, or to use the negative temperatures we have seen, we need a different model, the lattice gas, in which we have a set of fixed mesh sites and have vorticities which vary.

We have therefore split our fluid model into two branches: a vortex gas model in which points are of fixed strength, but may move, and which as a result require for Monte Carlo that all vorticities be the same sign and inverse temperature be positive; and a lattice model in which points are of fixed position but variable strength, and in which we may have positive and negative vorticities and still explore negative temperatures. This division of models will continue through the remainder of this book.

7

Spin-Lattice Models

7.1 Introduction

We built a model of fluid flow by representing the vorticity field of the fluid as a set of movable point vortices. But we know we cannot use this to represent every interesting phenomenon. We cannot consider vortices of mixed sign. Our inability to consider negative temperatures is another drawback.

In this chapter we consider a second model, inspired by finite element methods, in which we set a fixed mesh on our domain and allow changes in site strength to represent the evolution of the vorticity field. This removes the problems which limited the usefulness of the vortex gas problem. We call these generally **spin-lattice** models, lattice because we try placing the mesh sites in reasonably uniform patterns across the domain, and spin because these problems draw from the study of magnetism, in which each site value represents the spin of a small magnet.

We begin the detailed analysis of spin-lattice models, finite dimensional approximations to the statistics of ideal fluids on two dimensional surfaces. We explore the important topic of the relationship between these finite dimensional models and the continuum fluid statistics, and will review the types of continuum limits relevant to the continuum vortex statistics.

Finite dimensional Monte Carlo simulations cannot exactly capture phase transitions in a continuum system, but these transitions are of considerable interest. So we must study how to increase the size of the spin-lattice models to extract information about phase transitions.

For a fixed compact flow domain, such as the surface of a sphere or a bounded region of the plane, we focus on the **non-extensive continuum limit** to approximate the statistical properties of the continuum problem. We will see this issue of a thermodynamic limit is closely related to the particular first integrals, the conserved quantities, in the statistical mechanics model.

In two-dimensional vortex statistics, these first integrals are the flow kinetic energy, the total circulation and the enstrophy. Our spatial lattice decomposition of the vorticity field represents the kinetic energy by a finite dimensional Hamiltonian function with a logarithmic kernel, while quantities such as total circulation and enstrophy become respectively the sum and sum of squares of the lattice site values. It was observed by the first author in [264] that the discrete form of the enstrophy is just the spherical constraint in a spin-lattice model known as the **spherical model**.

The spherical model was introduced by T H Berlin and Mark Kac[1] as a model for ferromagnetism [41]. We will refer to this statistical mechanics model of two-dimensional inviscid macroscopic fluid flows as the **energy-enstrophy-circulation theory** since these are exactly the three conserved quantities.

7.2 Statistical Mechanics of Vortex Spin-Lattice Models

In their landmark 1952 paper *The Spherical Model of a Ferromagnet* T H Berlin and M Kac [41] described several different exactly-solvable Ising-type models of ferromagnetism. Their basis continues to inform us; while our fluid interactions do not use the same force law as magnets do, the form of the interactions is exactly the same – a set of discrete points, each with some given strength, with a pairwise interaction. We will take a few pages to review the Berlin and Kac models.

Fundamental is the **Ising model**. The model originally used a regular mesh of points on the plane, and on each site assigned a value of 1 or -1. Each site interacts with its nearest neighbors, with an interaction energy of $-Js_j s_k$ with J the interaction strength and s_j and s_k the site values. There may also be an an external magnetic field with strength hs_j for each site j.

The partition function of the one-dimensional, uniform-mesh Ising model is easy to solve. More complex is the two-dimensional problem for regular square meshes as well as for triangular meshes (it was on triangular meshes that the problem was first solved). Interestingly despite the simplicity of the model it can be a genuinely hard problem: Francisco Barahona and Sorin Istrail have demonstrated the general Ising model, with an irregular mesh and not necessarily on the plane, is an **NP-complete**[2] problem to solve.

The Ising model is well-suited to Monte Carlo methods, as the obvious experiment to conduct on each step – changing the strength of site s_j to $-s_j$ – is straightforward and the limitation to nearest-neighbor interactions means changes of energy can be quickly computed. It can also be extended easily,

[1] Mark Kac, 1914 - 1984, felt his best work was examining why some phenomena, such as phase transitions, happen only in certain numbers of spatial dimensions. His paper *Can one hear the shape of a drum?* neatly posed a central question of spectral theory. [288]

[2] The problem can be solved in non-deterministic polynomial time, and is equivalent to NP-hard problems; less formally, it is as difficult to find a solution for as the travelling salesman problem is.

for example into the **Potts**[3] **model**, which allows sites to take on one of a discrete set of strengths.

We use the term Ising-type models to describe any models with a mesh of points, site values, and pairwise interaction energies, even though in general these are non-uniform meshes, and may have interactions between pairs over the entire mesh and not only nearest neighbors. We will generally restrict the values which may be assigned site strengths; the interaction energy is dictated by the Hamiltonian of the problem we study.

The **Gaussian model** allows each site strength s_j to take any value from the whole real number line. Its name is because the distribution of site strengths is a Gaussian distribution: the probability of finding that site j has strength between s_j and $s_j + ds$ is $\frac{1}{\sqrt{2\pi}} \exp(-\frac{1}{2}s_j^2)ds$. In its Monte Carlo representation the Gaussian model is equivalent, as shown by Lim, to the **energy-enstrophy** model, in which energy and enstrophy are both constrained canonically. One problem noted by Berlin and Kac is the thermodynamic free energy can become a complex value if the temperature is below a critical value. This does not limit its use while the free energy is real.

The **spherical model** on the other hand, places only a limit on the enstrophy, the sum of squares of site strengths:

$$\sum_{j=1}^{N} s_j^2 = N \tag{7.1}$$

for the N sites. This is a constraint we will consider and re-examine several places in this book.

7.3 The Lattice Model

The basis of our lattice model is a set of N fixed mesh sites $\mathbf{x}_j$, each with an associated strength s_j. Our discretized lattice model approximates the vorticity field of the original fluid, which is by assumption a continuous function $\omega(\mathbf{x})$ defined on a domain D. There are two compelling alternate ways to define s_j – how to understand what we mean by this strength – which therefore need distinct set-up and interpretation and which require different thermodynamic limits when we try to understand the continuous fluid from our discrete version.

Common to both interpretations of s_j is our need for the mesh sites $\mathbf{x}_j$, and for the **Voronoi**[4] **diagram**. Given a set of N points $\mathbf{x}_j$ we construct the

[3] Named for Renfrey B Potts.

[4] Georgy Fedoseevich Voronoy (or Voronoi), 1868 - 1908, was particularly interested in algebraic integers – the roots of polynomials which have exclusively rational coefficients – and continued fractions. Voronoi diagrams he explored only shortly before his death. [288]

Voronoi diagram by partitioning the domain into N convex polygons, on the criteria that every polygon contains exactly one of the points $\mathbf{x}_j$, and every point in polygon j is closer to $\mathbf{x}_j$ than it is to any other point $\mathbf{x}_k$. The **Voronoi cell** D_j is the polygon containing point $\mathbf{x}_j$ and its interior.

We will also need the characteristic function for a set A, which as introduced in chapter 2 is a function (here) $H_A(x)$, equal to one if x is an element of A, and zero otherwise. This notion holds whether A is a set of real numbers, vectors, or more abstract elements; we use it only for the Voronoi cells D_j which are subsets of the domain.

We make our discrete approximation $q(\mathbf{x})$ to the vorticity field $\omega(\mathbf{x})$, which we assume to be continuous in space, in this way. Given N fixed mesh points $\mathbf{x}_1, \mathbf{x}_2, \mathbf{x}_3, \cdots, \mathbf{x}_N$ we assign site strengths $s_1, s_2, s_3, \cdots, s_N$, and

$$\omega(\mathbf{x}) \approx q(\mathbf{x}) = \sum_{j=1}^{N} s_j H_j(\mathbf{x}) \tag{7.2}$$

where H_j is the characteristic function for the domain D_j.

The choice of mesh sites $\mathbf{x}_j$ is arbitrary, but we will add one constraint, that the areas A_j of each domain D_j are approximately equal. This property is built into our mesh when we use the vortex gas algorithm to generate a mesh. This discretization allows us to construct the Hamiltonian energy of the set of particles by evaluating the kinetic energy of the fluid as represented here [264] [273] [324].

When we decide how to interpret the strengths s_j we become able to prove our vortex discretization in equation 7.2 does approximate the continuous vorticity field as the number of mesh sites grows infinitely large and the mesh sites remain approximately uniformly spaced.

7.3.1 Point Strength

Our first interpretation of the site strengths in equation 7.2 is for each j we choose $s_j = \omega(\mathbf{x}_j)$. Each site strength represents the vorticity of the continuous flow sampled just at that mesh site. The magnitude of s_j is therefore independent of the number of mesh sites N, which may be numerically convenient.

The fluid flow kinetic energy of the system is, modulo a singular self-energy term, given by

$$H(\mathbf{s}, \mathbf{x}) = -\frac{1}{8\pi} \sum_{j=1}^{N} \sum_{k \neq j}^{N} s_j s_k J_{j,k} \tag{7.3}$$

with $\mathbf{s} = (s_1, s_2, s_3, \cdots, s_N)$, $\mathbf{x} = (\mathbf{x}_1, \mathbf{x}_2, \mathbf{x}_3, \cdots, \mathbf{x}_N)$ and $J_{j,k}$ equal to $\frac{16\pi}{N^2}$ times the logarithm of the distance between $\mathbf{x}_j$ (the Green's function $G(\mathbf{x}_j, \mathbf{x}_k)$ which represents the fundamental solution to the inverse Laplace-Beltrami operator on the domain) and $\mathbf{x}_k$ in the domain D. The multiplication by 16π allows the elimination of constants from the energy expression; the division by N^2 means that as the number of points N increases, the total energy will remain relatively constant. As seen in the chapter on mesh generation the energy without any scaling would be proportional roughly to N^2, and this allows us to factor out these effects.

Of particular interest amongst other possible conserved quantities are the total circulation and the enstrophy. Both quantities are conserved as the system evolves in real time [88].

As the number of mesh sites N approaches infinity and the maximum value of A_j approaches zero, the sum of $s_j A_j$ will approach the fluid circulation. Consider the circulation of the discretized vorticity under this limit:

$$\lim_{N\to\infty} \int_D q(\mathbf{x})dA = \lim_{N\to\infty} \int_D \sum_{j=1}^{N} s_j H_j(\mathbf{x})dA \tag{7.4}$$

$$= \lim_{N\to\infty} \sum_{j=1}^{N} s_j \int_D H_j(\mathbf{x})dA \tag{7.5}$$

$$= \lim_{N\to\infty} \sum_{j=1}^{N} s_j A_j \tag{7.6}$$

$$= \lim_{N\to\infty} \sum_{j=1}^{N} \omega(\mathbf{x}_j) A_j \tag{7.7}$$

$$= \int_D \omega(\mathbf{x})dA \tag{7.8}$$

where the last lines we recognize is a Riemann[5] sum. To have the limit of this sum be numerical quadrature, an approximation to the Riemann-Stieltjes definition of the integral, is what motivated us to make each A_j approximately equal. More generally, Lebesgue integrals could be used in the place of Riemann integrals.

Similarly this discretized approximation to the fluid flow approximates the enstrophy of the fluid again as N approaches infinity and the maximum value of A_j approaches zero.

$$\lim_{N\to\infty} \int_D q^2(\mathbf{x})dA = \lim_{N\to\infty} \int_D \left(\sum_{j=1}^{N} s_j H_j(\mathbf{x}) \right)^2 dA \tag{7.9}$$

[5] Georg Friedrich Bernhard Riemann, 1826 - 1866, enjoyed a career of great work in many fields including non-Euclidean geometry, which laid foundations for the theory of relativity. He was praised for "gloriously fertile originality" by his thesis supervisor, Gauss. [288]

$$= \lim_{N \to \infty} \int_D \sum_{j=1}^{N} s_j^2 H_j^2(\mathbf{x}) dA \tag{7.10}$$

$$= \lim_{N \to \infty} \sum_{j=1}^{N} s_j^2 \int_D H_j^2(\mathbf{x}) dA \tag{7.11}$$

$$= \lim_{N \to \infty} \sum_{j=1}^{N} s_j^2 A_j \tag{7.12}$$

$$= \lim_{N \to \infty} \sum_{j=1}^{N} \omega^2(\mathbf{x}_j) A_j \tag{7.13}$$

$$= \int_D \omega^2(\mathbf{x}) dA \tag{7.14}$$

which is the enstrophy. Note the move from equation 7.9 to 7.10 is simplified because the characteristic functions $H_j(\mathbf{x})$ are an orthogonal set. For any j and k, the product

$$H_j(\mathbf{x}) H_k(\mathbf{x}) = \begin{Bmatrix} 1 \text{ if } j = k \\ 0 \text{ if } j \neq k \end{Bmatrix} \tag{7.15}$$

so all but the "perfect square" terms in $q(\mathbf{x})^2$ are zero.

It is numerically convenient to work with a simulated enstrophy equal to the sum of squares of site values. Provided A_j is approximately the area of the domain divided by the number of mesh sites then normalizing this sum to the true enstrophy is just multiplication by a constant.

Finally the energy of this discretized approximation converges to the actual energy of the fluid. Taking again the sort of Riemann sum limit used above, and letting $G(\mathbf{x}, \mathbf{y})$ be the strength of the interaction between points $\mathbf{x}$ and $\mathbf{y}$

$$E = -\frac{1}{8\pi} \lim_{N \to \infty} \iint q(\mathbf{x}) G(\mathbf{x}, \mathbf{y}) q(\mathbf{y}) dy dx \tag{7.16}$$

$$= -\frac{1}{8\pi} \lim_{N \to \infty} \iint \left(\sum_{j=1}^{N} s_j H_j(\mathbf{x}) \right) \left(\sum_{k=1}^{N} s_k H_k(\mathbf{y}) \right) G(\mathbf{x}, \mathbf{y}) dy dx \tag{7.17}$$

$$= -\frac{1}{8\pi} \lim_{N \to \infty} \iint \sum_{j=1}^{N} \sum_{k=1}^{N} s_j s_k H_j(\mathbf{x}) H_k(\mathbf{y}) G(\mathbf{x}, \mathbf{y}) dy dx \tag{7.18}$$

$$= -\frac{1}{8\pi} \lim_{N \to \infty} \sum_{j=1}^{N} \sum_{k=1}^{N} s_j s_k \iint H_j(\mathbf{x}) H_k(\mathbf{y}) G(\mathbf{x}, \mathbf{y}) dy dx \tag{7.19}$$

$$= -\frac{1}{8\pi} \lim_{N \to \infty} \sum_{j=1}^{N} \sum_{k=1}^{N} s_j s_k G(\mathbf{x}_j, \mathbf{x}_k) A_k A_j \tag{7.20}$$

$$= -\frac{1}{8\pi} \int_D \int_D \omega(\mathbf{x}) G(\mathbf{x}, \mathbf{y}) \omega(\mathbf{y}) dy dx \tag{7.21}$$

In calculating the energy we take the sum

$$E_N = -\frac{1}{8\pi} \sum_{j=1}^{N} \sum_{k=1}^{N} s_j s_k J_{j,k}$$

where

$$J_{j,k} = \left\{ \begin{matrix} \frac{16\pi^2}{N^2} G(\mathbf{x}_j, \mathbf{x}_k) & j \neq k \\ 0 & j = k \end{matrix} \right\} \tag{7.22}$$

and where the factor $\frac{16\pi^2}{N^2}$ represents (approximately) the area elements $A_j A_k$ over the sphere.

Since the minimum distance between any pair of mesh sites will be proportional to the inverse of $\sqrt{N}$ assuming the mesh sites are uniformly (or approximately uniformly) distributed on a finite domain, it may be expected the energy will diverge, because $J_{j,k}$ is proportional to $\log(N)$. This is not the case; while the minimum distance between any pair of points decreases this way the values of $J_{j,k}$ are not uniformly logarithmic in the number of mesh points. Because of this the energy converges.

The fact of this convergence can be made convincing by considering the case in which s_j uniformly equals 1 and the domain is the surface of the unit sphere. The energy as the number of points goes to infinity, with the scalings above, converges to $-\frac{1}{2}(\log(2)-1)$ [273]. From this it is straightforward to find upper and lower bounds for the energy based on the maximum and minimum of $\omega(\mathbf{x})$. More, as we have a maximum and a minimum energy for a given maximum and minimum vorticity we believe negative temperature may be observed.

7.3.2 Normalized Strength

The obvious alternative to interpreting the site strength s_j as the vorticity at a single point $\mathbf{x}_j$ is for it to be the scaled or normalized circulation of Voronoi cell D_j,

$$s_j = \frac{1}{A_j} \int_{D_j} \omega(\mathbf{x}) d\mathbf{x} \tag{7.23}$$

(To be precise this is the circulation of the boundary ∂D_j as the "internal" vorticity subtracts out.) The difference initially appears slight as the proofs the discretized vorticity field equation 7.2 with these s_j resemble those of the point-strength model closely. Consider the circulation of our new discretized vorticity as the number of mesh sites grows infinitely large:

$$\lim_{N \to \infty} \int_D q(\mathbf{x}) dA = \lim_{N \to \infty} \int_D \sum_{j=1}^{N} s_j H_j(\mathbf{x}) dA \tag{7.24}$$

$$= \lim_{N \to \infty} \int_D \sum_{j=1}^{N} H_j(\mathbf{x}) \frac{1}{A_j} \int_{D_j} \omega(\mathbf{y}) d\mathbf{y} d\mathbf{x} \quad (7.25)$$

$$= \lim_{N \to \infty} \sum_{j=1}^{N} \frac{1}{A_j} \int_D H_j(\mathbf{x}) d\mathbf{x} \int_{D_j} \omega(\mathbf{y}) d\mathbf{y} \quad (7.26)$$

$$= \lim_{N \to \infty} \sum_{j=1}^{N} \int_{D_j} \omega(\mathbf{y}) d\mathbf{y} = \int_D \omega(\mathbf{y}) d\mathbf{y} \quad (7.27)$$

which is the circulation of the continuous vorticity field.

The enstrophy of the discretized vorticity is a constant (if A_j is approximately uniform across the domain) times a lower bound to that of the continuous vorticity.

$$\int q^2(\mathbf{x}) d\mathbf{x} = \int \left(\sum_{j=1}^{N} s H_j(\mathbf{x}) \right)^2 d\mathbf{x} \quad (7.28)$$

$$= \int \sum_{j=1}^{N} s_j^2 H_j^2(\mathbf{x}) d\mathbf{x} \quad (7.29)$$

$$= \sum_{j=1}^{N} s_j^2 \int H_j^2(\mathbf{x}) d\mathbf{x} \quad (7.30)$$

$$= \sum_{j=1}^{N} s_j^2 A_j \quad (7.31)$$

Even in the limit as the number of mesh points grows infinitely large our discrete vortex approximation in this view provides only a lower bound of the actual enstrophy. But since we assume the vorticity $\omega(\mathbf{x})$ is continuous it is provable the difference between the enstrophy of the discretization and the enstrophy of the continuous fluid can be made arbitrarily small.

We next find the energy. Let $G^0(\mathbf{x}, \mathbf{y})$ be the Green's function on the sphere modified so that $G^0(\mathbf{x}, \mathbf{x})$ equals zero. Then the energy of the continuous vorticity is

$$E = -\frac{1}{8\pi} \iint \omega(\mathbf{x}) G^0(\mathbf{x}, \mathbf{y}) \omega(\mathbf{y}) d\mathbf{y} d\mathbf{x} \quad (7.32)$$

while that of our discretized version is

$$E = -\frac{1}{8\pi} \iint q(\mathbf{x}) G^0(\mathbf{x}, \mathbf{y}) q(\mathbf{y}) d\mathbf{y} d\mathbf{x}$$

$$= -\frac{1}{8\pi} \iint \left(\sum_{j=1}^{N} s_j H_j(\mathbf{x}) \right) G^0(\mathbf{x}, \mathbf{y}) \left(\sum_{k=1}^{N} s_k H_k(\mathbf{y}) \right) d\mathbf{y} d\mathbf{x} \quad (7.33)$$

$$= -\frac{1}{8\pi}\sum_{j=1}^{N}\sum_{k=1}^{N}\iint s_j s_k G^0(\mathbf{x},\mathbf{y}) H_j(\mathbf{x}) H_k(\mathbf{y}) d\mathbf{y} d\mathbf{x} \tag{7.34}$$

$$= -\frac{1}{8\pi}\sum_{j=1}^{N}\sum_{k=1}^{N} s_j s_k G^0(\mathbf{x}_j,\mathbf{x}_k) \iint H_j(\mathbf{x}) H_k(\mathbf{y}) d\mathbf{y} d\mathbf{x} \tag{7.35}$$

$$= -\frac{1}{8\pi}\sum_{j=1}^{N}\sum_{k=1}^{N} s_j s_k G^0(\mathbf{x}_j,\mathbf{x}_k) A_j A_k \tag{7.36}$$

if we accept that $G^0(\mathbf{x},\mathbf{y})$ may be approximated, assuming $\mathbf{x}$ to be in D_j and $\mathbf{y}$ to be in D_k as $G^0(\mathbf{x}_j,\mathbf{x}_k)$.

While the point-strength interpretation found the magnitude of s_j to be independent of the number of mesh sites N, in the mean-strength formulation there is a dependence. As N increases each s_j will naturally decrease. This may present numerical problems: floating point arithmetic becomes less reliable as we perform pairwise operations with numbers of very different magnitudes. As N increases each s_j decreases while the interaction strengths $J_{j,k}$ do not. In the range of mesh sizes we find interesting the magnitudes of numbers will not be so large as to make floating point errors likely.

However, this normalized-strength interpretation is both a physically sensible interpretation and a mathematically interesting model. In this construction any particular s_j is a **coarse-grained variable**, representing the theoretically complicated vorticity within a Voronoi cell by a single data point. We have only an average vorticity or circulation to consider. Averaging a quantity's value over a cell is a familiar step in modelling the quantity.

The intriguing question raised is, does a particular arrangement of site values $(s_1, s_2, s_3, \cdots, s_N)$ represent a microstate? While the specific assignment of values is unique, and therefore is a microstate, obviously one can imagine there being two or more arrangements of fluid flow which have the same averages and which therefore are both represented by the same state vector, making this a macrostate. Which is it? And, in hindsight, the same question can be asked about the representative-point model; so does that find a microstate or macrostate?

Obviously whether it is a micro- or a macro-state depends on context. With either interpretation of s_j the state $\mathbf{s} = (s_1, s_2, s_3, \cdots, s_N)$ is a microstate if we regard the site values as the limits of what can be measured in our fluid. If we consider our model as approximating a more fundamental level of information – one with vorticity values that can be measured at infinitely many sites arbitrarily close together, for example – then obviously $\mathbf{s}$ is a macrostate of all these.

7.4 Negative Temperatures

We have discussed briefly the notion of negative statistical mechanics temperatures and asserted that these will exist when a system has a maximum possible energy. It is worth taking a moment to sketch the argument that these negative temperatures will occur in the vortex problems we study.

Theorem 9. *Negative temperature states exist for point vortex problems provided the vortex problem has a phase space of bounded volume.*

The phase space for our vortex problem is the set of site strengths and positions $(s_1, \mathbf{x}_1, s_2, \mathbf{x}_2, \cdots, s_N, \mathbf{x}_N)$. Our argument derives from Onsager.

Let dV be the differential volume element of phase space, so

$$dV = ds_1 d\mathbf{x}_1 ds_2 d\mathbf{x}_2 \cdots ds_N d\mathbf{x}_N \tag{7.37}$$

By assumption the total volume is finite, so

$$\int dV = \left(\iint ds_j d\mathbf{x}_j \right)^N = A^N \tag{7.38}$$

We then also define a function $\phi(E)$ to be the volume of phase space which energy is less than E, so

$$\phi(E) \equiv \int_{H<E} dV = \int_{-\infty}^{E} \phi'(\tilde{E}) d\tilde{E} \tag{7.39}$$

with boundary conditions $\phi(-\infty) = 0$ and $\phi(\infty) = A^N$.

The derivative of the volume function, $\phi'(E)$, must be nonnegative for all energies E, and in fact we expect it to be positive. In order for $\phi(\infty)$ to be finite, therefore, $\phi'(E)$ must decrease to zero as E increases – and, in fact, there must be some maximum E_c, the critical energy at which $\phi''(E_c) = 0$.

The entropy of a system at a particular energy, $S(E)$, is the natural log of the area of the surface with energy E; and the inverse temperature β of a system is the rate of change of the entropy with respect to the energy. Therefore

$$\beta = \frac{d}{dE} \log\left(\phi'(E)\right) \tag{7.40}$$

$$= \frac{\phi''(E)}{\phi'(E)} \tag{7.41}$$

which, as $\phi'(E)$ is always positive, must therefore be positive when $E < E_c$, and negative when $E > E_c$.

The last point which must be proven is that our vortex models have a phase space of finite volume. For points on the surface of the sphere this is easy: the volume of phase space of each spatial coordinate $\mathbf{x}$ is just the area of the sphere on which the points lay (remember the phase space is the range of values which may be taken on; since the points are constrained to the surface of the sphere they are in two dimensions). The phase space of site strengths is constrained by the requirement we draw that enstrophy, the sum of squares of site values, be fixed; with that there are obviously maximum and minimum possible values for any s_j. Therefore we have negative temperatures. □

This argument shows we have negative temperatures both for the problem in which positive and negative vortices may occur, and for the problem in which all vortices are of the same strength. When we consider the vortex gas, with particles of fixed strength free to move, this justifies using negative temperatures, though we will find in the Monte Carlo simulation this causes points to cluster together around a few sites. There are no spatially uniform distributions in negative temperatures. This is to be expected from consideration of how to maximize a vortex gas's energy; it is also not unexpected given Onsager began studying negative temperatures to explain the accumulation of vortices into large structures.

This argument also requires our mesh be distributed over a compact space. There is no objection to this if we want to study the problem on the sphere, or inside a bounded region of the plane, or on finite but unbounded surfaces such as the torus. It does mean we do not at present have a justification for considering negative temperatures on the unbounded plane. This should be expected: given any collection of site values $(s_1, s_2, s_3 \cdots, s_N)$ and of mesh sites $(\mathbf{x}_1, \mathbf{x}_2, \mathbf{x}_3, \cdots, \mathbf{x}_N)$ the Hamiltonian

$$H = -\frac{1}{8\pi}\frac{16\pi^2}{N^2}\sum_{j=1}^{N}\sum_{k \neq j}^{N} s_j s_k \log|\mathbf{x}_j - \mathbf{x}_k| \tag{7.42}$$

may always be increased by the simple expedient of multiplying the coordinates of each $\mathbf{x}_j$ by a constant greater than one. Since the energy can invariably be increased there is no maximum energy and so we do not expect negative temperatures.

7.5 Phase Transitions

What is a phase transition? We all have a good intuitive picture of a phase transition, the change of water vapor into liquid water, or of liquid water into ice. We derive the name phase transition from these substantial changes in the physical properties of a system; that they should have some representation in statistical mechanics is almost implicit in our view that these transitions are caused by changes in temperature.

(They can also be caused by changes in pressure, or changes in chemical composition or even mechanical agitation, which can also be described as statistical mechanics processes.)

We know from physical experiments that the heat capacity of water – the amount by which the temperature of a body of water will rise given a certain quantity of additional heat – changes enormously as we approach the change from ice to liquid water, or from liquid water to steam. At the transitions there is an interval in which adding more heat only changes the state of the water, and does nothing to increase its temperature. We have effectively a discontinuity in the ratio between the heat added and the temperature change.

An **intensive variable** is a property inherent to the pieces of the model – physically these are properties like the conductivity or the density of a substance, or the inverse temperature at which phase transitions occur. An **extensive variable** depends on the amount of material present – the total mass, or charge, or in our model the net vorticity of a fluid.

Given a set of n extensive variables $Q_1, Q_2, Q_3, \cdots, Q_n$, we can define $n-1$ densities as

$$\rho_j = \frac{Q_j}{Q_n} \text{ for } j = 1, 2, 3, \cdots, n-1 \tag{7.43}$$

(with, often, Q_n chosen to be a quantity like the volume or the number of particles in the system, making each ρ_i explicitly a density) [242]. A **phase transition of the first kind** is a discontinuity in one or more of these densities. A **second-order phase transition** is one in which each ρ_i is continuous, but the first derivative (and higher order derivatives) have a finite discontinuity. A **third-order phase transition** similarly is continuous and has a continuous first derivative, but the second and higher-order derivatives are discontinuous. (This is the definition of Ehrenfest[6]. Fisher[7] recommends instead dividing just into first-order and second- or higher-order phase transitions.)

A classic statistical mechanics problem, such as an Ising model of ferromagnetism on a two-dimensional grid, can reveal phase transitions; they show themselves as discontinuities in (for example) the derivative of the mean magnetism per unit site as the temperature (or equivalently – and more usefully for us – inverse temperature β) varies.

In Monte Carlo experiments we do not see discontinuities. We can make numerical estimates of the derivatives, but cannot find the exact derivative of (for example) energy with respect to β and so do not see divergences in any quantity. What we see instead is **critical slowing down**, mentioned in chapter 4.

While measuring quantity f after some number N of Monte Carlo states, the size of fluctuations in the quantity should decrease. When the fluctuations

---

[6] Paul Ehrenfest, 1880 - 1933, became excited by mathematics and physics after attending a series of classes on the mechanical theory of heat which Ludwig Boltzmann taught. He was a pioneer of quantum mechanics and of non-equilibrium statistical mechanics. [288]

[7] Michael Ellis Fisher, born 1931, has studied at length the problem of phase transitions and Monte Carlo simulations. [130]

are sufficiently small we accept our results as near the equilibrium value. We take f after N sweeps and then again t sweeps later on. Then

$$\frac{\langle f(N+t)f(N)\rangle}{\langle f^2\rangle - \langle f\rangle^2} \sim \exp\left(-\frac{t}{\tau}\right) \tag{7.44}$$

with τ the correlation time [166] [353]. Around phase transitions is a significant increase in the correlation time. This does not prove a phase transition exists, but it suggests one and provides points to study by more rigorous methods. Numerical and analytical work have always complemented and inspired one another; fluid modelling is a particularly rich field for such approaches.

7.6 Energy-Enstrophy-Circulation Model

Our **energy-enstrophy-circulation model**, is derived from the Euler equation for inviscid fluid flow on the surface of the sphere. It examines the fluid flow by the spin-lattice approximation, representing the vorticity field by mesh sizes and strengths. In the Monte Carlo algorithm site strengths s_j are allowed to vary subject to particular constraints. The fluid flow kinetic energy of the system is, modulo a singular self-energy term, given by

$$H(\mathbf{s}, \mathbf{x}) = -\frac{1}{8\pi} \sum_{j=1}^{N} \sum_{k \neq j}^{N} s_j s_k J_{j,k} \tag{7.45}$$

with $\mathbf{s} = (s_1, s_2, s_3, \cdots, s_N)$, $\mathbf{x} = (\mathbf{x}_1, \mathbf{x}_2, \mathbf{x}_3, \cdots, \mathbf{x}_N)$ and $J_{j,k}$ equal to $\frac{16\pi^2}{N^2}$ times the logarithm of the distance between $\mathbf{x}_j$ and $\mathbf{x}_k$, which on the surface of a unit sphere is $1 - \mathbf{x}_j \cdot \mathbf{x}_k$. Of particular interest in this vortex model are the discretized total circulation, $\sum_{j=1}^{N} s_j = \Gamma$, and enstrophy, $\sum_{j=1}^{N} s_j^2 = \Omega$.

We microcanonically conserve both the discretized total circulation and the enstrophy, and seek to find statistical equilibria of the lattice vortex system for a wide range of values of the inverse temperature β. This sets the numerical simulations to be a version of Kac's spherical model discussed in the previous section. Fixing the temperature of a statistical simulation is indicative of a method based on Gibbs canonical statistics; the lattice system interacts with a virtual energy bath to reach equilibrium. However, because we also hold the total circulation and enstrophy fixed, these constraints are modelled microcanonically.

The Gibbsian statistics of the spherical model or the energy-enstrophy-total circulation theory is completely specified by sampling the system according to the probability measure

$$P(\mathbf{s}; \beta, \Omega, \Gamma) = \frac{1}{Z_N(\beta, \Omega, \Gamma)} \exp\left(-\beta H_N(\mathbf{s}, \mathbf{x})\right) \delta\left(\sum_{j=1}^{N} s_j^2 - N\right) \delta\left(\sum_{j=1}^{N} s_j\right) \tag{7.46}$$

where the sum in the partition function

$$Z_N(\beta, \Omega, \Gamma) = \sum_{\mathbf{s}} \exp\left(-\beta H_N(\mathbf{s}, \mathbf{x})\right) \delta\left(\sum_{j=1}^{N} s_j^2 - N\right) \delta\left(\sum_{j=1}^{N} s_j\right) \quad (7.47)$$

is taken over all lattice vortex vectors $\mathbf{s}$. We produce a sequence of states by the Metropolis-Hastings rule to explore the phase space of all possible site strengths subject to the circulation and enstrophy constraints. The larger the size N of the lattice, the more sharply peaked is the probability distribution around the most probable state $\mathbf{s}_*$ [410]. Thus, for fixed β and fixed N reasonably large, it will be a simple matter to run the algorithm and find the statistical equilibrium $\mathbf{s}_*$ with overwhelming probability $P(\mathbf{s}; \beta, \Omega, \Gamma)$.

7.7 Solution of the Spherical Ising Model for $\Gamma = 0$

We now briefly review the exact solution of the spherical model for barotropic flows in the inertial frame and refer the reader to the literature for details [258]. The partition function for the spherical Ising model has the form

$$Z_N \propto \int D(\mathbf{s}) \exp\left(-\beta H_N(\mathbf{s})\right) \delta\left(\Omega \frac{N}{4\pi} - \sum_{j=1}^{N} \mathbf{s}_j \cdot \mathbf{s}_j\right) \quad (7.48)$$

where the integral is a path integral taken over all the microstates $\mathbf{s}$ with zero circulation. In the **thermodynamic** or **continuum limit** considering the integral as $N \to \infty$, this partition function can be calculated using Laplace's integral form,

$$Z_N \propto \int D(\mathbf{s}) \exp\left(-\beta H_N(\mathbf{s})\right) \delta\left(\Omega \frac{N}{4\pi} - \sum_{j=1}^{N} \mathbf{s}_j \cdot \mathbf{s}_j\right) \quad (7.49)$$

$$= \int D(\mathbf{s}) \exp\left(-\beta H_N(\mathbf{s})\right) \left(\frac{1}{2\pi \mathrm{i}} \int_{a-\mathrm{i}\infty}^{a+\mathrm{i}\infty} d\eta \exp\left(\eta \left(\Omega \frac{N}{4\pi} - \sum_{j=1}^{N} \mathbf{s}_j \cdot \mathbf{s}_j\right)\right)\right) \quad (7.50)$$

Solving these Gaussian integrals requires first writing the site vorticities $\mathbf{s}_j$ in terms of the spherical harmonics $\{\psi_{l,m}\}_{l=1}^{\infty}$, which are the natural Fourier modes for Laplacian eigenvalue problems on S^2 with zero circulation, and using this to diagonalize the interaction in H_N. This process is known as diagonalization because we can view the calculation of the Hamiltonian as evaluating the product $\mathbf{s}^T H \mathbf{s}$ for the appropriate matrix of interactions H and the appropriate state vector $\mathbf{s}$, and as with all matrix problems matters become analytically simpler if we can reduce the problem to considering the diagonal of the matrix. Since we rewrite site values in terms of orthonormal basis functions, we can write

$$Z_N \propto \int D(\alpha) \exp\left(-\frac{\beta}{2} \sum_{l=1} \sum_{m=-l}^{l} \lambda_{lm} \alpha_{lm}^2\right) \times \left(\frac{1}{2\pi \mathrm{i}} \int_{a-\mathrm{i}\infty}^{a+\mathrm{i}\infty} d\eta \exp\left(\eta N \left(1 - \frac{4\pi}{\Omega} \sum_{l=1} \sum_{m=-l}^{l} \alpha_{lm}^2\right)\right)\right) \quad (7.51)$$

We have that the eigenvalues of the Green's function for the Laplace-Beltrami operator on S^2 are

$$\lambda_{l,m} = \frac{1}{l(l+1)}, \quad l = 1, 2, \cdots, \sqrt{N}, m = -l, -l+1, \cdots, 0, \cdots, l \quad (7.52)$$

and $\alpha_{l,m}$ are the corresponding amplitudes so that

$$s(\mathbf{x}_j) = \sum_{l=0}^{\sqrt{N}} \sum_{m=-l}^{l} \alpha_{l,m} \psi_{l,m}(\mathbf{x}_j) \quad (7.53)$$

for each of the mesh sites x_j.

Next we exchange the order of integration in equation 7.51. This is allowed provided that a is positive and is chosen large enough so that the integrand is absolutely convergent. Rescaling the inverse temperature by setting $\beta' N \equiv \beta$ yields

$$Z_N \propto \frac{1}{2\pi \mathrm{i}} \int_{a-\mathrm{i}\infty}^{a+\mathrm{i}\infty} d\eta \exp\left(\eta N \left(1 - \frac{4\pi}{\Omega} \sum_{l=1} \sum_{m=-l}^{l} \alpha_{l,m}^2\right) - \frac{\beta' N}{2} \sum_{m=-1}^{1} \lambda_{1,m} \alpha_{1,m}^2\right) \times \int_{l \geq 2} D(\alpha) \exp\left(-\sum_{l=2} \sum_{m=-l}^{l} \left(\frac{\beta' N \lambda_{l,m}}{2} + N\eta \frac{4\pi}{\Omega}\right) \alpha_{l,m}^2\right). \quad (7.54)$$

We can explicitly solve this inner integral because it is the product of a collection of Gaussian integrals. So

$$\int_{l \geq 2} D(\alpha) \exp\left(-\sum_{l=2} \sum_{m=-l}^{l} \left(\frac{\beta' N \lambda_{l,m}}{2} + N\eta \frac{4\pi}{\Omega}\right) \alpha_{l,m}^2\right) = \prod_{l=2}^{\sqrt{N}} \prod_{m=-l}^{l} \left(\frac{\pi}{N\eta \frac{4\pi}{\Omega} + \frac{\beta' N}{2} \lambda_{l,m}}\right)^{1/2} \quad (7.55)$$

provided this physically important conditions holds:

$$\frac{\beta' \lambda_{l,m}}{2} + \eta \frac{4\pi}{\Omega} > 0, \quad l = 2, 3, \cdots, \sqrt{N}, \quad m = -l, -l+1, \cdots, 0, \cdots, l \quad (7.56)$$

So we can now simplify equation 7.54 as

$$Z_N \propto \int_{a-\mathrm{i}\infty}^{a+\mathrm{i}\infty} d\eta \exp\left(N \begin{bmatrix} \eta\left(1 - \frac{4\pi}{\Omega}\sum_{m=-1}^{1}\alpha_{1,m}^2\right) \\ -\frac{\beta'}{2}\sum_{m=-1}^{1}\lambda_{1,m}\alpha_{1,m}^2 \\ -\frac{1}{2N}\sum_{l=2}\sum_{m}\log\left(N\eta\frac{4\pi}{\Omega} + \frac{\beta' N}{2}\lambda_{l,m}\right) \end{bmatrix}\right) \tag{7.57}$$

which we can cast in a form suitable for the saddle point method or method of steepest descent,

$$Z \propto \lim_{N\to\infty} \frac{1}{2\pi\mathrm{i}} \int_{a-\mathrm{i}\infty}^{a+\mathrm{i}\infty} d\eta \exp\left(NF(\eta, \Omega, \beta)\right) \tag{7.58}$$

in the thermodynamic limit as $N \to \infty$, where the free energy per site – after separating out the threefold degenerate ground states $\psi_{1,0}, \psi_{l,1}$, and $\psi_{l,-1}$ – is, modulo a factor of $-\beta'$, given by

$$\begin{aligned} F(\eta, \Omega, \beta') = \eta\left(1 - \frac{4\pi}{\Omega}\sum_{m=-1}^{1}\alpha_{1,m}^2\right) - \frac{\beta'}{2}\sum_{m=-1}^{1}\lambda_{1,m}\alpha_{1,m}^2 \\ -\frac{1}{2N}\sum_{l=2}\sum_{m}\log\left(N\eta\frac{4\pi}{\Omega} + \frac{\beta' N}{2}\lambda_{l,m}\right) \end{aligned} \tag{7.59}$$

The saddle point condition is

$$0 = \frac{\partial F}{\partial \eta} = \left(1 - \frac{4\pi}{\Omega}\sum_{m=-1}^{1}\alpha_{1,m}^2\right) - \frac{2\pi}{\Omega}\sum_{l=2}^{\sqrt{N}}\sum_{m=-l}^{l}\left(N\eta\frac{4\pi}{\Omega} + \frac{\beta' N}{2}\lambda_{l,m}\right)^{-1}. \tag{7.60}$$

To close the system we need a set of three additional constraints. These are given by the equations of state for $m = -1$, 0, and 1:

$$0 = \frac{\partial F}{\partial \alpha_{1,m}} = \left(\frac{8\pi\eta}{\Omega} + \beta'\lambda_{1,m}\right)\alpha_{1,m} \tag{7.61}$$

The last three equations have as solutions

$$\alpha_{1,m} = 0 \quad \text{or} \quad \frac{8\pi\eta}{\Omega} + \beta'\lambda_{1,m} = 0, \text{ for each } m. \tag{7.62}$$

This means that in order to have nonzero amplitudes in at least one of the ground or condensed states, which are the only ones to have angular momentum,

$$\frac{4\pi\eta}{\Omega} = -\frac{\beta'}{4} \tag{7.63}$$

which implies that the inverse temperature must be negative,

$$\beta' < 0. \tag{7.64}$$

The Gaussian condition, equation 7.56, on the modes with $l = 2$

$$\frac{\beta'}{1,2} - \frac{\beta'}{2} > 0 \tag{7.65}$$

can only be satisfied by a negative temperature, $\beta' < 0$, when there is any energy in the angular momentum containing ground modes.

When we substitute this nonzero solution into the saddle point equation it yields

$$0 = \left(1 - \frac{4\pi}{\Omega} \sum_{m=-1}^{1} \alpha_{1,m}^2\right) - \frac{4\pi}{\Omega} \frac{T}{N} \sum_{l=2}^{\sqrt{N}} \sum_{m=-l}^{l} \left(\lambda_{l,m} - \frac{1}{2}\right)^{-1} \tag{7.66}$$

$$= \left(1 - \frac{4\pi}{\Omega} \sum_{m=-1}^{1} \alpha_{1,m}^2\right) - \frac{T}{T_c} \tag{7.67}$$

where the critical inverse temperature is negative, has a finite large N limit, and is inversely proportional to the relative enstrophy Ω,

$$-\infty < \beta_c' = \frac{4\pi}{\Omega N} \sum_{l=2}^{\sqrt{N}} \sum_{m=-l}^{l} \left(\lambda_{l,m} - \frac{1}{2}\right)^{-1} < 0 \tag{7.68}$$

The saddle point equation gives us a way to compute the equilibrium amplitudes of the ground modes for temperatures hotter than the negative critical temperature T_c. (Remember that this is an extraordinarily high-energy state; a temperature of -1 is hotter than that of -2.) For temperatures T so that $T_c < T < 0$,

$$\sum_{m=-1}^{1} \alpha_{1,m}^2(T) = \frac{\Omega}{4\pi} \left(1 - \frac{T}{T_c}\right) \tag{7.69}$$

This argument shows that at positive temperatures (these correspond to a low barotropic energy; we examine this more explicitly in a few chapters), there cannot be any energy in the solid-body rotating modes. In other words, there is no phase transition at positive temperatures. This is the spin-lattice representation of the self-organization of barotropic energy into a large-scale coherent flow at very high energies in the form of symmetry-breaking Goldstone modes.

Moreover, these extremely high energy ground modes carry a nonzero angular momentum, which can be directed along an arbitrary axis, despite the fact that this problem is formulated in the inertial frame. This underscores the fact, further elaborated in chapter 13, that the spin-lattice Hamiltonians in this and the last chapter of the book model generalized barotropic flows, rather than the standard version of the Barotropic Vorticity Model in which angular momentum is conserved. The reason for doing this is explained in some detail in chapter 13.

These predictions for the spherical model of these barotropic vortex statistics should be compared with the results of Monte Carlo simulations, which we do in the next chapter. In particular, the linear dependence of the negative critical temperature $T_c = T_c(\Omega) < 0$ on the relative enstrophy should be noticed in the spherical model solution as well as the Monte-Carlo simulations.

8

Monte Carlo Simulations of Spin-Lattice Models on the Sphere

8.1 Introduction

We have now introduced most of the components we need, and with an understanding of how to conduct each experiment we can set up a Monte Carlo simulator. Pick a mesh size N, pick an inverse temperature β, and conduct a large number of sweeps. Each sweep consists of N experiments. Each experiment consists of picking three distinct sites and attempting to vary their strength as described in the above section. The change in enthalpy ΔH from an experiment is calculated, and a random number r is drawn from the interval $[0, 1]$; the change in site values is accepted if r is less than $\exp(-\beta \Delta H)$ and is rejected otherwise.

We have not yet established what mesh sites $\mathbf{x}_1, \mathbf{x}_2, \mathbf{x}_3, \cdots, \mathbf{x}_N$ to use. We will make a more thoughtful choice later, but for now will use mesh sizes N that are perfect squares, and place a "square grid" of points on the sphere. There will be $\sqrt{N}$ lines of latitude on which points are placed, and as many lines of longitude. Along each line of latitude the vortices will be uniformly spaced in longitude; the angle between one longitudinal line and the next is $\frac{2\pi}{\sqrt{N}}$.

We would like latitudinal lines evenly spaced, which brings up an interesting condition. We remember from chapter 6 the conjugate variables for vortex dynamics on the unit sphere are the longitude and the cosine of the co-latitude of a point. If we placed a square grid uniformly on the plane, we would have the lines be uniformly spaced in the horizontal and vertical directions. For vortex problems on the plane the horizontal and vertical directions are the conjugate variables. By analogy then we will put the lines of latitude uniformly spaced in the cosine of the co-latitude of their positions. The cosine of the co-latitude of the point on a unit sphere is just the Cartesian z-coordinate; so we place our rows to be uniformly spaced in z and in longitude.

It is worth exploring whether there are observable effects on the simulation results from choosing a mesh that is uniform in latitude, rather than in z, and we will look for those effects later.

For now though we will consider the relationship between any one mesh site and its nearest neighbor. There are three obviously likely relationships between the site values.

Antiparallel alignment. Suppose the two sites are likely to have strength the opposite sign; that is, the vortices point in antiparallel directions. The interaction energy of the pair is $-s_j s_k J_{j,k}$, with J_{jk} proportional to the logarithm of the distance between the two. For sites near one another J_{jk} is negative. Thus pairwise interactions have negative energy and if the majority of mesh sites are antiparallel to their neighbors then the system will have a negative energy. We expect positive temperatures β to produce this. With positive temperature changes which decrease enthalpy are always accepted and those increasing enthalpy are rarely accepted.

It is unlikely *every* site will be antiparallel to its neighbors, even if the inverse temperature is very large. To align every site opposite its neighbors is equivalent to showing the graph consisting of mesh sites and the connections between nearest neighbors is **bipartite**. A bipartite graph is a collection of **vertices** (points) and **edges** (curves which start at a vertex and end at a vertex, and which do not pass through any vertices besides its ends) which can divided into two sets, so that every edge connects a vertex in the first set to one in the second.

Still we can expect (and will see) a majority or more of sites aligned antiparallel to neighbors in these positive temperature regions. This state corresponds to the shortest wavelengths, that is, the largest wave-numbers, as our analysis of the partition function suggests we should see for positive β.

The plot in figure 8.1 shows an example of the antiparallel alignment as calculated by this Monte Carlo Metropolis-Hastings algorithm. Mesh sites are marked with a + symbol; the colors (or shading) correspond to the strength of the mesh site within the colored region. The colorings are assigned by a Voronoi diagram logic – the color of a point on the sphere matches the strength of the mesh point nearest it. The plot of site strengths shows a great peak and smaller peaks of smaller magnitude surrounding it, and an approximately checkerboard pattern around that which shows sites do tend to have opposite the sign of their nearest neighbors.

The second part of the figure presents the energy of the site values and positions after a thousand Metropolis-Hastings sweeps. Each sweep is N attempted experiments, where N is the number of points in the numerical mesh. We find the energy very quickly drops to a level following which the energy fluctuates but remains relatively constant. This is typical for systems that have reached a statistical equilibrium.

Chaotic alignment. Suppose one site is no more likely to have strength the opposite or the same sign as its nearest neighbor. The interaction energy between the two sites is as likely to be negative as positive and we cannot guess whether the overall energy will be low or high. An unpredictable arrangement suggests the system is disorganized. If our intuition from real temperature – in which low temperatures give solids, medium temperatures liquids and high

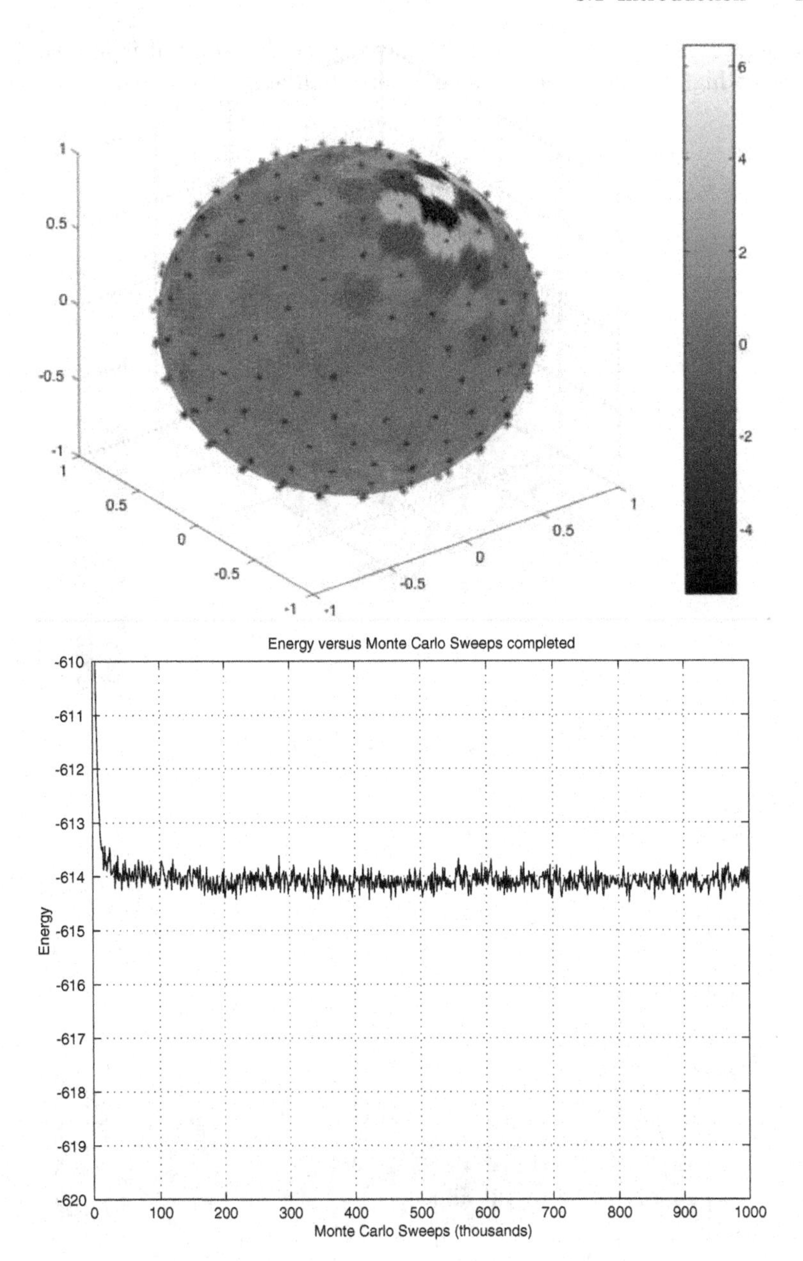

Fig. 8.1. An example of antiparallel alignment of site strengths. For this example there are 256 points, with $\beta = 10$ and run for $1,000,000$ sweeps.

temperatures gases – hold then we may expect this state, if it appears, will do so at high temperatures. This is inverse temperatures β near zero.

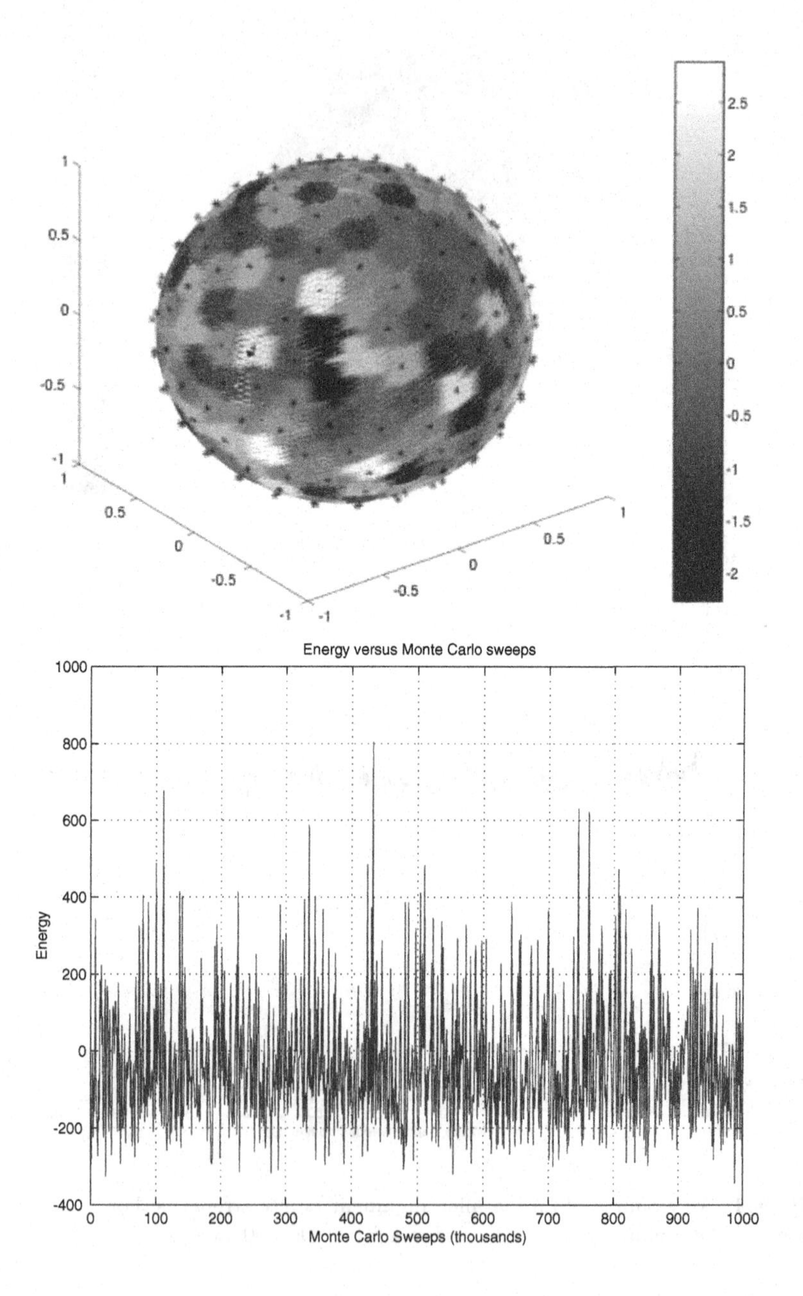

Fig. 8.2. An example of chaotic alignment of site strengths. For this example there are 256 points, with $\beta = 0$ and run for $1,000,000$ sweeps.

If β equals zero every proposed change in site strengths is approved, and site values will be randomly distributed. A small nonzero β may show bias towards some organization, but if the inverse temperature is small enough a great number of "counterproductive" moves – which increase the product of β and enthalpy – are approved and neighboring sites will be almost randomly aligned.

While we expect this to appear when β is zero, it is obviously of little importance. It appears where the magnitude of β is small because in the region near a phase transition there is a critical slowing down, a region of β in which the system takes very many Metropolis-Hastings experiment to settle into a single phase, if it ever should settle. Detecting such a phase near $\beta = 0$ suggests there is a phase transition around this temperature.

The plot of site values in figure 8.2 shows no particular organization between sites, and the energy plot in the second part of the figure shows wild fluctuations in the system energy as the number of sweeps increases. From this it is not hard to conclude all we see here are random fluctuations as sites vary independently of their neighbors.

Parallel alignment. Suppose the two are likely to have strength the same sign; that is, the vortices are parallel. The interaction energy is therefore positive. If many sites are to be parallel then this must be a high-energy state. This reflects a system dominated by the longest wavelength, that is, when the wave-number is smallest, and just as the partition function suggests for negative temperatures the system will be dominated by the smallest wave-number or largest wavelength.

It would be remarkable to see this state for positive inverse temperature. But in negative temperatures the Metropolis-Hastings algorithm approves steps that increase the enthalpy. This is done when the sites are given strengths approximately equal in magnitude and direction.

The plot of site values in figure 8.3 shows one hemisphere of positive and one of negative vorticity, with the strength decreasing from a "north vorticity pole" – the site of greatest vorticity – to a south – the least, generally negative. The energy after a thousand sweeps shows the system reaching very close to a maximum and approximately staying there; fluctuations are much smaller than are seen for the antiparallel state.

The results of our preliminary considerations are not surprising; we will want experiments at negative, near-zero, and positive β and we do not yet know what mesh sizes are most likely productive. We do know to consider negative temperatures, and expect that near zero the system will be chaotic.

8.2 Correlation Functions

It is obviously interesting whether there are structures being built of the site strengths. In the problem of site magnetism, invariably a reference point for spin-lattice models, whether a material in general is ferromagnetic or not

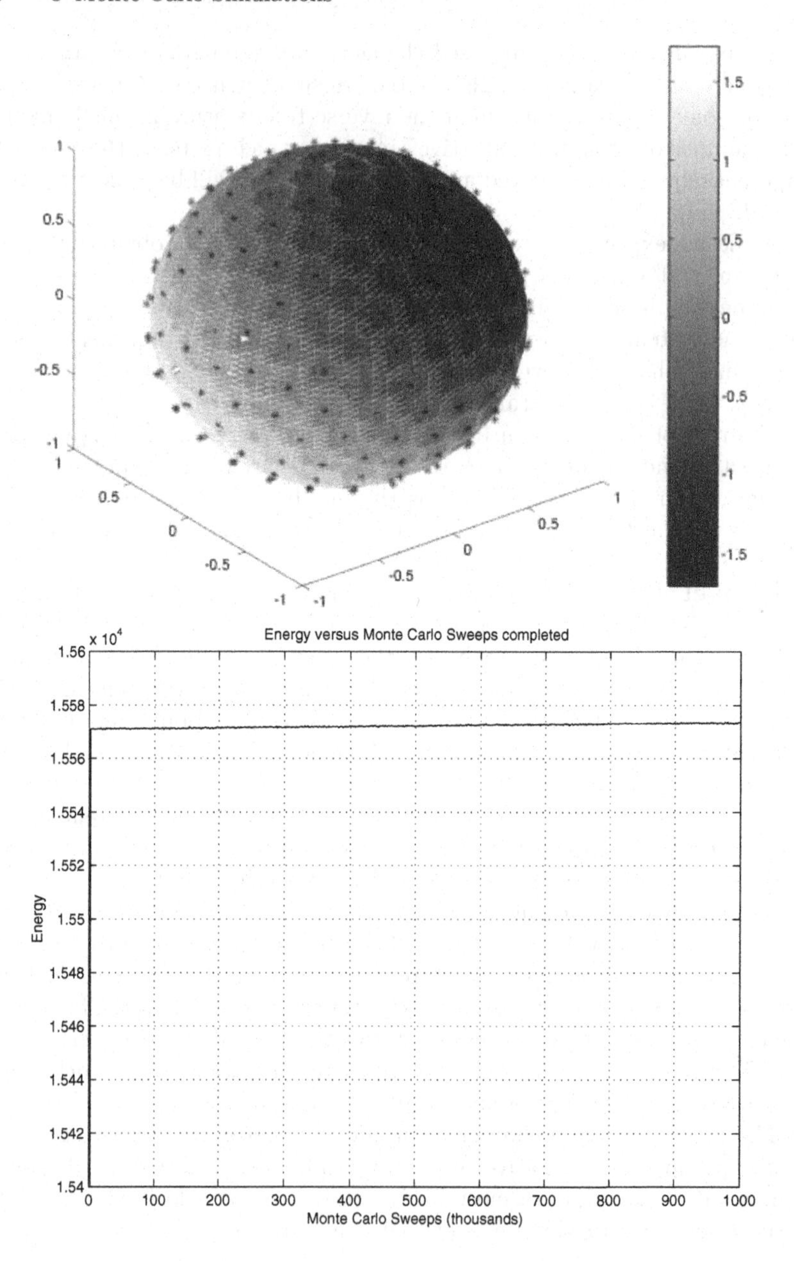

Fig. 8.3. An example of parallel alignment of site strengths. For this example there are 256 points, with $\beta = -10$ and run for $1,000,000$ sweeps.

depends on the building of structures, of whether many sites have magnetisms parallel or antiparallel to one another, or whether there is no relationship at all among sites. This extends naturally to our spin-lattice model; even the parallel alignment that leads to a large magnetic moment obviously corresponds to our parallel state.

The **correlation** r between any two random variables X and Y is

$$r(X,Y) = \frac{\langle (X - \langle X \rangle)(Y - \langle Y \rangle) \rangle}{\sigma_X \sigma_Y} \tag{8.1}$$

where σ_X and σ_Y are the standard deviations of X and Y. This is exactly the correlation coefficient found from attempting to perform a least mean squares linear approximation to the relationship between X and Y, and it is obviously related to the correlation time discussed in previous chapters regarding whether two elements of an ensemble were independent.

The closer the correlation is to 1 or to -1 the stronger the relationship between the two variables. Positive numbers correspond to the two variables increasing linearly together, negative numbers to one linearly decreasing as the other increases. Non-linear correlations are not necessarily detected by this function, but independent variabes have correlations near zero.

While this is generally useful for comparing multiple variables this does not right away give us a new tool to use in studying the site strengths. Obviously the correlation of site strengths with themselves will be one; the correlation between distance and site strength – if we measure distance from the "north vorticity pole" – is non-linear; the site strengths vary as a cosine from the axis of rotation, in the parallel case, and vary as a much more complex function in the antiparallel case.

So we define now the **spin correlation function**, a function of distance d from a reference point we will label s_j:

$$S_j(d) = \langle s(d) s_j \rangle - \langle s(d) \rangle \langle s_j \rangle \tag{8.2}$$

where $s(d)$ is the collection of site strengths for sites a distance d from s_j, and similarly the expectation values are taken over all the points a distance d from s_j. The **spin-spin correlation** function $S(d)$ is the mean of $S_j(d)$ taken over all sites j.

This function, properly, is nonzero only for the handful of distances representing any pairwise distance, and even in a uniform and regular mesh there may be very few points going into the "expectation value" for any pairwise correlation. The function becomes more nearly continuous as the number of points in the mesh grows, to the point that when spin-spin correlation functions are examined for physical models (where each site is a molecule in a crystal) we ignore the discontinuities. It is also not rare to take the sites and the expectation values over sites over the annulus with inner radius $d - \frac{\delta}{2}$ and outer radius $d + \frac{\delta}{2}$ for some small δ.

Obviously the strength of any one site can in principle affect any other site, but in practice there is a maximum length between which any two sites have a measurable correlation. This is known as the **correlation length** ξ. In statistical mechanics terms we find that for inverse temperatures β near a critical point – a phase transition temperature β_c – the correlation function becomes approximately

$$S(d) = \frac{1}{d^p} \exp\left(-\frac{d}{\xi}\right) \tag{8.3}$$

for a critical power exponent p. The correlation length is not constant; it depends on β and near the inverse temperature it will be approximately

$$\xi \approx \left|1 - \frac{\beta}{\beta_c}\right|^{-\nu} \tag{8.4}$$

for a critical exponent ν. Note the parallel to the correlation time, the slowing-down of the convergence of a Monte Carlo Metropolis-Hastings algorithm to a particular state as inverse temperature comes nearer a transition point; they are aspects of the same phenomenon.

Still despite the general utility of spin-spin correlation functions we find they have a few limitations that restrict their use in our problems. The most significant limitation is the averages needed for the system are most accurate when there are many points of the same distance apart. For problems with uniform meshes this is fine, but we want to use meshes generated by Monte Carlo algorithms, in which there will not be regular grids. While (we will find) this produces meshes that are nearly uniform, there will be few exact duplications of pairwise lengths. We may "blur out" the distances by the annulus method above, at the cost of decreasing the accuracy of our results. More important is that the spin-spin correlation function is tailored to linear correlations between distance and spins, and we have little reason to expect linear relationships.

Therefore we will seek a function which provides generally less information, but which is better at detecting the point we are most interested in, whether sites are aligned parallel or antiparallel to their neighbors.

8.3 The Mean Nearest Neighbor Parity

It is not difficult to interpolate the site strengths s_j produced by a Monte Carlo Metropolis-Hastings simulation in the parallel state to a continuous curve $\omega(\mathbf{x})$ describing the vorticity. The parallel state, also called the solid-body rotation state because its vorticity distribution varies just as the site vorticity of a solid-body rotating on a single axis, has a vorticity described by equation $\omega(\mathbf{x}) = A \cos(\mathbf{x} \cdot \boldsymbol{\theta})$ for some amplitude A and some axis of rotation θ.

This axis is easily approximated from the site strengths s_j and positions $\mathbf{x}_j$. Letting max be the index of the greatest site strength and min the index

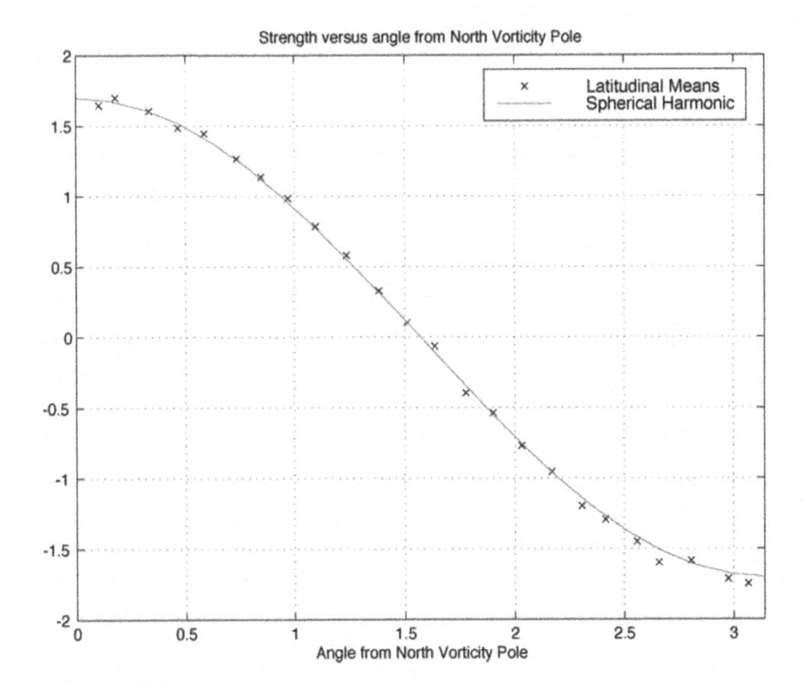

Fig. 8.4. The site strengths found in a parallel state compared to the vorticity as determined by a spherical harmonic.

of the smallest then the axis of rotation is approximately $\mathbf{x}_{\max} - \mathbf{x}_{\min}$. Let A be the greater of $|s_{\max}|$ and $|s_{\min}|$. This is the comparison made in figure 8.4, in which a parallel state was found from the Metropolis-Hastings simulation, and then its axis of rotation and amplitude A calculated from the maximum and minimum site strengths.

Compared in figure 8.5 is the Hamiltonian energy of the parallel configuration for meshes of various sizes (we have used meshes from 32 through 1024 points, a convenient range) compared to the energy of the first spherical harmonic with the matching amplitude and axis of rotation.

One may wonder why the energy of this parallel state increases as the number of sites increases, and why it is always less than that the of the matching spherical harmonic. The cause is obvious from considering the energy of N interacting particles

$$H(\mathbf{s}, \mathbf{x}) = -\frac{1}{8\pi} \sum_{j=1}^{N} \sum_{k \neq j}^{N} s_j s_k \log(d_{j,k}) \nu_N \tag{8.5}$$

and the energy of the spherical harmonic

$$H = -\frac{1}{8\pi} \int_{S^2} \int_{S^2} s(\mathbf{x}) s(\mathbf{x}') \log(1 - \mathbf{x} \cdot \mathbf{x}') d\mathbf{x}' d\mathbf{x} \tag{8.6}$$

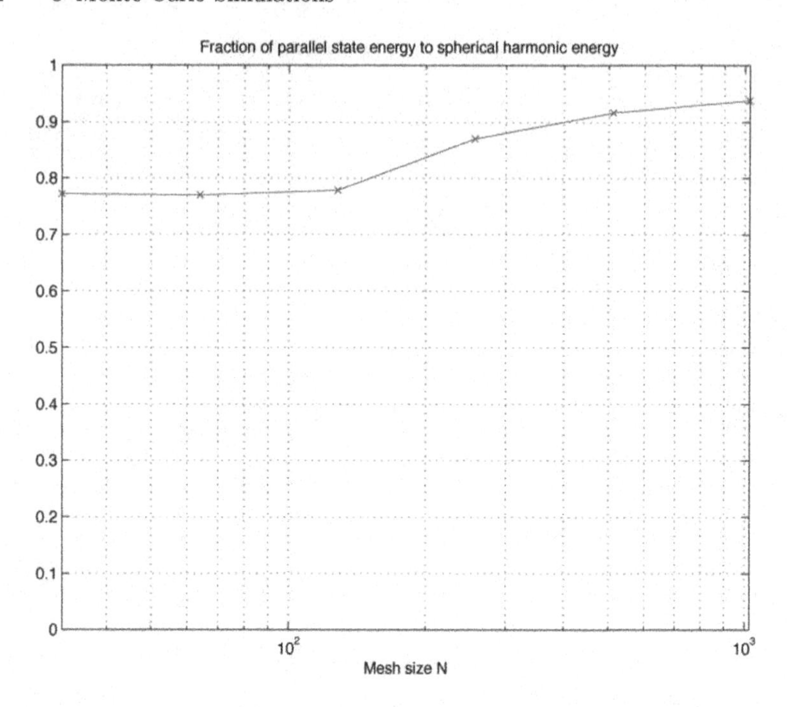

Fig. 8.5. The site strengths found in a parallel state compared to the vorticity as determined by a spherical harmonic.

$$s(\mathbf{x}) = A\boldsymbol{\theta} \cdot \mathbf{x} \tag{8.7}$$

where $\boldsymbol{\theta}$ is the axis of rotation and A the maximum amplitude. With a mesh of N discrete points the energy of the interactions of sites nearer to $\mathbf{x}_j$ than to any other mesh site are omitted from the energy sum. Thus the energy of our discretized approximation to the vorticity field will be less than the energy of the continuous field.

While curve-fitting works well for the parallel state which we expect to be the simplest spherical harmonic, it is less efficient to find a curve fitting for the antiparallel state. The number of longitudinal and latitudinal zones – and the axis of rotation to be used – is unclear; and while it is possible to identify all these parameters it is hard to be completely convinced by them.

And yet we should be able to determine given the site strengths and mesh positions whether a system is in the relatively organized, antiparallel, state, or if it is in the chaotic state. To find some estimate of what state systems are in we introduce the **mean nearest neighbor parity**.

Definition 3. *Let the parity of a pair of sites be the product of the sign of one site of the pair's vorticity with the sign of the vorticity of the other site. If the sites are both positive or both negative this is* $+1$*. If one is positive and one is negative this is* -1*. If either site is zero this is* 0*.*

For site j, the nearest neighbor k is the site closest to it. The nearest neighbor parity is then the parity of the pair of points j and k.

Definition 4. *Let the mean nearest neighbor parity be the arithmetic mean of the nearest neighbor parities for each nearest neighbor pair.*

(This slightly circumlocutious definition is used to avoid ambiguity in the case one site has multiple nearest neighbors. That condition seems unlikely given a mesh generated by Monte Carlo Metropolis-Hastings algorithms as outlined in chapter 10 but remains possible numerically and analytically.)

This property provides an excellent measure for the structure of site vorticities. When sites are generally aligned parallel to their neighbors the mean nearest neighbor parity is close to $+1$. (It is not likely to be exactly one, as the constraints of this problem require total circulation to be zero, so there must be both positive and negative vorticities, and it is quite likely at least one pair of nearest neighbors will have opposite signs.) While it is possible the mesh sites may be arranged so that the positive region and the negative region are so separated no positive site is the nearest neighbor of any negative site, this is an improbable configuration.

When sites are typically antiparallel to one another site mean nearest neighbor parity tends to be close to -1. As with the parallel state it is improbable that the nearest neighbor of each and every site will be one of the opposite sign.

In the chaotic state, in which no particular correlation exists between one site's values and another's, the mean nearest neighbor parity will average closer to 0. Rarely will it be exactly zero, as above.

While the mean nearest neighbor parity provides a measure of a system's organization which fits well an intuitive understanding of the parallel, antiparallel, and chaotic states, it is not infallible. One can design point meshes and vorticity assignments so the mean nearest neighbor parity misrepresents the system. This does not invalidate the utility of the mean nearest neighbor parity and it is argued in proposition 14 that such misleading results are improbable, but does remind one that these numerical tools require some caution.

Proposition 13. *It is possible for the mean nearest neighbor parity to provide a misleading understanding of the structure of a system.*

Proof. Let the vorticity of a system $u(\theta, z)$ (where θ is the longitude and z the cosine of co-latitude of a point on the unit sphere) equal 1 if $z \geq 0$ and -1 if $z < 0$. This is the parallel state, a high-energy state common to negative inverse temperatures β. Let the numerical mesh be located at the points (θ_j, z_j) for $j = 1, 2, 3, \cdots, 2n$, where $\theta_j = 2\pi \frac{(j+1)}{2n}$ for j odd, $\theta_j = 2\pi \frac{j}{2n}$ for j even, $z_j = \cos\left(\frac{\pi}{8n}\right)$ for j odd, and $z_j = -\cos\left(\frac{\pi}{8n}\right)$ for j even. and $z_j = (-1)^j \left(\frac{1}{2n+1}\right)$ for all j. Though the system is by construction in the parallel configuration, with all the positive vorticity in one region and all the negative vorticity in another, the mean nearest neighbor parity for this case is -1. □

In practice, the mean nearest neighbor parity is measured only for vorticity assignments which have reached a statistical equilibrium, and the tendency is for this to prevent such counterintuitive conclusions.

Proposition 14. *The state described in the proof of Proposition 13 does not maximize the Gibbs factor for negative temperatures β, in that vorticity assignments on the mesh sites could be rearranged to provide a system with higher Gibbs factor, and therefore is unlikely to be the result of the Monte Carlo algorithm used in this chapter.*

Proof. The example of $n = 2$ demonstrates the argument. Given the mesh sites $\left(0, \cos\left(\frac{\pi}{16}\right)\right)$ and $\left(\pi, \cos\left(\frac{\pi}{16}\right)\right)$ with the vorticity $+1$, and given the mesh sites $\left(0, -\cos\left(\frac{\pi}{16}\right)\right)$ and $\left(\pi, -\cos\left(\frac{\pi}{16}\right)\right)$ with the vorticity -1, the Gibbs factor for any negative β is

$$\exp\left(-2\beta\left(\log\left(2\cos^2\left(\frac{\pi}{16}\right)\right) - \log\left(2 - 2\cos^2\left(\frac{\pi}{16}\right)\right) - 2\log(2)\right)\right) \quad (8.8)$$

By rearranging the vorticities of the sites – so that the mesh sites with locations $\left(0, \cos\left(\frac{\pi}{16}\right)\right)$ and $\left(0, -\cos\left(\frac{\pi}{16}\right)\right)$ have vorticity 1, and so that the sites with locations $\left(\pi, -\cos\left(\frac{\pi}{16}\right)\right)$ and $\left(\pi, -\cos\left(\frac{\pi}{16}\right)\right)$ have vorticity -1, the Gibbs factor is

$$\exp\left(-2\beta\left(-\log\left(2\cos^2\left(\frac{\pi}{16}\right)\right) + \log\left(2 - 2\cos^2\left(\frac{\pi}{16}\right)\right) - 2\log(2)\right)\right) \quad (8.9)$$

which is larger and therefore more likely as the final state of the Monte Carlo chain. □

The result of a Monte Carlo run on the above mesh and with the above vorticities will most likely be to have the sphere divided into the two regions, and the mean nearest neighbor parity for that outcome will be $+1$. This is what is expected from the system when β is negative. So while it is possible for the mean nearest neighbor parity to misrepresent the system, it is not likely to.

The mean nearest neighbor parity has difficulty detecting modes which have medium-range structure, with regions larger than one mesh site but smaller than a hemisphere alternating values. But since such medium-range structures are not expected and not observed numerically this is not a practical limitation.

With a program capable of finding a statistical equilibrium for a given mesh size N, enstrophy Ω, at an arbitrarily chosen inverse temperature β, we can experiment to search for evidence suggesting a phase transition as the temperature changes. The first and simplest experiment is to try a fixed $N = 256$, and to allow the enstrophy to vary; picking a numerically convenient range we try from $\Omega = 32$ to $\Omega = 1024$. Results of plotting the mean nearest neighbor parity at equilibrium versus β are plotted in figures 8.6 and 8.7.

The first suggestion of a phase transition comes from looking at the negative temperature and the positive temperature plots for the same enstrophy. In each of the plots from figure 8.6 (negative temperatures, the extremely high-energy states) the mean nearest neighbor parity begins relatively near 1, corresponding to the parallel state – one hemisphere is all positive vorticities and the other negative. As β approaches zero (the temperature approaches negative infinity), the system becomes less energetic and at some temperature the mean nearest neighbor parity drops to zero.

In positive temperatures as β increases from zero (so temperature decreases from infinitely large – quite hot, but not as hot as possible – to zero) the mean nearest neighbor parity decrease from zero. In the range of temperatures examined it never drops completely to zero, but it does steadily decrease, and if we examine a plot of one of these equilibria it appears to be in the antiparallel state. This fact highlights a weakness of the mean nearest neighbor parity as a tool.

Whatever the equilibrium it is not unreasonable for a few of the mesh sites to be assigned an "incorrect" vorticity, a bit off of the value the site would have if we solved for the exact, continuous fluid flow, equilibrium. This does not change the mean nearest neighbor parity, however, unless the site value is already near zero, so the small error changes the site's parity with its neighbors. For negative temperatures one hemisphere is positive values and one negative, approximately; it is only along the rim that site values are small enough that they might be changed. This is very few points of the whole, and the fraction of such rim sites decreases as N increases.

In positive temperatures, however, there is a regular oscillation between positive and negative values; nearly any mesh site is near the change of sign of the vorticity. Therefore proportionately more sites have near-zero vorticities and there are more chances for a "misplaced" sign to change the parity of a nearest-neighbor pair. Given that, the parity for positive temperatures being more nearly zero is to be expected, and we can accept the flatter curves of mean nearest neighbor parity versus β.

(There is also a greater fluctuation around the curves that appear to be present in smaller meshes than in larger ones. That too is to be expected, from reasoning like that above.)

The asymmetry in how parity drops as one recedes from zero in negative and in positive β suggests there may be another interesting phase transition in the negative region. So let us consider the region of negative β where the mean nearest neighbor parity drops from the large near-one number down to zero. We look at a specific example with $N = 256$, and enstrophy $\Omega = 256$, and a finer placement of sampling temperatures β.

The curve of data points is chosen to fit the typical curve we see in a phase transition,

$$m(\beta) = C\left(1 - \frac{\beta_c}{\beta}\right)^{\alpha} \tag{8.10}$$

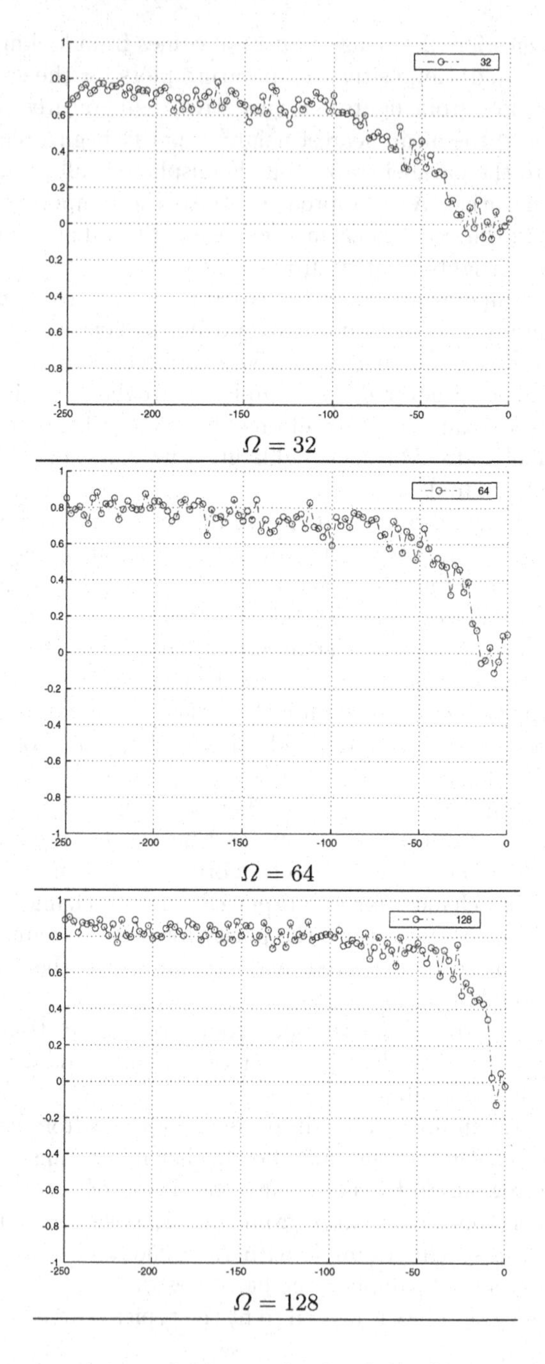

Fig. 8.6. The mean nearest neighbor parity for the problem on the sphere versus inverse temperatures on a range from $\beta = -250$ to $\beta = 0$ with 256 points but varying enstrophy.

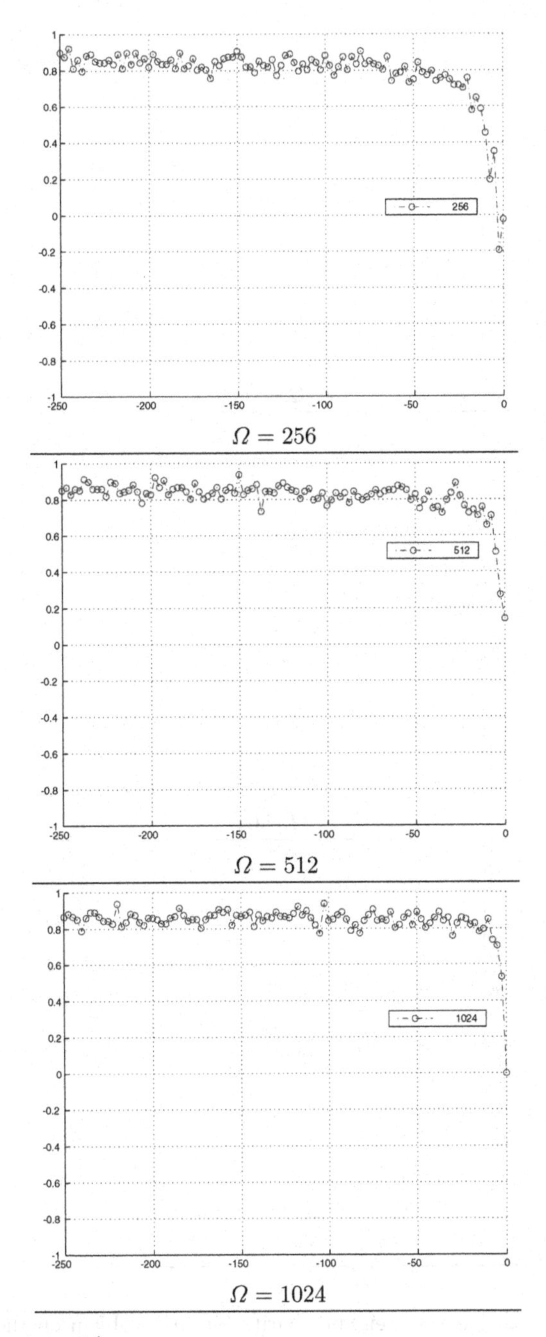

$\Omega = 256$

$\Omega = 512$

$\Omega = 1024$

Fig. 8.6. (Continued.)

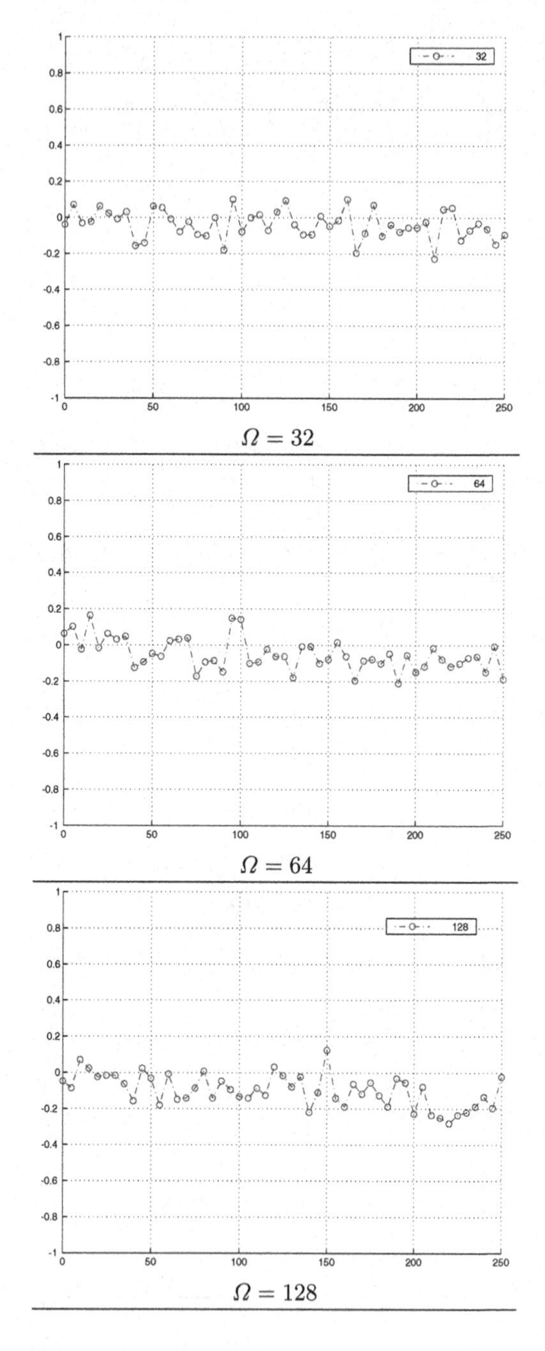

Fig. 8.7. The mean nearest neighbor parity for the problem on the sphere versus inverse temperatures on a range from $\beta = 0$ to $\beta = 250$ with 256 points but varying enstrophy.

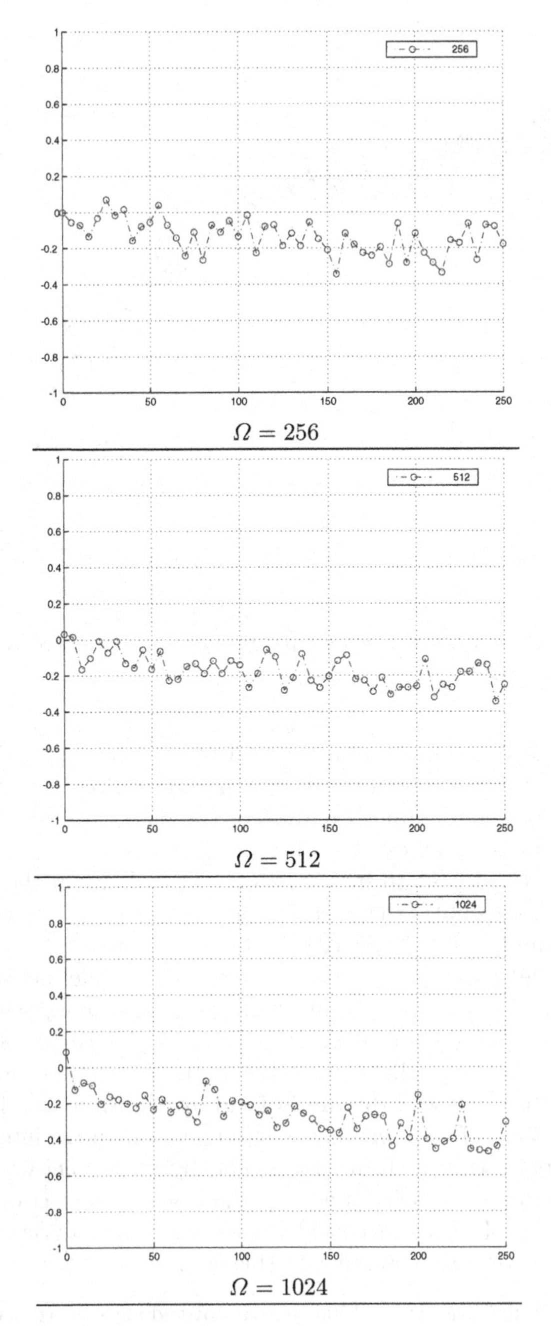

Fig. 8.7. (Continued.)

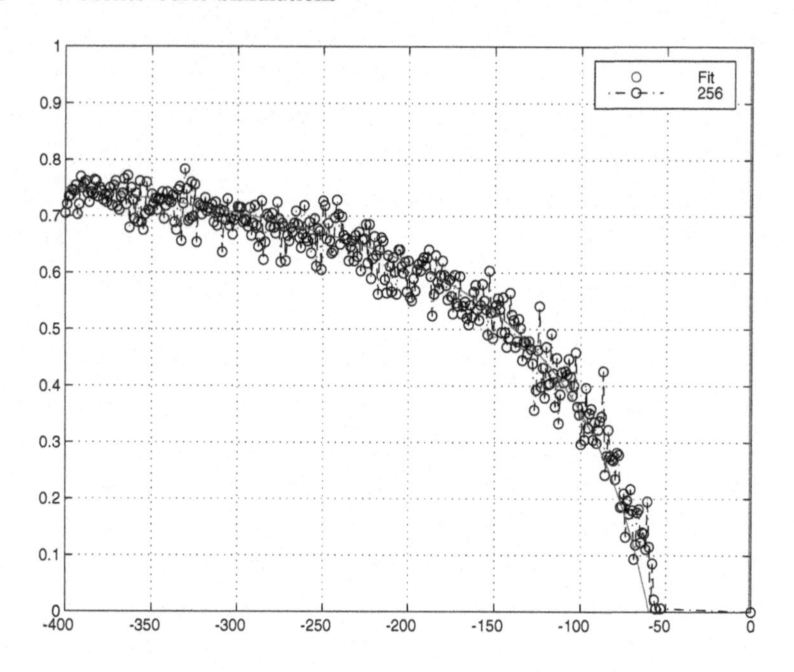

Fig. 8.8. The mean nearest neighbor parity and a fitting curve across a range of inverse temperatures.

for some scaling constant C, some critical temperature β_c and some scaling constant α. In this case, by a touch of experimentation backed up by linear regression analysis on a transformation of the data, we find that $C = 0.75$, $\beta_c = -60$ and $\alpha = 1$ is an excellent fit.

This is a surprise and an insightful one – the decay of the mean nearest neighbor parity resembles the decrease in magnetism in a ferromagnet with temperature and the way magnetism will drop to zero when the temperature passes the Curie[1] point. This is far from the first parallel between ferromagnetism and our vorticity models, but it raises an intriguing question.

The Curie point temperature and the transition from a magnetic to a disorganized system can be predicted analytically from the properties of point ferromagnets by making use of the **mean field approximation**. In this scheme we treat the energy of a system as depending not on all the interactions of all the particles, but rather as the interaction of any given point with an averaged out approximation to the overall field, perhaps with a correction given for the nearest neighbors. This allows considerable savings in analytic work but is not valid for every problem. Is it valid for this one?

[1] Named for Pierre Curie, 1859 - 1906, who discovered piezoelectricity and with wife Marie Sklodowska Curie, 1867 - 1934, discovered the radioactivity of thorium and uranium, and isolated uranium, radium, and polonium.

We do not know, although we suspect it likely is. We have the numerical experiments outlined in this chapter which are consistent with a mean field theory, which proves nothing but suggests the conclusion. The real question is whether the mean field theory becomes exact – whether its derivations of the partition function, for example, match the original – when the number of mesh sites N grows infinitely large. Research in the similar Onsager vortex gas problem [67] [68] [293] suggests these N-vortex spin-lattice models with a spherical model constraint on the site values should be a mean field theory, but a complete proof of this has not been made.

The next thought comes from considering the behavior as the enstrophy increases from 32 to 1,024. In negative temperatures there is a clear falling of the mean nearest neighbor parity from 1 to 0 as β nears zero. How sharp a fall this is increases as enstrophy does, and the value of β at which the parity stops being nearly constant and decreases grows nearer to zero as enstrophy increases. The phase transition is happening as β grows closer to zero; that is, at temperatures which are becoming increasing in magnitude while staying negative, which is consistent with the exact solutions of the spherical model obtained by Lim [258] and reported in chapter 7.

The next obvious thing to consider is whether a phase transition can be found if the enstrophy is kept constant but the number of mesh sites is allowed to vary. This appears to be supported by the plots of figure 8.9 where we find that the critical inverse temperature becomes more negative with increasing N in nearly a linear fashion. This is consistent with the scaling of the inverse temperature by the number of lattice points that is required in the exact solution of the spherical model mentioned in the previous chapter [258].

The appearance of the variation of mean nearest neighbor parity versus β for positive β, with different mesh sizes N, as seen in figure 8.10, is interesting. We know from experience with a constant mesh size and varying enstrophy that the mean nearest neighbor parity drops towards -1 as enstrophy increases; here we see it sticking near zero as N increases. From these plots it seems likely the mean nearest neighbor parity at a given β depends on the mean enstrophy per mesh site; certainly it is easy to suppose the parity would be close to zero if most of the mesh sites are near zero and therefore their sign being rather arbitrarily distributed.

As a check, though, we can examine a lot of one of these apparently-zero mean nearest neighbor parity states to see if it is more fairly described as an antiparallel state. Figure 8.11 is an example of this with $N = 256$ and $\Omega = 256$; and it does seem the sphere is divided into checkerboard-like regions of relatively large mesh sizes, and then relatively flat regions where average mesh values are close to zero. It seems this analysis of the mean nearest neighbor parity is supported by the numerical results, and that it is consistent with this phase transition.

Considered now is the effect on the mean nearest neighbor parity of the number of sweeps completed. This is needed to establish in part how long the experimental runs need to be in order to have an estimate of the system's

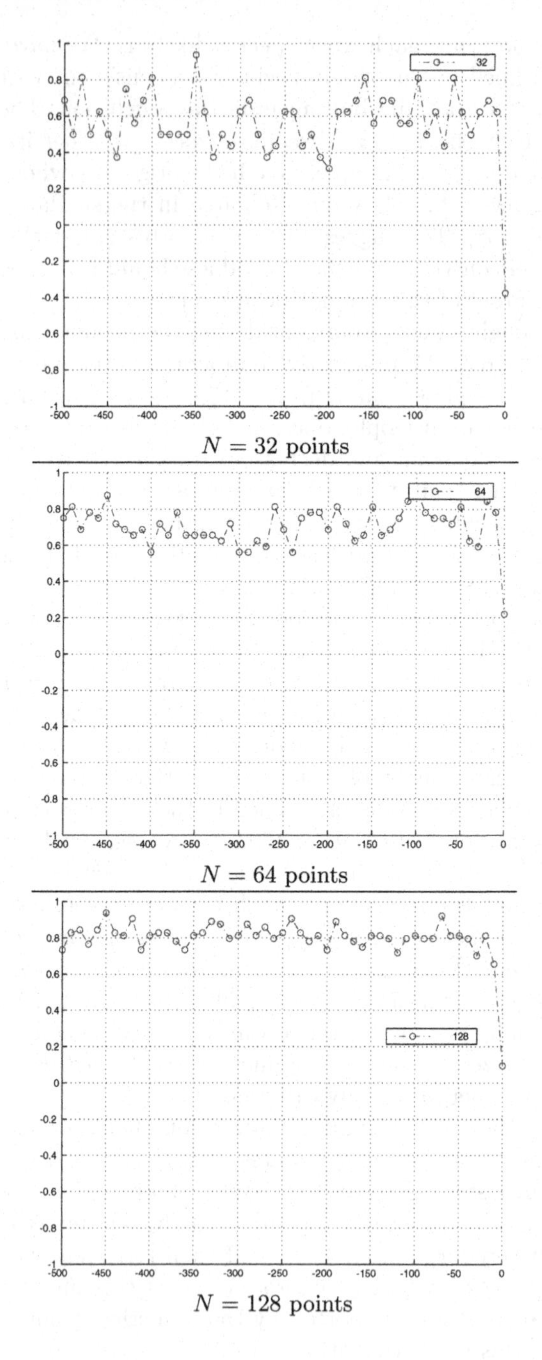

Fig. 8.9. The mean nearest neighbor parity for the problem on the sphere versus inverse temperatures on a range from $\beta = -500$ to $\beta = 0$ with constant enstrophy $\Omega = 256$ but varying N.

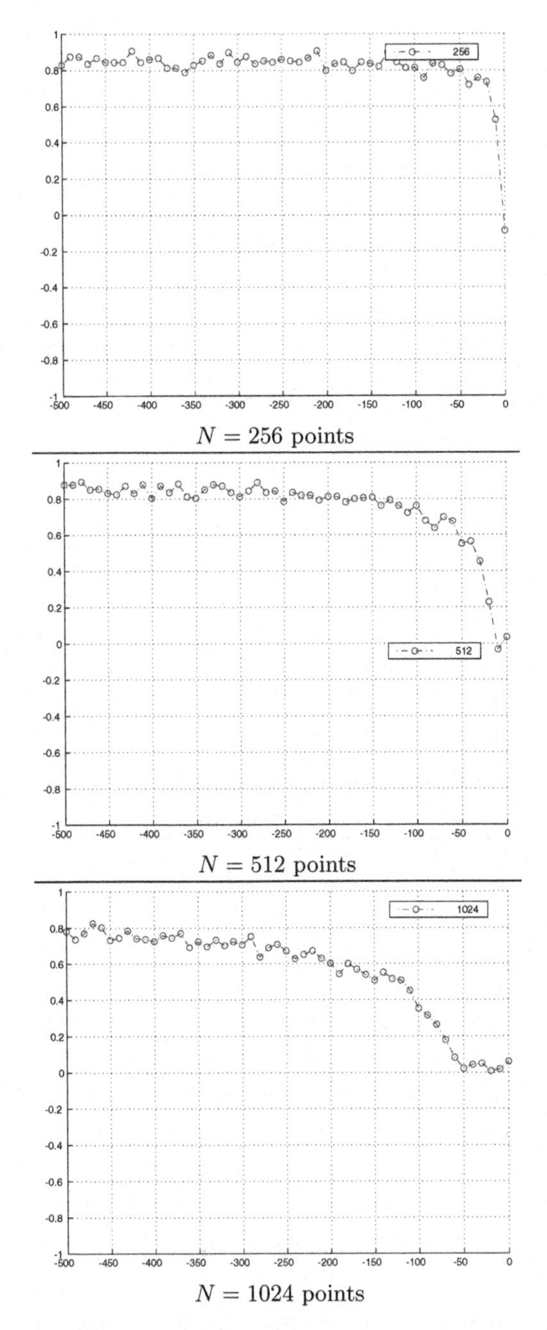

Fig. 8.9. (Continued.)

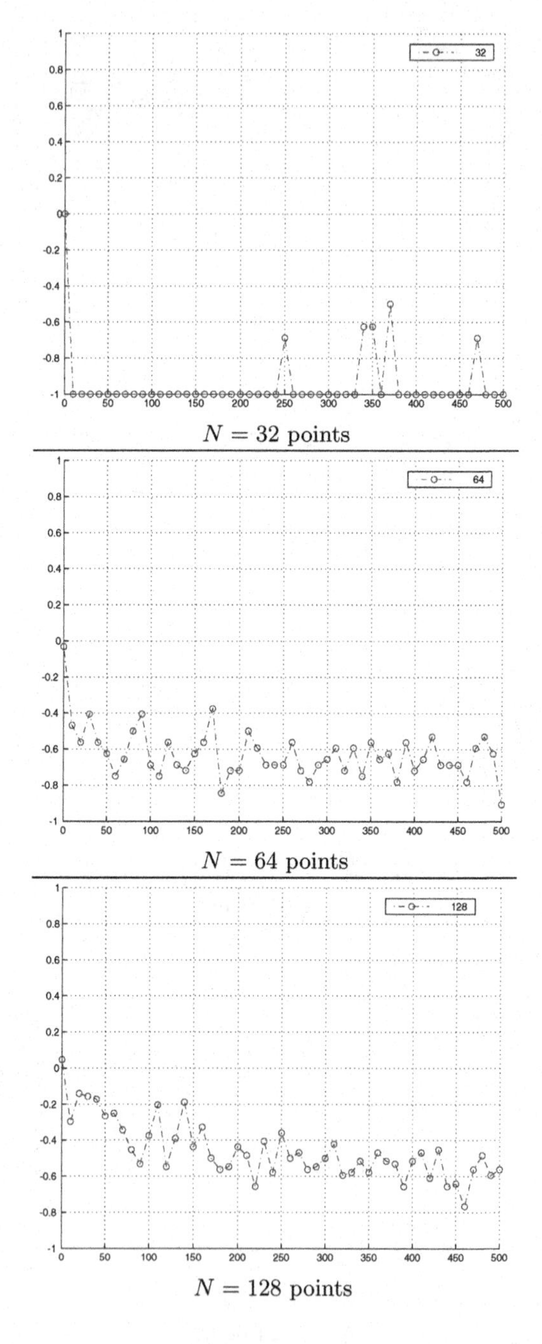

Fig. 8.10. The mean nearest neighbor parity for the problem on the sphere versus inverse temperatures on a range from $\beta = 0$ to $\beta = 500$ with constant enstrophy $\Omega = 256$ but varying N.

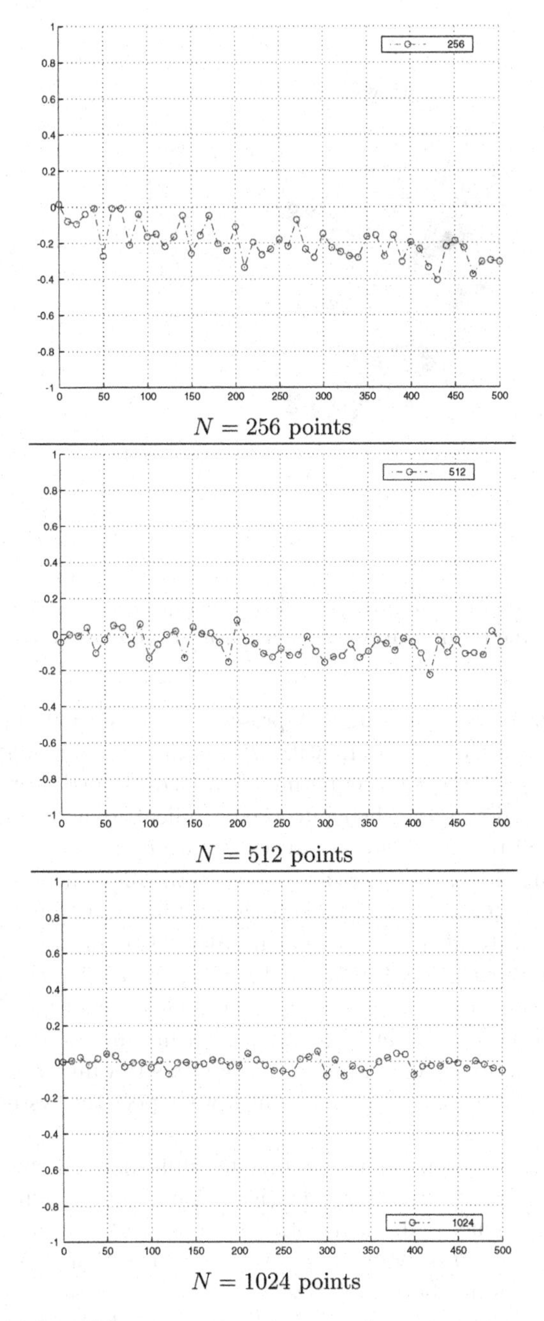

Fig. 8.10. (Continued.)

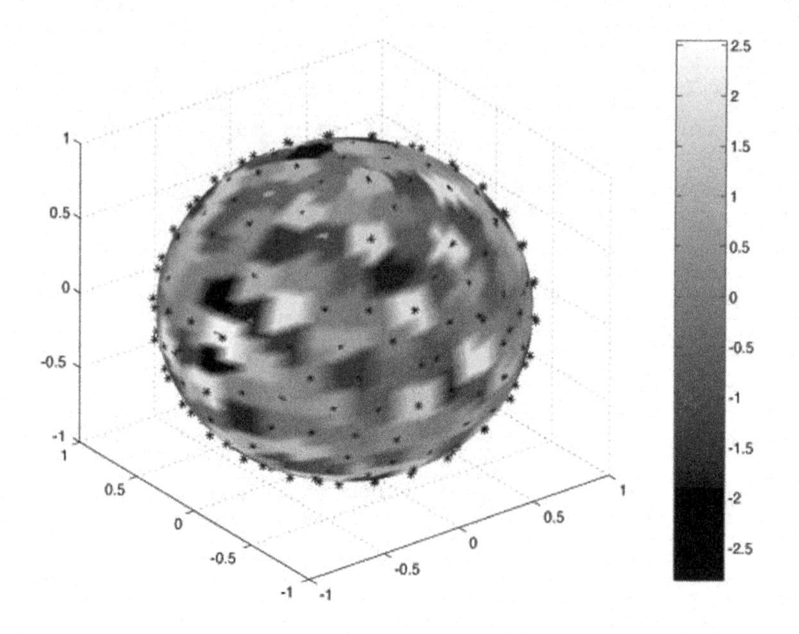

Fig. 8.11. An equilibrium with $\beta = 500$ at mesh size $N = 256$ and enstrophy $\Omega = 256$, where the mean nearest neighbor parity is around -0.25.

state measured correctly. Figure 8.12 presents the evolution of the mean nearest neighbor parity of a system with 256 points and circulation 0 run for 1,000,000 sweeps. Only a few experiments are considered, but they are fairly representative and so serve their purpose of indicating how long Monte Carlo simulations need to run before useful results are found.

The first question is whether this change in the mean nearest neighbor parity as the sweep count increases reflects the system settling into its final state, and what that state ultimately is. For negative β the mean nearest neighbor parity over time (excluding the first data point, which was the initial random assignment) has a mean of 0.86 with a standard deviation of 0.04. (These numbers are rather consistent across multiple runs and different meshes.) The mean nearest neighbor parity for that mesh with site values assigned to match those of the first spherical harmonic (which we expect the system approaches) is 0.8359.

Another interesting point indicating the uniformity of the meshes generated by the Metropolis-Hastings algorithm here is that multiple meshes of 256 points were generated, each with different seed values. The "spherical harmonic" mean nearest neighbor parity for each of them was close to 0.836 despite the meshes not being identical.

In positive β the mean nearest neighbor parity fluctuates more, and one may question whether the value settles at all. Here statistical examination of

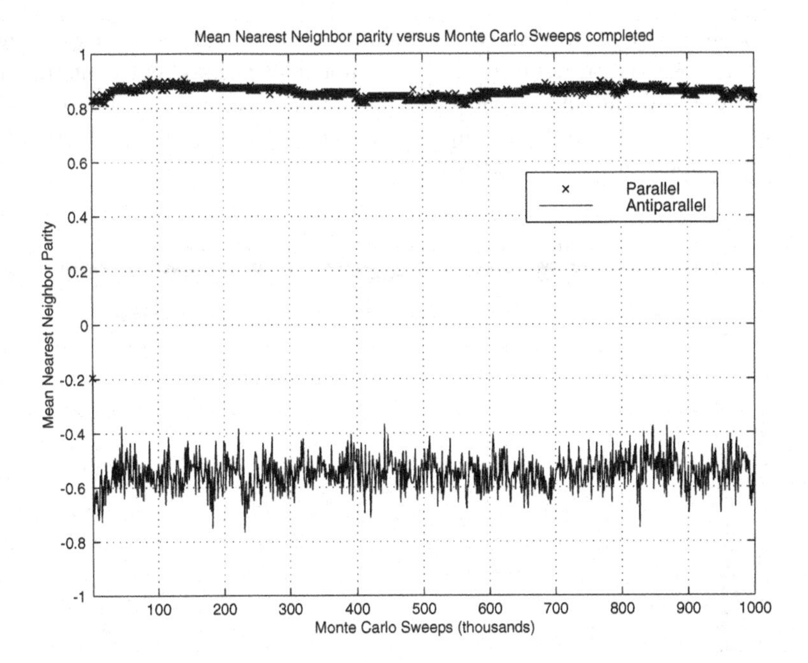

Fig. 8.12. The mean nearest neighbor parity for the problem on the sphere versus Monte Carlo sweeps completed.

the mean nearest neighbor parity at different times is useful: a histogram of its values over the sweeps reveals their distribution is reasonably close to a bell curve. For this particular run at positive β the mean of the mean nearest neighbor parities after the first data point is -0.55 with a standard deviation of 0.06. (As above the mean and standard deviation are rather consistent across multiple runs and different meshes.)

One method which could improve the mean nearest neighbor parity as a measure would be to identify all the neighbors of each site, and take the mean neighbor parity over all its adjacent sites, which is more consistent at least with the mean field theory we expect will apply here.. Nevertheless the mean nearest neighbor parity approximates its long-term mean value after as few as $20,000$ sweeps. We may take its value, with the recognition that there must be an error margin whose size decreases as the number of mesh sites N increases, to be approximately the equilibrated mean nearest neighbor parity.

8.4 Distances

Another property which can be measured and which we may expect to have some significance in describing a system is the distance between the north and south vorticity poles. In the parallel state we expect the north and south poles to be nearly antipodal, so the distance between them is close to two, and this is observed in figure 8.13. The minor fluctuations represent the fact near the pole all the sites have nearly the same size, a consequence of the parallel state being a discretized approximation of the solid-body rotation the negative temperature state represents.

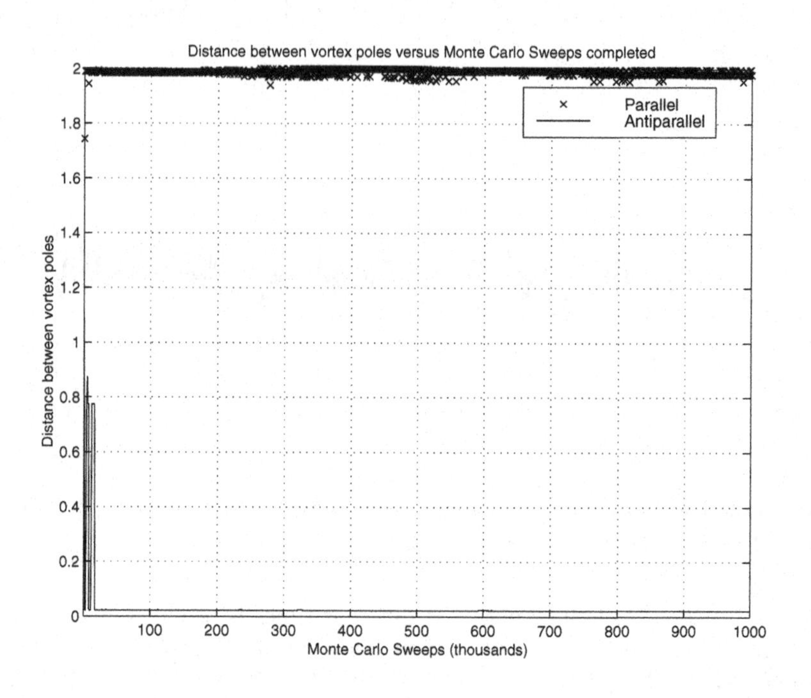

Fig. 8.13. The mean nearest neighbor parity for the problem on the sphere versus Monte Carlo sweeps completed.

The behavior of the distance between north and south vorticity poles is more interesting in the antiparallel case than it is for the parallel: it often shows a pattern of small fluctuations interrupted by large jumps (though they are obscured in the simulation used to generate figure 8.13). The process here is that each site will tend to have a nearest neighbor of opposite sign, and the Metropolis-Hastings algorithm's tendency to reduce the energy when β is positive makes states in which $s_i s_j$ is minimized more probable. This pairwise product can be maximized, when the enstrophy has to stay constant, by having s_i and s_j of roughly equal magnitude and opposite sign. The result is the antiparallel state tends to create "dipoles," neighboring pairs of approximately equal magnitude and opposite sign.

Generally, multiple dipole pairs will tend to be created, and they will tend to have approximately equal sizes. Often there will even be a chain of alternating signs and roughly equal magnitude over a good segment of the sphere. (The example plotted in figure 8.1 lacks a long chain, but it does have a checkerboard-like pattern of positive and negative in the upper right of the picture, showing these dipoles collected into a quite organized island.) The result is the minor fluctuations in site values along these multiple dipole pairs will make either vorticity pole jump among several nearly equal sites. The result is a progression of times when the same two sites (not necessarily in a single dipole pair) are the vorticity poles, with sudden jumps when another site's magnitude grows enough to become one of them.

The expectation is that nearest neighbors will have opposite sign, and the drive to minimize energy while holding enstrophy constant encourages the nearest neighbors to have roughly equal magnitude. This indicates that we may expect – if we assume the north and south poles will eventually settle around one dipole of nearest neighbors – the distance to approximate $\frac{2\pi}{N}$, roughly the distance of any neighboring pair of vortices. In the example used for figure 8.1 the minimum distance is 0.022; as $\frac{2\pi}{256}$ is 0.025 this reinforces our supposition that the mesh generated is close to uniform and the north and south vorticity poles are being drawn as close together as possible.

Experiments show after just a few hundred Monte Carlo sweeps both the mean nearest neighbor parity and the distance between vorticity poles will be quite close to their "ultimate" values, the numbers reached after considerably longer runs. This is the behavior Monte Carlo experiments are designed to produce.

8.5 Remarks

We have finally assembled the various components of the spin-lattice model representing an inviscid fluid by a discrete vorticity field, with a bit of analytic modelling to predict the phases and phase transitions we expect, and to a Monte Carlo Metropolis-Hastings experiment which produces statistical equilibria. And it has given us reason to suspect we have a mean field model; whether it is will be examined later in the book.

While our simulations can be set up and run quickly and we can understand much of inviscid fluid flow through them, we gain much of this ease and speed by the simplicity of the problems we simulate. These Monte Carlo methods particularly are tailored to statistical equilibria; we could not represent a problem of obvious interest such as the progression of an igniting flame through a medium, or the course of air around an airplane or rocket. We have not introduced something as simple as boundaries to problems; nor have we introduced viscosity or internal forces.

The rapid convergence on a statistical equilibrium has several fortunate side effects; one is that considerable research can be done on personal computers. Advanced hardware is needed only if one wants to consider extremely large meshes – which, since we understand the scaling laws involved, are not much needed. It means also one can write programs running these simulations without much effort on optimizing the programs. Software optimization is a challenging task, worth a separate course of study. Since generally code becomes faster at the cost of making it harder to understand (and debug) avoiding the need for optimization is an advantage.

There is one bit of software optimization which is straightforward and beneficial enough to be used. Calculating the site interaction energy requires the natural logarithm of distances between sites. On any computer this will be computationally expensive, as the functions which approximate the logarithm are involved. It is worthwhile, if program memory allows, to make an array of interaction energies $J_{j,k}$ instead.

This is a classic trade-off: the program becomes dramatically faster but at the cost of keeping a potentially quite large array in memory. As usual one can make code faster at the cost of increasing code complexity or increasing the memory required (and incidentally also increasing the start-up time before any results are available). In this case the trade is almost certainly worthwhile. Program implementation choices like this will be necessary in writing most simulations.

9

Polyhedra and Ground States

9.1 Introduction

The general problem of how to place point vortices on the sphere to minimize energy is unsolved. There are some known results, though, for example that we can minimize the energy of four point vortices by placing all four at approximately 109.4 degrees apart from the other points. That number is suggestive: we know those are the internal angles of the vertices of a tetrahedron. Were we to draw lines between the equilibrium points we would get the tetrahedron.

This gives us a new question to explore and a very fruitful field to study. If we take the vertices of a polyhedron, do we find an equilibrium for the vortex gas problem? Right away we realize we must consider only polyhedra which have vertices that can fit on the surface of a sphere, which eliminates some interesting shapes including nearly all non-convex polyhedra, but we have many to explore.

Many polyhedra have been developed such as by Coxeter[1] [96] or Sutton [412]; their coordinates have a preferred direction, typically symmetry around the z-axis. Metropolis-Hastings has no preferred direction, but the **radial distribution function** lets us convince ourselves some of our equilibria are polyhedron vertices. We will define the radial distribution function properly in section 10.3; for now it is enough that it is a function describing how the frequency of finding a pair of points separated by a given distance depends on the distance. The meshes generated by polyhedra may take us away from the goal of uniform meshes, but they will inspire new questions.

It is convenient to refer to the number of vertices of a polyhedron, which we will call the **size** of the polyhedron. (We do not have much need to describe the volume inside a polyhedron.) Thus to say one polyhedron is smaller than

[1] Harold Scott MacDonald Coxeter, 1907 - 2003, was a master of geometry and group theory. Some of his geometric discoveries were used architecturally by Richard Buckminster Fuller, 1895 - 1983. [288]

another is a comparison of vertex count. It is also convenient shorthand to say, for example, "the energy of the polyhedron" when we mean "the energy of the mesh of points placed at the vertices of the polyhedron"; when we are interested in a property of the shape rather than a mesh of points of uniform strength at the vertices of the shape we will make that distinction explicit.

This topic also brings us more into graph theory, so we will need some of its terminology. A **graph** is a set of points, the **vertices**, and the lines connecting pairs of vertices, the **edges**. The **degree** of a vertex is the number of edges which contain it.

The topics of this book indicate what we expect the approach to be. The question is why should we be able to use Monte Carlo methods with low but finite temperatures to find vertices of polyhedra. The polyhedra we find correspond to minimal-energy vortex states on the sphere; why?

Planck's theorem (Theorem 5) tells us that a statistical equilibrium at a given temperature corresponds to a minimizer of the free energy,

$$F = \langle E \rangle - \frac{1}{\beta} S \tag{9.1}$$

for energy E and entropy S. States of dynamic equilibrium, however, are associated with minimizers of the energy E. But when the temperature is very low, β is extremely large, and so the free energy is approximately $\langle E \rangle$. This limit is well-documented on the plane and one naturally expects it to follow on the sphere.

From that argument then the ground state at temperature zero, or the dynamic equilibrium, should be approximated reasonably well by finite-temperature minimizers of the free energy F. These free-energy minimizers are exactly the most probable vortex configurations that we find by the Metropolis-Hastings Monte Carlo algorithm.

The most surprising thing we will find is that the ground state is approximated very well by minimizers of the free energy for a wide range of finite temperatures, including for temperatures on the order of 1.

9.2 Face-Splitting Operations

There is no limit to the polyhedra we could examine. We must narrow our focus to simple set and a handful of rules to have a manageable problem. We start with a convex polyhedron with faces composed entirely of triangles. Though the faces need not be regular, the Platonic solids of the tetrahedron, the octahedron, and the icosahedron form a useful basic set. All represent ground states for the vortex problem with the corresponding number of vortices.

Daud Sutton considers [412] shapes constructed by performing several operations on polyhedra of fewer vertices. Such operations include taking duals to existing polyhedra, truncating the corners of a shape, compounding several shapes together, twisting edges, and the like. We will pick just two operations. This set – the centroid and the geodesic split – construct robust families of polyhedra. These are defined for spherical triangles, but extend naturally to other shapes.

9.2.1 Centroid Splitting

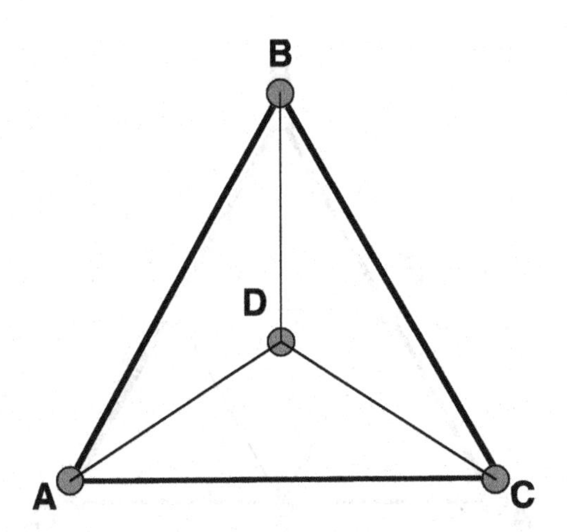

Fig. 9.1. An example of the centroid split. The centroid of the planar triangle ABC is added, and then the point at which the ray from the center of the sphere passing through this centroid is added as D. The edges AD, BD, and CD are added.

Definition 5. *The* **centroid** *split of a triangle ABC is to add to it a point D where the ray beginning at the center of the sphere and passing through the centroid of ABC intersects the sphere, and adding to the polyhedron the edges AD, BD, and CD.*

This construction does not preserve the degrees of the original vertices A, B, or C, but degree counts are arbitrary and do not require attention here. This splitting increases the number of vertices in the polyhedron by the number of faces, triples the number of faces, and adds three times the number of faces to the number of edges. The count of vertices, edges, and faces in the new polyhedron is:

$$\begin{aligned} v' &= v + f \\ e' &= e + 3f \\ f' &= 3f, \end{aligned} \tag{9.2}$$

An apparent anomaly is that certain configurations appear to create a square face from a polyhedron with only triangular faces – notably the centroid split of a regular tetrahedron gives the square cube. But by keeping the edges this effect will not propagate – splittings after this shape restore the triangular faces.

9.2.2 Geodesic Splitting

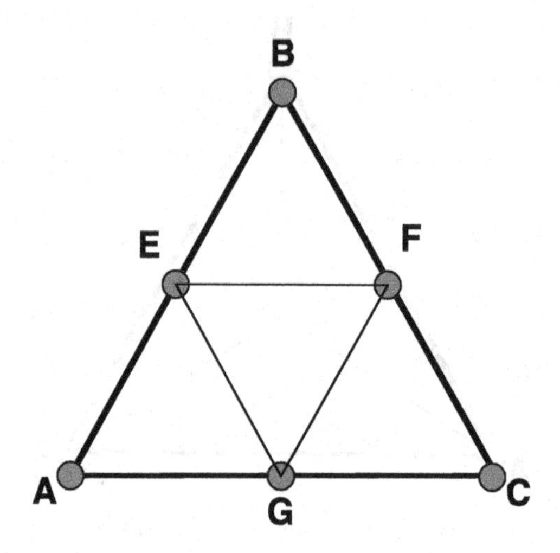

Fig. 9.2. An example of the geodesic split. The midpoints of geodesics AB, BC, and CA are added as the points E, F, and G respectively, and then the edges EF, FG, and GE are added.

Definition 6. *The* ***geodesic*** *split of a triangle ABC is to add to it a point E where the ray from the center of the sphere and passing through the midpoint of AB intersects the sphere, a point F similarly based on the bisection of BC, and a point G similarly based on the bisection of CA, and adding to the polyhedron the edges EF, FG, and GE.*

This splitting preserves the degree of the original vertices, and adds vertices of degree six. It increases the number of vertices in the polyhedron by the number of edges, quadruples the number of faces, and doubles and then adds three times the number of faces to the number of edges. The count of vertices, edges, and faces in the new polyhedron is:

$$\begin{aligned} v' &= v + e \\ e' &= 2e + 3f \\ f' &= 4f, \end{aligned} \tag{9.3}$$

9.2.3 Noncommutivity

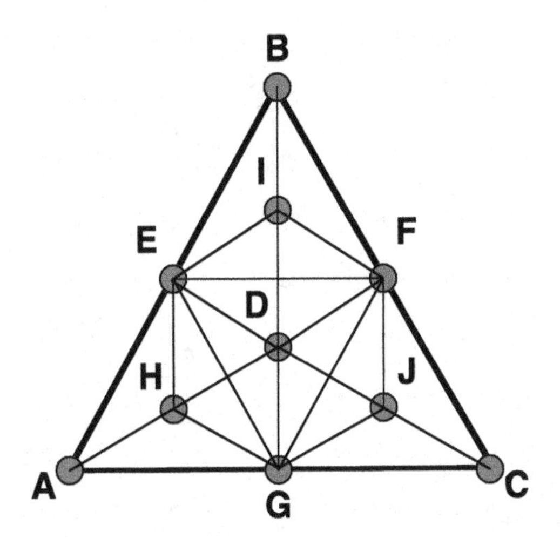

Fig. 9.3. An illustration of how the centroid and geodesic splits may be expected to commute. As constructed on the plane the triangle AEG must be similar to ABC (and similarly EBF and GFC are also similar to ABC), making it possible to show the centroid and geodesic operations may be performed in either order. However, similar triangles do not exist on the surface of the sphere and so the proof breaks down. Figure 9.4 provides an example of a centroid-geodesic split resulting in different points than the geodesic-centroid order does.

If we split a triangle first by centroid and then by geodesic cuts, and compare it to the results of a geodesic then centroid cut, we may suspect the splits commute. In either order the original face is split into ten points, and when we draw the results of both operations we see the vertices produced – though not the edges between triangles – are congruent.

Theorem 10. *For a triangle resting in the plane, the centroid split and the geodesic split of that triangle are commutative in their placement of vertices.*

Proof. Let triangle ABC rest in the plane. A centroid split creates the point D and the edges AD, BD, and CD. The geodesic split of this creates point E bisecting AB, point F bisecting BC, point G bisecting CA, point H bisecting AD, point I bisecting BD, and point J bisecting CD.

Now consider the geodesic split followed by the centroid split. Again on triangle ABC, let point E' bisect AB, F' bisect BC, and G' bisect CA. By construction $E' \sim E$, $F' \sim F$, and $G' \sim G$. In applying the centroid operation let D' be the centroid of triangle $E'F'G'$, H' centroid to $AE'G'$, I' centroid to $E'BF'$, and J' centroid to $G'F'C$.

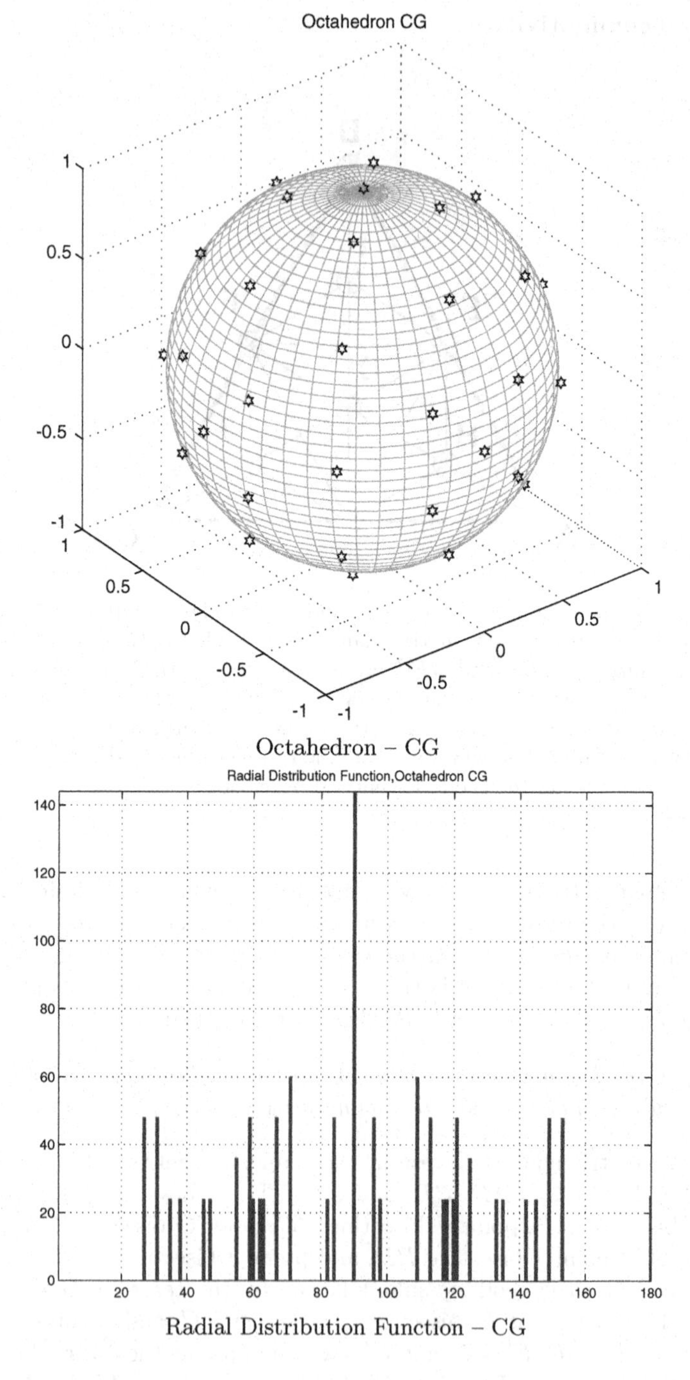

Fig. 9.4. The octahedron split by the geodesic word GC.

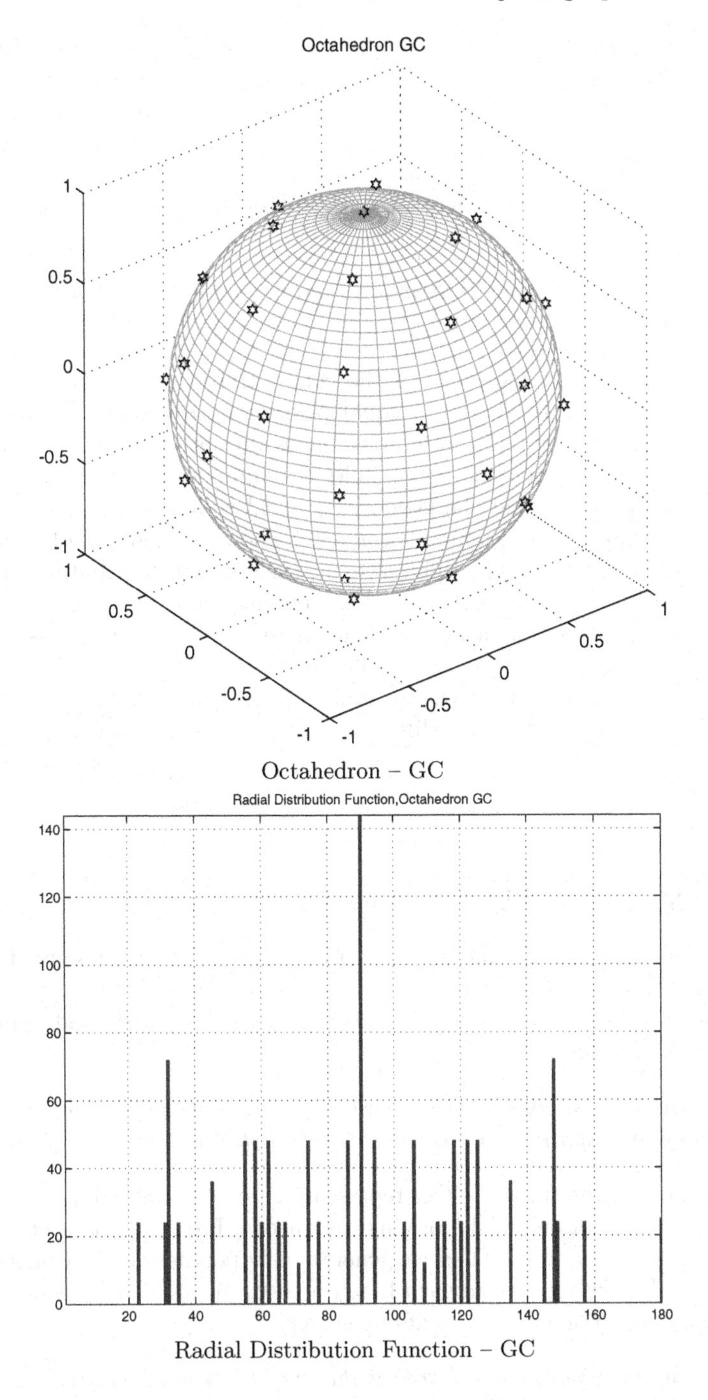

Fig. 9.5. The octahedron split by the geodesic word CG, showing the final polyhedra to not be identical to that split by the word GC.

$D' \sim D$ as the triangle $E'F'G'$ is similar to ABC, with edge $E'F'$ parallel to AB, $F'G'$ parallel to BC, and $G'E'$ parallel CA. $H' \sim H$ because $AE'G'$ is similar to ABC, with the length AE' half that of AB (and so on), therefore the distance from A to the centroid H' is half the distance from A to the centroid D'. By similar arguments $I' \sim I$ and $J' \sim J$.

Thus on the plane the centroid and geodesic splittings of a triangle commute. □

Corollary 1. *For a spherical triangle the centroid and geodesic splittings do not commute.*

Proof. Critical to the proof of the commutivity in the plane is the use of similar triangles to show the equal positioning of points H and H', of I and I', and of J and J'. Similar triangles do not exist on the surface of the sphere, so the proof does not hold. Examining the difference between the centroid-geodesic and the geodesic-centroid split starting from a simple polyhedron shows this. Figures 9.4 and 9.5 show an example of these non-commuting operations. The differences in their radial distribution functions make it obvious these are not equivalent configurations. □

As the area of a spherical triangle decreases – as the triangle is closer to planar – the centroid and geodesic splits come closer to commuting. This we will see as the energies of polyhedra split by long chains of operations varying only slightly when the chains differ only in the last operations – and these chains need not be long. Sequences as short as three operations can see their last two letters come near commuting.

9.3 Polyhedral Families

By applying the centroid and geodesic splittings repeatedly a family of related polyhedra may be constructed. Letting C stand for a centroid split and G stand for a geodesic split, one can construct a geodesic word and apply it to a basic polyhedron.

Definition 7. *A* ***geodesic word*** *is a string of C and G operations representing the repeated splitting of faces of a basic polyhedron.*

As example the word $CCCG$ represents three centroid splits and then a geodesic split from one of the original polyhedra. From any base polyhedron which has only triangular faces we generate new polyhedra. We can see these families as binary trees from each base polyhedron, with each geodesic word corresponding to one node or leaf of the tree.

Definition 8. *A* ***polyhedron tree*** *is the tree constructed by taking a polyhedron base and applying to it all geodesic words.*

It is possible to start from the same base and by different geodesic words reach polyhedra of the same size. Given that our operations do not commute this gives us several distinct polyhedra at a given size. But it is not guaranteed we can reach any arbitrary size: we know geodesic words and bases to give us a polyhedron of size 258, but 259 may evade us.

Several polyhedra of interest are generated by the branches of our polyhedron trees: the tetrahedron split by C is the cube, and the icosahedron split by C is the truncated icosahedron (and its own dual). Extremely few of these polyhedra are known "interesting" configurations such as Platonic or Archimedean solids, but those which are will receive specific attention.

By symmetry we expect the tetrahedron, octahedron, and icosahedron are equilibria. This and the fact they are made of triangular faces make them our first choices for base polyhedra. From these shapes we expect any chain of purely geodesic splits, or a chain of geodesics followed by one centroid split, also produces an equilibrium.

Counter-intuitively, configurations built with several centroid splits produce very asymmetric polyhedra, but still have energies very close to the low energy configuration of nearly uniform points. This is worth further examination.

9.4 Polyhedral Families Versus Vortex Gas

As the number of mesh sites increases we recognize the number of pairs of points grows quadratically, and therefore we may expect the energy of the mesh will also grow quadratically. It does [39], whether we place points at the vertices of polyhedra or at the equilibria found by the Metropolis-Hastings algorithm, as illustrated in figure 9.6.

These tables present the first few levels of a polyhedron tree based on the tetrahedron, the octahedron, and the icosahedron. These bases are Platonic solids with triangular faces, with vertices uniformly of degree three, four, and five, respectively. In each table the energy of the generated polyhedron is compared to the energy of the Metropolis-Hastings problem for a mesh of the same size.

Several of these polyhedra have energies close to configurations from the Metropolis-Hastings algorithm, and one case – the icosahedron split by CG – the polyhedron energy is below the Metropolis-Hastings energy. This reminds us that our Markov chain Monte Carlo can get quickly close to a minimum energy, but may have difficulty getting to exactly the lowest-energy configuration. This is the degeneracy of states coming into play; the set of states close to the minimum energy is much larger than the set of minimum energies.

The difference between positions system energies as the final terms in the geodesic word are reversed decreases words grow longer. From the tetrahedron energies for GCG and GGC are nearly identical, and even CCG and CGC differ by only about one percent.

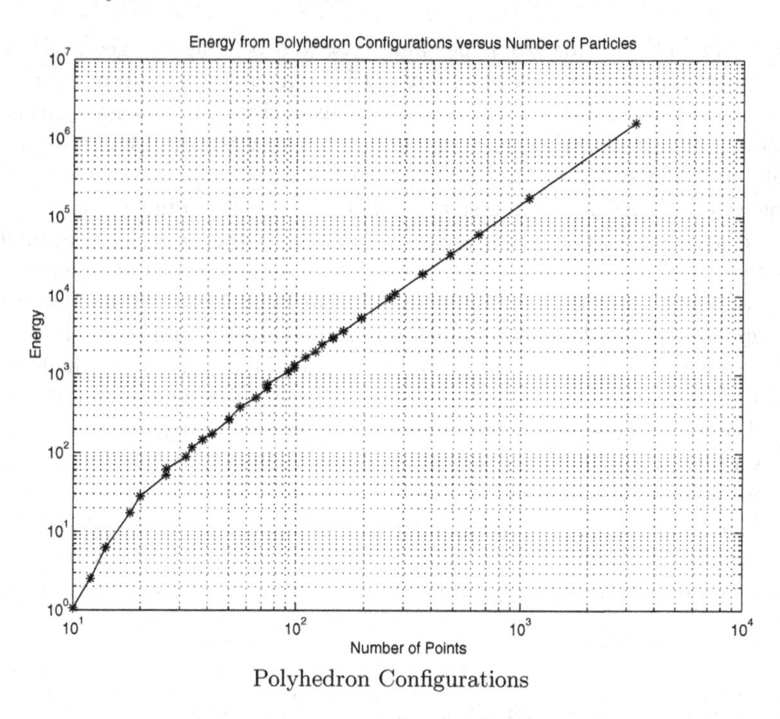

Fig. 9.6. The system energy as a function of the number of particles for the examined polyhedra.

Polyhedra grown from the octahedron show even closer agreements between energies as final geodesic letters. We do not encounter any cases where the polyhedron energy is below the Metropolis-Hastings energy, though. Similarly the polyhedra grown from the icosahedron find quite good fits between energy and minimum energies.

Though the icosahedron split by *GGG* and the Metropolis-Hastings mesh with 642 free points agree closely in energy, the shapes are considerably different, as is revealed by their radial distribution functions. The Metropolis-Hastings mesh's radial distribution function is close to a sinusoidal curve, what we would expect from a uniform distribution, while that for the polyhedron has fewer and higher peaks, a crystal-like configuration.

9.4.1 Pairwise Interaction Energies

We know the energy of the system grows as the square of the number of particles [39]) so it is reasonable to ask for the average energy of interaction of a pair. Figure 9.8 and the tables in this section show energy divided by $N(N-1)$, both for the polyhedron trees generated from the tetrahedron,

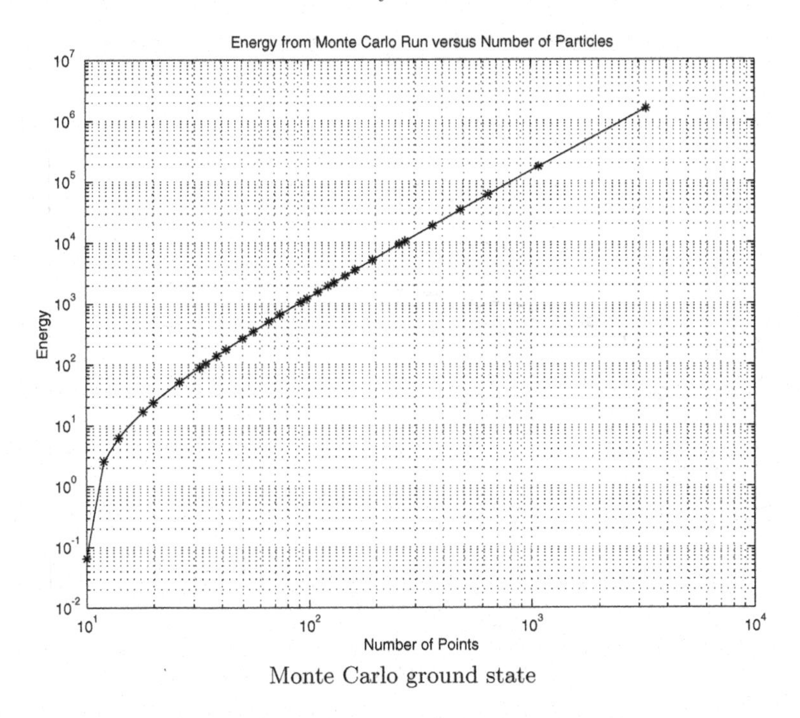

Monte Carlo ground state

Fig. 9.7. The system energy as a function of the number of particles for the ground state as found by a Monte Carlo algorithm.

octahedron, and icosahedron, and for the Metropolis-Hastings configurations for the corresponding number of points.

Once the number of points grows moderately large the pairwise energy appears to approach a constant near approximately 0.15. As may be expected from the near-commutivity of centroid and geodesic splits the pairwise energy depends only slightly on the geodesic word and more on mesh size.

The pairwise Metropolis-Hastings energy tends to be slightly lower than the polyhedron-based configurations. The calculations of Bergersen et at [39] and reiterated here and in [273] indicate the energy per pair of particles should approach a limiting mean value of $-\frac{1}{2}(\log(2) - 1)$ as the number of particles grows without bound, and this seems to be approached by both the Metropolis-Hastings and the polyhedron configurations.

We have produced a mesh that satisfies our objectives, a set of approximately uniformly spaced points which is almost but not exactly regular. For our Metropolis-Hastings work explored in later chapters this is enough. In general deterministic methods for solving differential equations will require not just the points but also edges. A typical finite-elements problem requires a **triangulation** of the domain, the dividing of space into **elements**, which are triangles (on the plane) or spherical triangles (on the sphere). Obviously

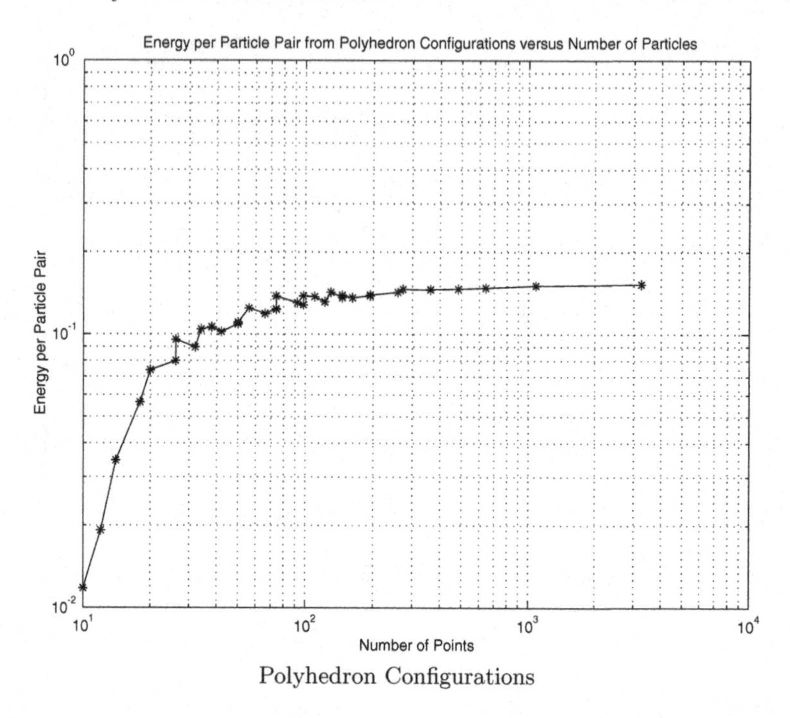

Fig. 9.8. The system energy per number of pairs as a function of the number of particles for the examined polyhedra.

our mesh sites will be the vertices of these elements – the **nodes** – but we do not know which possible edges to choose.

In finite-elements methods we measure the **isotropy** of an element; this is the smallest angle within the shape. An element with isotropy $\frac{\pi}{3}$ is an isosceles triangle and this tends to be numerically stable. The isotropy of a mesh is the smallest isotropy of any of its elements. We would therefore either like to draw edges so the isotropy of the mesh is close to $\frac{\pi}{3}$, or to draw edges so that most elements have that isotropy and accept a handful of elements that do not fit well. A quick estimate of the number of possible edges to draw makes clear we cannot hope to test all these combinations to find the optimum. It does, however, suggest an obvious new Monte Carlo problem, this one of adding or moving edges between sites in an effort to maximize the isotropy of the mesh.

Rather than take on the problem of finding the optimum edge arrangement, we will introduce a simple method which is likely good enough for most purposes. Given our mesh M with points at $\mathbf{x}_j$ for $j = 1, 2, 3, \cdots, N$ all inside the domain D, we divide the entire domain into Voronoi cells $D_1, D_2, D_3, \cdots, D_N$. A point $\mathbf{x}$ is placed in cell D_j if and only if $\mathbf{x}$ is closer to $\mathbf{x}_j$ than to any other $\mathbf{x}_k$.

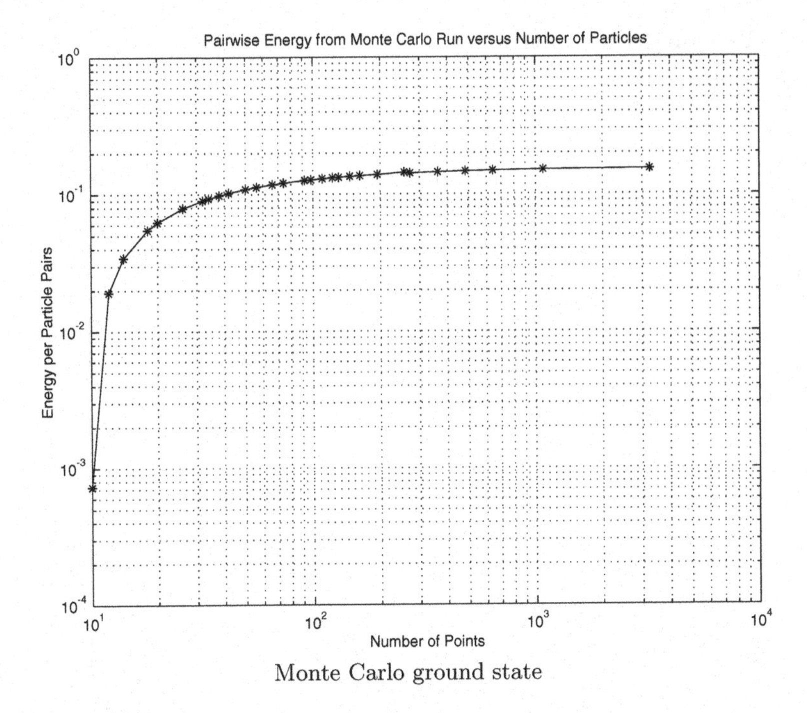

Monte Carlo ground state

Fig. 9.9. The system energy per number of pairs as a function of the number of particles for the ground state as found by a Monte Carlo algorithm.

This divides D into a map of convex polygonal regions. None of the edges of the polygon bounding D_j touches a mesh site, but any perpendicular dropped from $\mathbf{x}_j$ to a polygon edge will reach one of the neighboring mesh sites. It is these perpendicular lines which we take as the edges for our mesh.

9.5 Energy of Split Faces

The most interesting property possessed by polyhedra with faces split repeatedly by the rules we have established is that they create mesh point placements which have very low energies, energies quite near what we would have from a uniform distribution of vorticity across the sphere. This we expect from repeated geodesic splits on the faces, as they add to the polyhedron new nearly equilateral faces roughly uniformly distributed.

But something startling happens even with the centroid split, which creates near-isosceles faces very far from equilateral triangles. We examine the results of three face-splitting operations starting from the tetrahedron, the octahedron, and the icosahedron as examples. We compare the energies and radial distribution functions generated by these polyhedron configurations to those produced by the free Monte Carlo simulation.

9.5.1 Tetrahedron Splittings

CCC Split

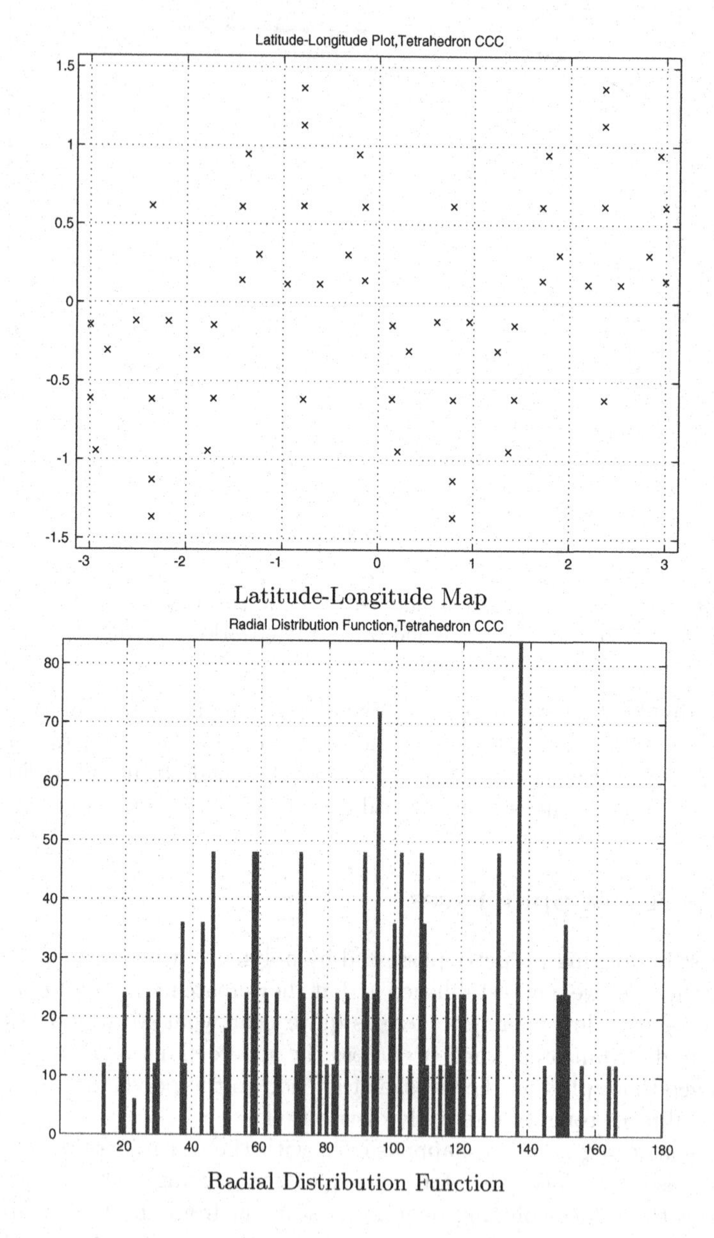

Fig. 9.10. State energy: 383.80063908

Free Particles

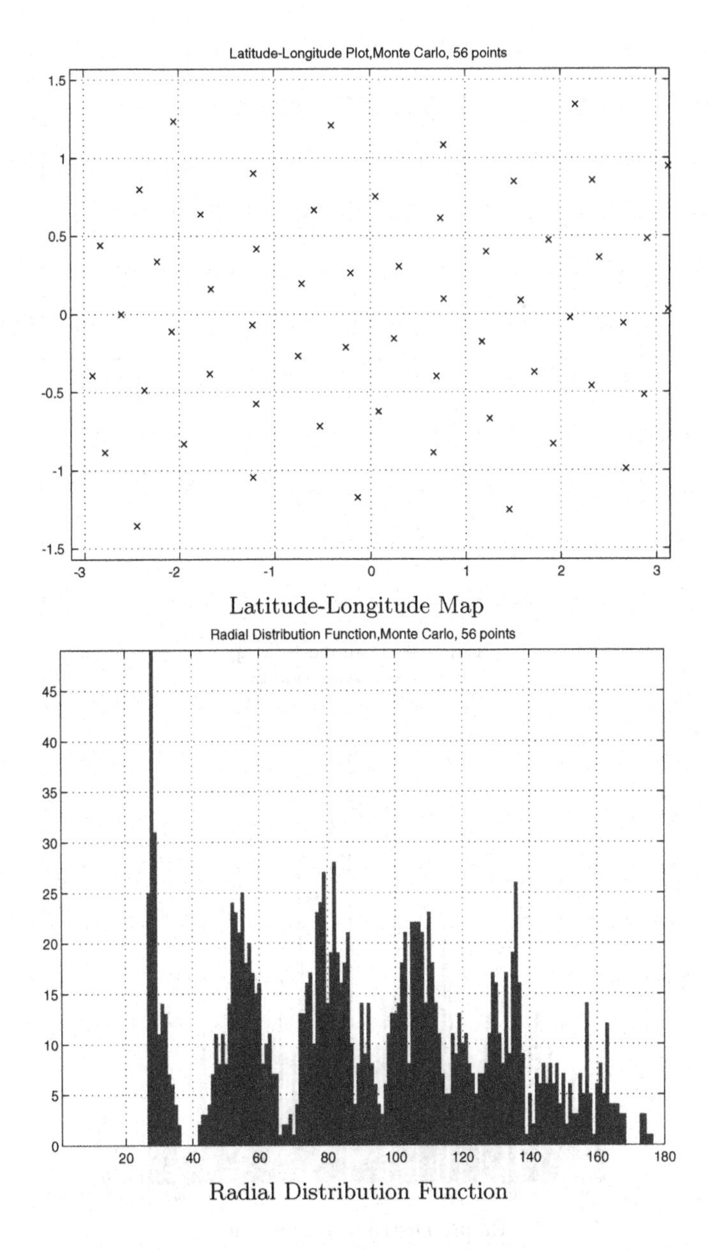

Fig. 9.11. State energy: 346.36114899

9.5.2 130 Vortices

GGG Split

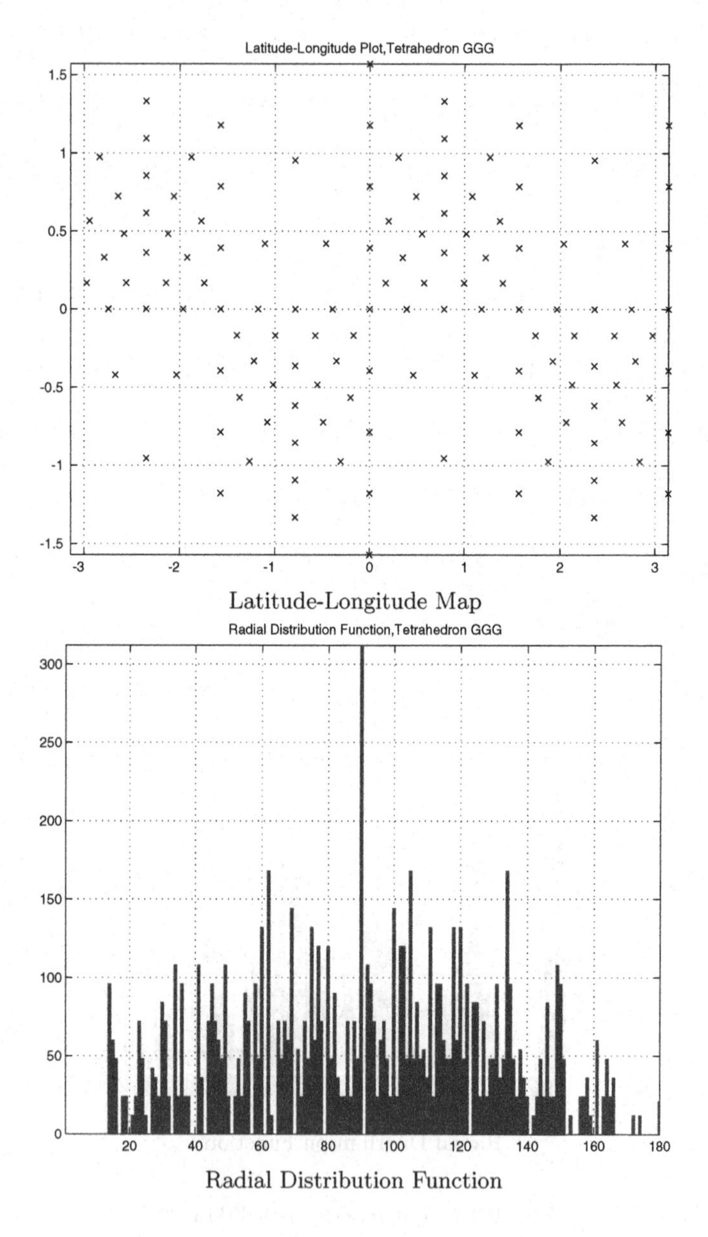

Fig. 9.12. State energy: 2390.85999192

Free Particles

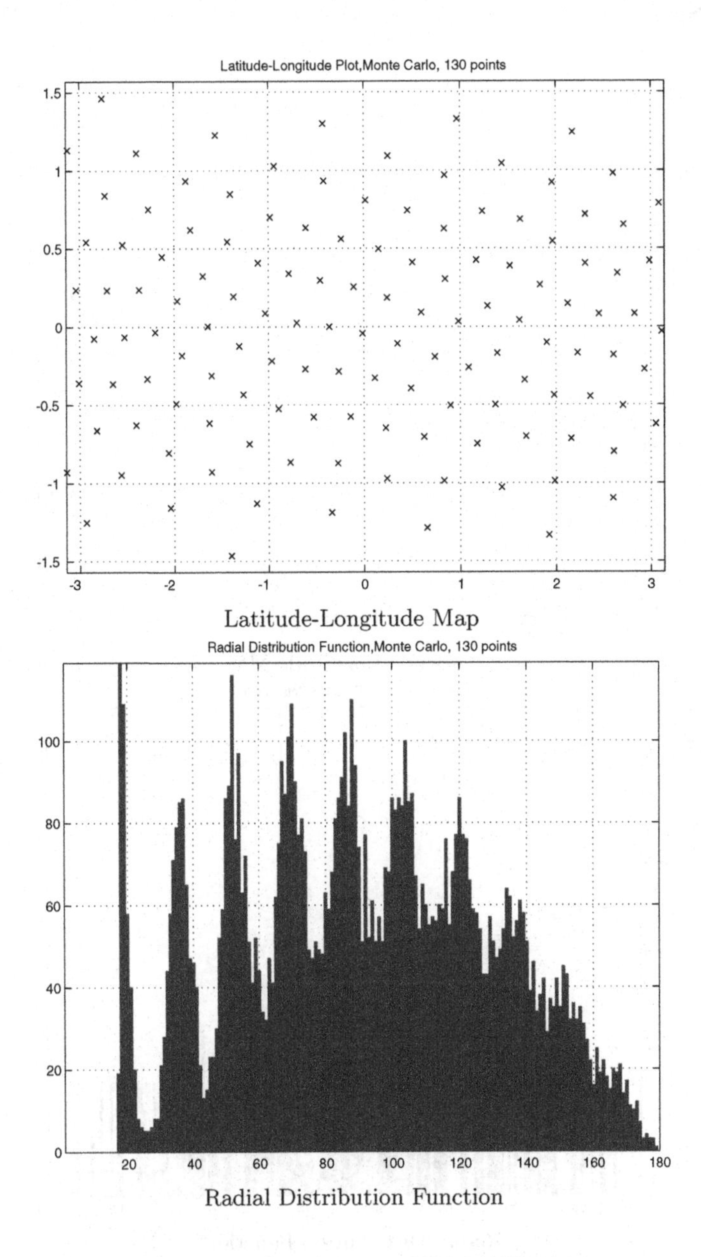

Fig. 9.13. State energy: 2225.11446834

9.5.3 Octahedron Splittings

CCC Split

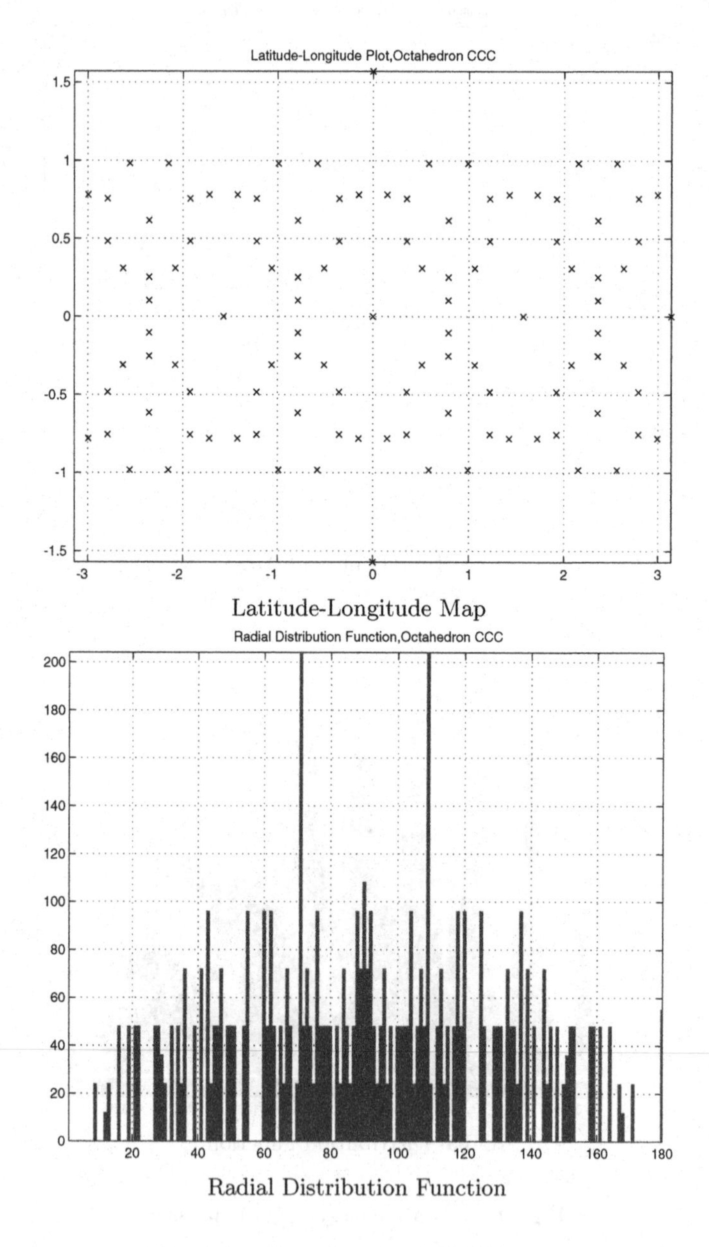

Fig. 9.14. State energy: 1648.07973705

Free Particles

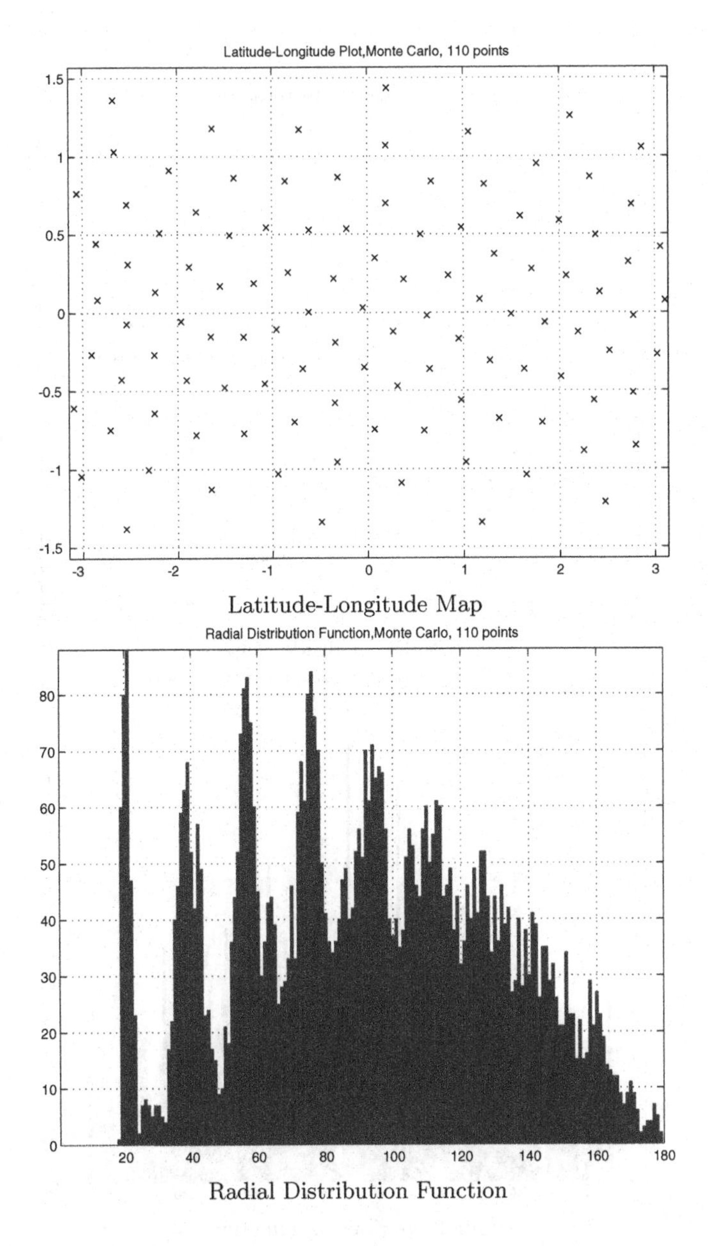

Fig. 9.15. State energy: 1554.46190646

9.5.4 258 Vortices

GGG Split

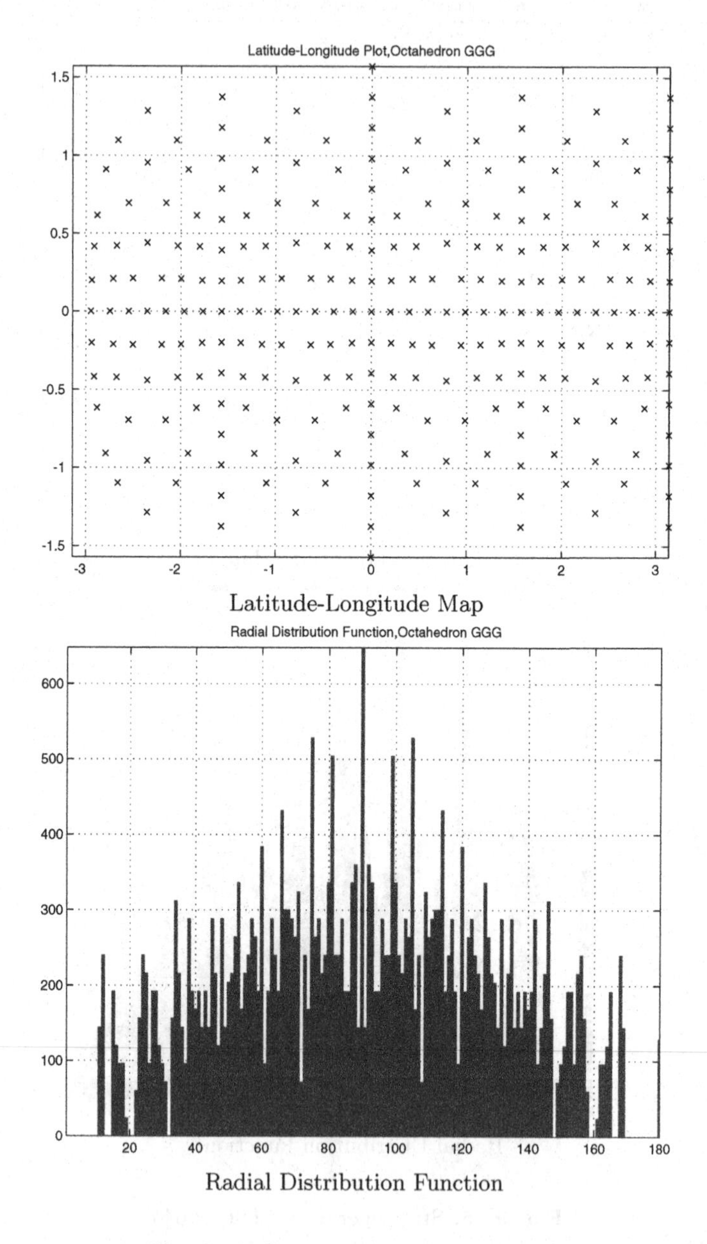

Fig. 9.16. State energy: 9455.00969009

Free Particles

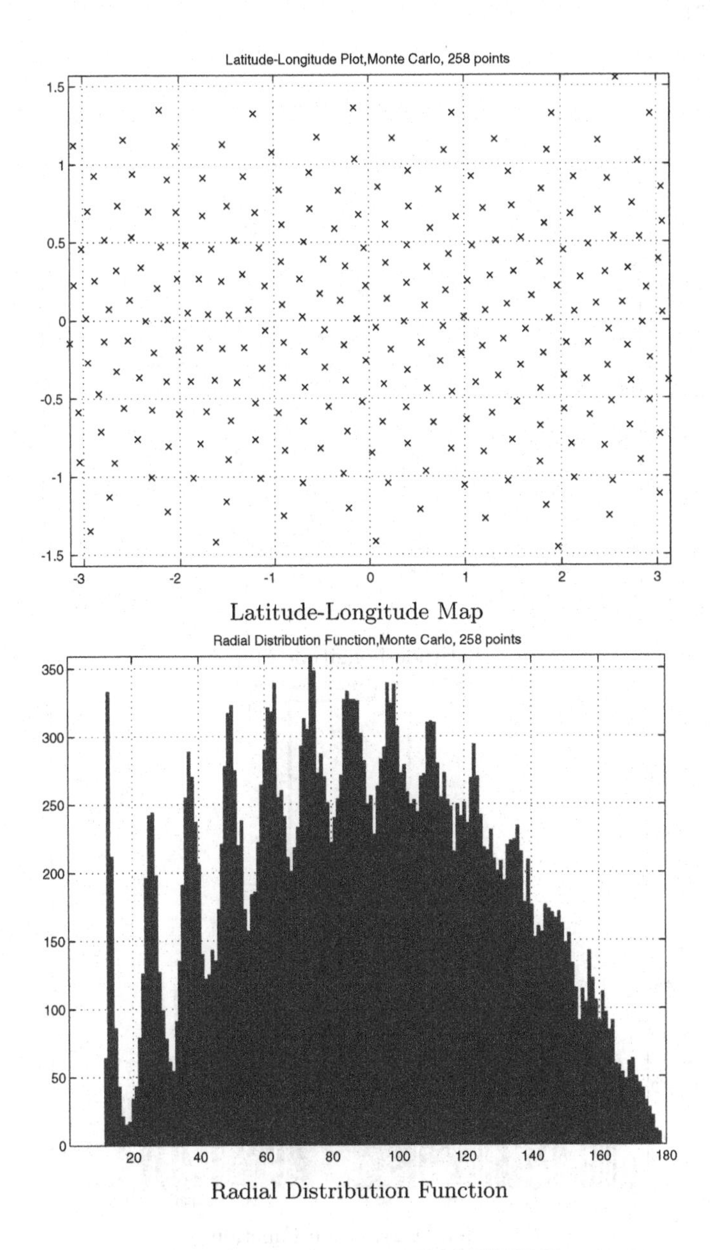

Fig. 9.17. State energy: 9394.27066383

9.5.5 Icosahedron Splittings

CCC Split

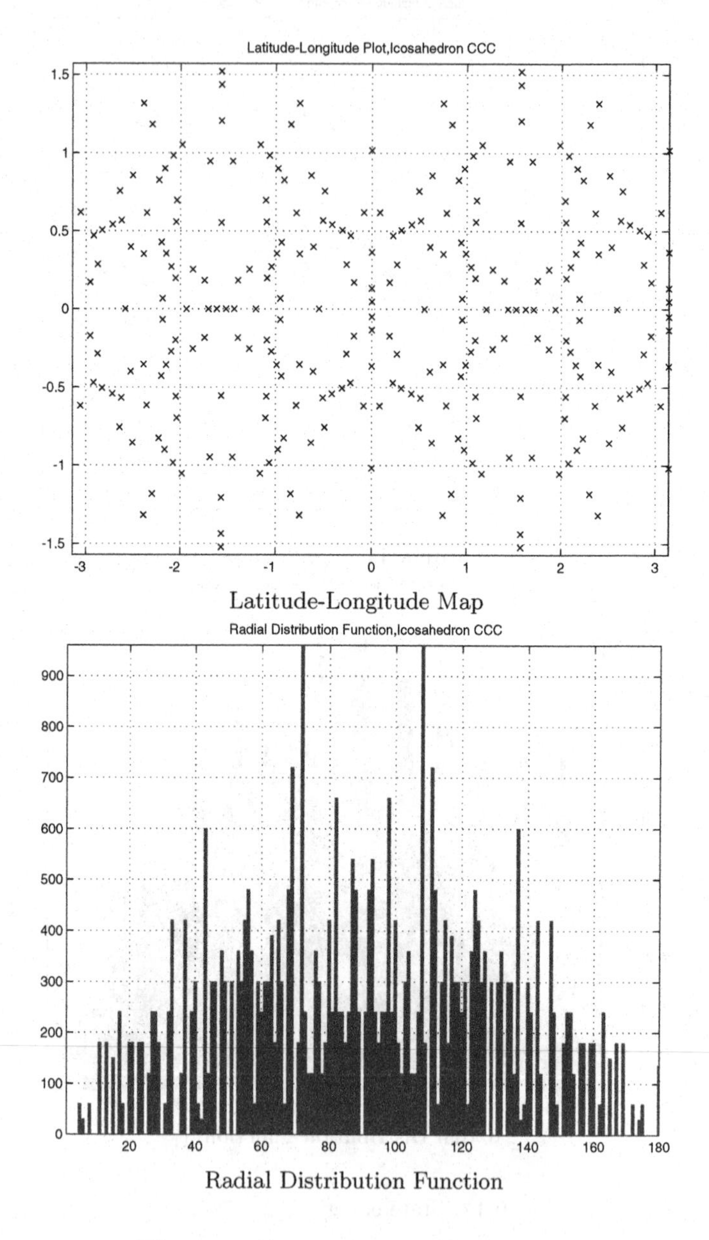

Fig. 9.18. State energy: 10774.41028121

Free Particles

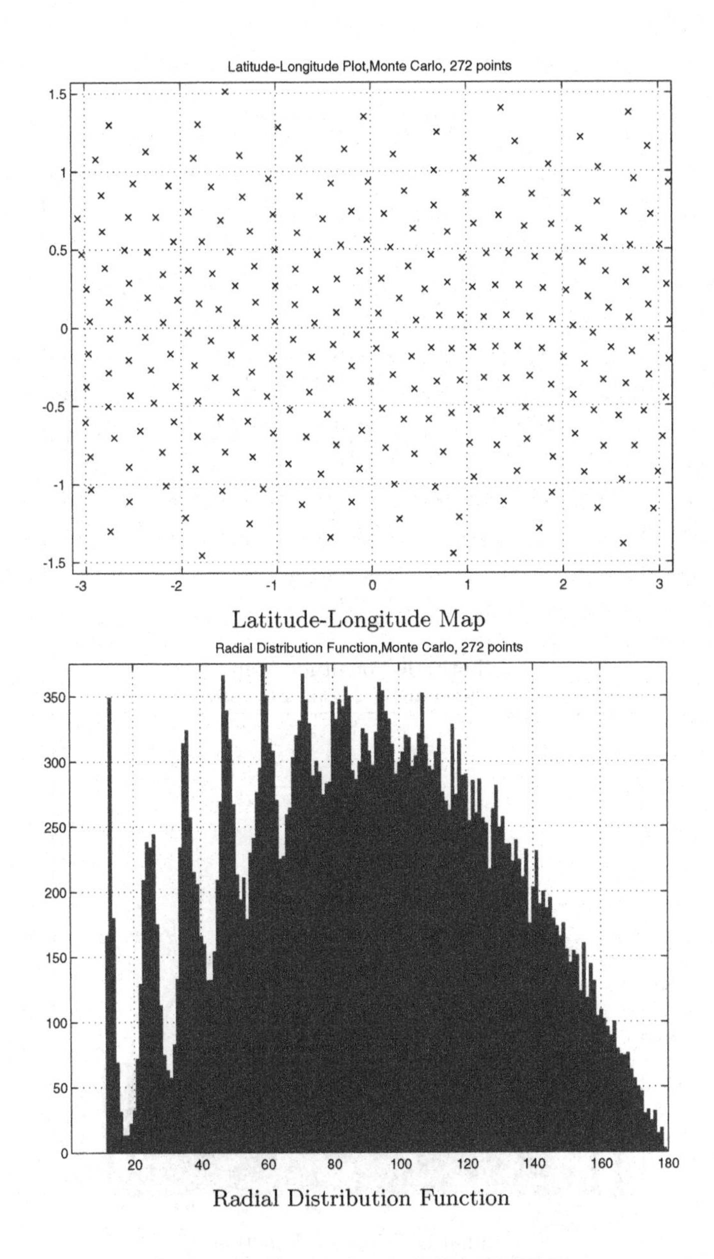

Fig. 9.19. State energy: 10481.07936965

9.5.6 642 Vortices

GGG Split

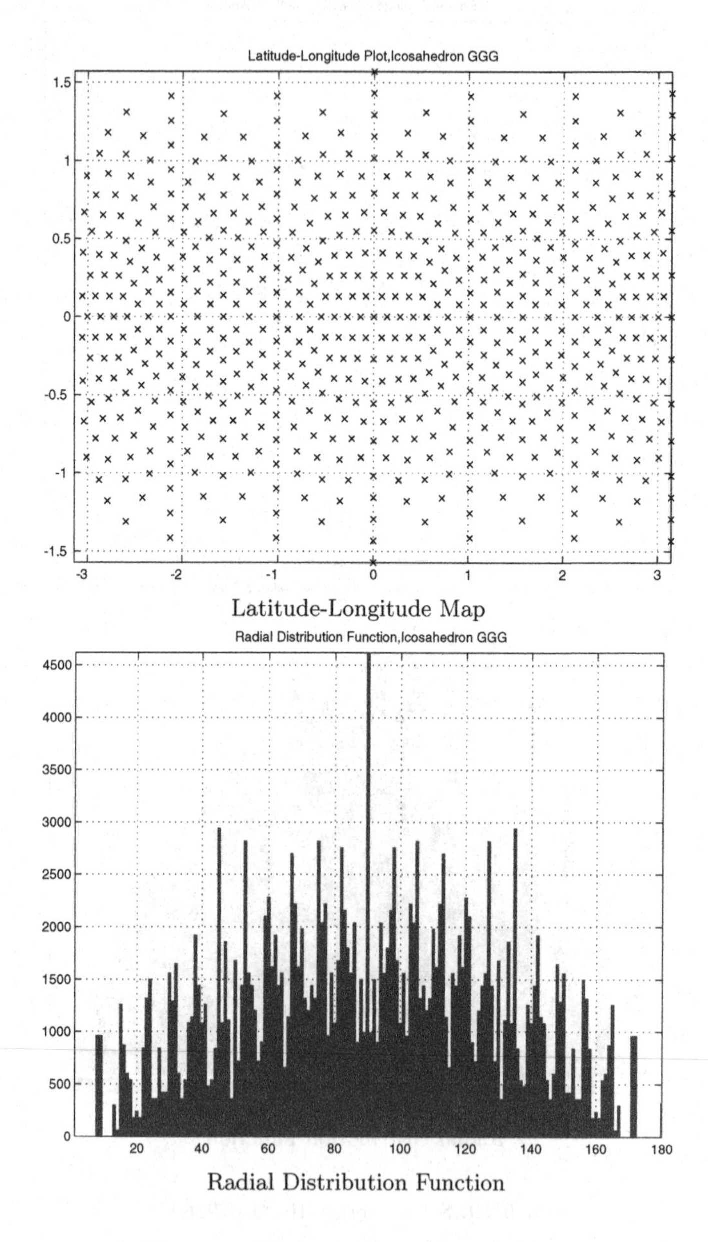

Fig. 9.20. State energy: 60926.02395237

Free Particles

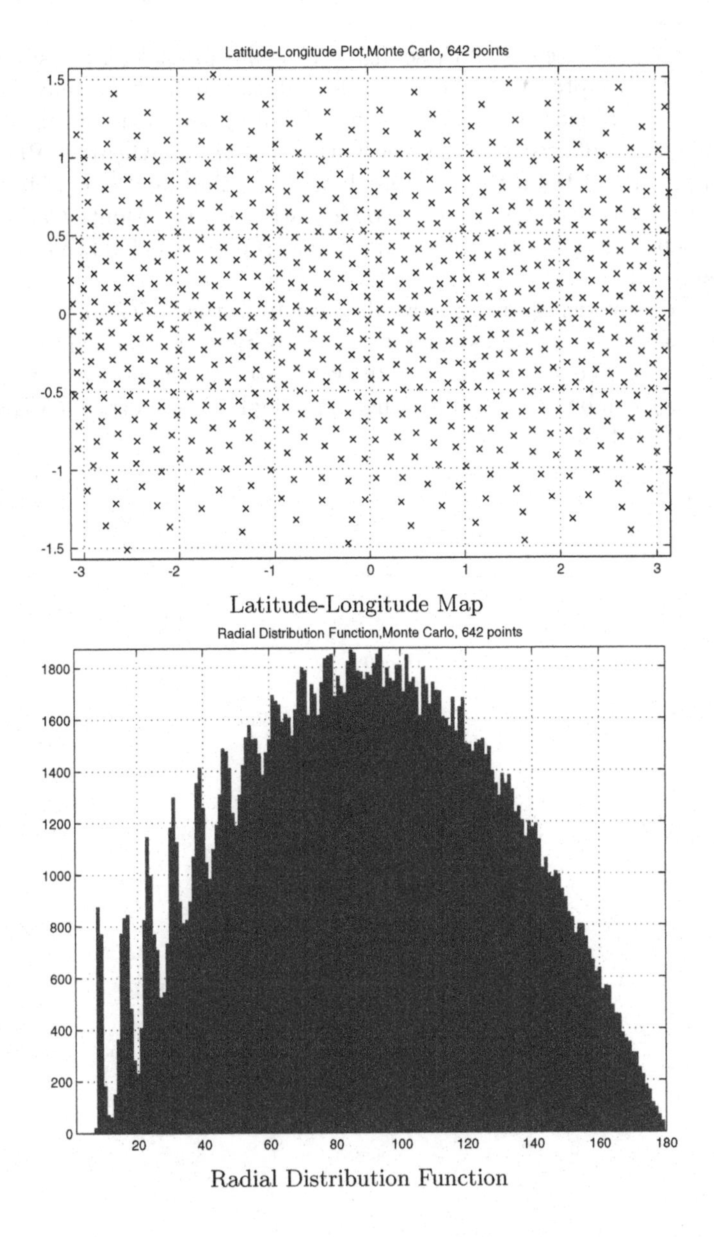

Fig. 9.21. State energy: 60907.99893598

The most remarkable result is that the very irregular shapes we find from CCC splittings have energies almost as low as those which are more uniformly distributed over the sphere. While we have many points close to one another we have also large regions free of any points, and remarkably these factors balance one another.

We would not use one of these highly irregular meshes for the spin-lattice work in other chapters, as we want meshes which roughly uniformly cover the sphere. It is still surprising to encounter. That these points are positions of an equilibrium follows from the construction of the polyhedron. But we do not expect any of them to be stable dynamic equilibria; a simple test using these mesh points as the starting configuration for the Monte Carlo vortex gas problem suggests that – these shapes very rapidly break down to the more uniform meshes instead.

Another surprising property we discover from the GGG splittings, which are extremely uniform and symmetric configurations, is that symmetry is *not* a guarantee of an energy minimum. In every case a configuration with slightly lower energy can be found by breaking up the highly symmetric GGG patterns into figures just slightly askew.

10

Mesh Generation

10.1 Introduction

In this chapter we continue exploring the vortex gas problem, partly for its dynamic interest, and partly to apply Monte Carlo methods to it. The vortex gas system has been analytically explored and quite a few relative equilibria are known by Lim, Montaldi, and Roberts to exist [270] and have been classified in shape and in dynamic stability. We will be interested here in the shape, and use the problem to easily generate meshes for other numerical computations.

What are the properties of a good numerical-computation mesh? We want the difference between a solution computed on that discrete mesh and the exact solution analytical methods would provide to be small. We also want computations on the mesh to be stable, so that the numerical errors – those introduced by representing exact and often complex functions with simpler approximations – and computational errors – those caused by using floating-point arithmetic and similar implementation details – stay limited. How well any mesh behaves depends on the problem we study and on the mesh we use. So what meshes are likely to be good?

We want mesh points to be approximately uniformly spaced over the domain, and want to avoid large regions of too-regular placings of points (which can lead to mesh artifacts). Computationally useful is also an algorithm which handles a wide range of numbers of points, and which can produce several distinct patterns for any given mesh size.

We turn to the Monte Carlo algorithm. By fixing vortex strengths – and we will for this chapter fix the strength of each vortex to $\sqrt{4\pi}$ to simplify the Hamiltonian used – and varying positions of vortices we will find a distribution of vortices which is a local minimum of the energy of interactions. This will be a relative equilibrium (at least) and we will see this distribution has the properties we want.

10.2 The Vortex Gas on the Sphere

Let us consider the vortex gas problem with vortices of uniform strength on the surface of the sphere. (We will consider the problem on the plane in a later section.) This is a mechanics problem which can be studied analytically [270] or by ordinary numerical integration tools like Runge-Kutta integrators. Though we will remember the dynamics we will study statistical equilibria generated by Monte Carlo methods.

The potential energy of this vortex gas on the sphere if we take, without loss of generality, each of the N vortices to have strength $\sqrt{4\pi}$ will be

$$H(\mathbf{z}_1, \mathbf{z}_2, \cdots, \mathbf{z}_N) = -\frac{1}{2}\sum_{j=1}^{N}\sum_{k\neq j}^{N} \log|1 - \mathbf{z}_j \cdot \mathbf{z}_k| \tag{10.1}$$

for particles on the surface of the sphere (with $\mathbf{z}_j = (x_j, y_j, z_k)$ the coordinates of vortex j, and the constraint that $|\mathbf{z}_j| \equiv 1$ for all j) [215] [265] [324].

A Monte Carlo sweep is a series of N experiments – matching the number N of vortices – in each of which one vortex is selected (randomly, although we could as well allow the selection to be done in a preselected order). On each vortex a proposed displacement is tested: Consider moving the chosen vortex by a random amount – a rotation around a randomly selected axis $\mathbf{r}$ by an angle θ uniformly distributed on $(0, \epsilon)$.

It is worth mentioning how we calculate the position of the rotated vortex. There are many ways to represent three-dimensional rotations; our preferred method is to use **quaternions**. Quaternions are an extension of complex variables originally developed by Sir William Rowan Hamilton, and represents a number

$$z = a + bi + cj + dk \tag{10.2}$$

where a, b, c, and d are all real numbers and where

$$i^2 = j^2 = k^2 = -1 \tag{10.3}$$

$$ij = k \quad ji = -k \tag{10.4}$$

$$jk = i \quad kj = -i \tag{10.5}$$

$$ki = j \quad ik = -j \tag{10.6}$$

and finally the conjugate of a quaternion z is

$$\bar{z} = a - bi - cj - dk \tag{10.7}$$

Quaternions are well-suited to representing rotations around a unit vector. If (x, y, z) are the Cartesian coordinates of a point on the unit sphere, (r_x, r_y, r_z) are the Cartesian coordinates of the unit vector $\mathbf{r}$, and θ is the angle of rotation then construct the quaternions

$$p = 0 + xi + yj + zk \tag{10.8}$$
$$q = \cos\left(\frac{\theta}{2}\right) + \sin\left(\frac{\theta}{2}\right) r_x i + \sin\left(\frac{\theta}{2}\right) r_y j + \sin\left(\frac{\theta}{2}\right) r_z k \tag{10.9}$$

then the coordinates of the rotated vector are the i, j, and k components of the quaternion

$$p' = qp\bar{q} \tag{10.10}$$

which is a form easy to represent and calculate numerically.

We choose the axis of rotation $\mathbf{r}$ by picking its z coordinate randomly from a uniform distribution between -1 and 1. We then pick a longitudinal angle ϕ uniformly from between $-\pi$ and π, and set the Cartesian coordinates of $\mathbf{r}$ to be $(\sqrt{1-z^2}\cos(\phi), \sqrt{1-z^2}\sin(\phi), z)$. We may also use this approach for picking unit vectors at random to be the initial placing of points.

With the proposed new position for this vortex we calculate the change in energy ΔH this move would cause. Based on a preselected inverse temperature β, and s chosen from a uniform random distribution on $(0, 1)$, the change is accepted if $s < \exp(-\beta \Delta H)$ and rejected otherwise. This is the Metropolis-Hastings algorithm [227].

Recall that if β is positive then this results in any change decreasing energy being accepted. This our key – with positive β, moves which on average increase the spread of vortices from one another are approved. The result is a covering of the whole domain. We cannot justify theoretically using negative β, since there is no maximum energy – the energy can always be made greater by moving points even closer together. If we attempt the Metropolis-Hastings algorithm anyway we see vortices clumping together, not necessarily at a single point.

We repeat sweeps until we reach a statistical equilibrium. This we can measure by observing the fluctuations in energy, or in the radial distribution function (discussed below) or by other exact measures; or we can run preliminary experiments to estimate how long a run is needed to get close to an equilibrium.

Once we have our equilibrium we take the positions vortices to be the mesh sites. Given positive β energy is generally lowered by spreading the points apart, so we will find reasonably uniform distributions of points. With β very large few moves which put vortices on average closer together are accepted. The result is a mesh that comes very close to the most uniform possible distribution of points over the sphere.

This algorithm does not require moving free particles. We can hold, for example, several vortices to be aligned along a little circle and make each experiment the moving of this entire circle as a rigid body.

10.3 Radial Distribution Function

We want an algorithm which can produce different meshes for the same number of points. So how can we tell whether two meshes are different? The fact the sphere is rotationally symmetric and all the vortices are of uniform strength indicates that if we generate two meshes M_1 and M_2 by the Metropolis-Hastings algorithm above, they are unlikely to have exactly the same set of coordinates for each point. If M_1 and M_2 differ only by a rotation and a relabelling of vortices they are not really different though no coordinates may match.

It is easy to suppose we might compare meshes by testing whether rotations (and reflections) of M_1 can make its points congruent to the points of M_2, but the number of possible rotations and reflections that would need to be considered make it clear this is not practical. We need a tool which does not consider the "absolute" positions of points or their labels.

We draw inspiration from the study of crystals and of real gases, in which structure represented by the **radial distribution function**. This function is the density of particles, relative to a uniform density, as a function of distance from a reference point [165]. In gas and crystal contexts this is the probability of finding a particle at a given distance; in a way it measures average distances between particles. We will tailor the definition to our needs.

What is unique about a mesh is the set of relative differences between points and their relative orientations. Our **radial distribution function** is defined for a given angular separation θ. (On the surface of the unit sphere, this is equivalent to the separation by a distance θ.) We will let the function for any given distance be the number of pairs of points separated by that distance.

Definition 9. *For a set of particles M define the radial distribution function to be the set of ordered pairs*

$$D = \{(\theta, n) |\ n \text{ is the number of pairs of points with separation } \theta\} \quad (10.11)$$

– that is, θ is a radial separation with n pairs of points with that separation.

Now we have a tool which we expect to distinguish between patterns. We can easily specify when two patterns are **equivalent**.

Definition 10. *Two patterns of points on the sphere are equivalent if a combination of rotations and reflections in $O(3)$ make the positions of one pattern coincident to the positions of the other.*

$O(3)$ is the set of orthogonal transformations in three dimensions, that is, the rotations and reflections that can be performed on a unit sphere. The radial distribution function lets us avoid the task of checking every possible rotation and reflection of M_1 to compare it to M_2.

Theorem 11. *The radial distribution function $D(M)$ is a property of the group orbit under $O(3)$ actions.*

Proof. The radial distribution function is a measure of the angles between points in a given configuration. $O(3)$ actions rotate and reflect these points on the surface of the sphere but do not change the relative separations of any of these points. Therefore the radial distribution functions will be identical for all elements in the orbit of a given configuration M. □

Corollary 2. *Two vortex configurations cannot be equivalent unless their radial distribution functions agree.*

Proof. If a configuration M_2 is equivalent to state M_1 there exists a combination of rotations and reflections – an action in $O(3)$ – which turns M_1 into M_2. Therefore the radial distribution function for configuration M_2 must be equal to that of configuration M_1. □

There is a caveat necessary to observe. While two equivalent configurations must have identical radial distribution functions, it is not proven that two configurations with identical radial distribution functions must be equivalent.

The radial distribution function also presents an interesting means of studying any pairwise interaction such as a chemical potential. Consider an energy-like potential function $U(M)$ for a collection of points M with uniform strength.

Theorem 12. *The value of $U(M)$, for a system in which all particles have uniform strength, is uniquely determined by the radial distribution function of M.*

Proof. The value of the energy-like potential function $U(M)$ is defined by

$$U(M) = \sum_{j=1}^{N} \sum_{k=j}^{N} s_j s_k f(\mathbf{z}_j, \mathbf{z}_k) \tag{10.12}$$

with $f(\mathbf{z}_j, \mathbf{z}_k)$ a pairwise distance-dependent function. Assuming each particle has strength $s_j = 1$, then the value of this function is

$$U(\mathbf{z}) = \sum_{j=1}^{N} \sum_{k=j}^{N} f(\mathbf{z}_j, \mathbf{z}_k) \tag{10.13}$$

$$= \sum_{j=1}^{N} \sum_{k=j}^{N} g(d(\mathbf{z}_j, \mathbf{z}_k)) \tag{10.14}$$

$$= \sum_{d_j \in D} n(d_j) g(d_j) \tag{10.15}$$

where the functions f and g are as above, d is the scalar distance function, and $n(d)$ is the number of pairs separated by the distance d.

From the construction of the energy-like potential function as the summation over the elements in the radial distribution function from equation 10.15 it is obvious this property of the points M uniquely defines the value of $U(M)$, regardless of whether any two configurations points M_1 and M_2 with the same radial distribution function are congruent placements of points. □

10.4 Vortex Gas Results

The tetrahedron is one of the Platonic solids [96] [412] and is a commonly recognized shape. In the vortex problem it is an equilibrium. It has a simple radial distribution function: every point is equally distant from its neighbors.

Four completely free particles find the tetrahedron configuration. As the Metropolis-Hastings algorithm has no means of detecting latitude the tetrahedron "points" in a randomly chosen direction. The energy and the radial distribution function indicate the tetrahedron has been found.

The octahedron is another Platonic solid. It can be viewed as a triangle of vortices in the northern hemisphere and another, staggered with respect to lines of longitude, triangular arrangement of vortices in the southern hemisphere. This bears some resemblance to the von Kármán trail, and so gives us more reason to consider what interesting properties may arise from considering rigid rings of vortices rather than solely free particles.

Six free particles finds the octahedron, as measured by energy and radial distribution function, as well as by simply looking at the points and drawing the obvious edges between neighbors.

The cube is another Platonic solid. We can consider it as a pair of rings of four points each, mirrored around the equator. This is unlike the von Kármán trail in not being staggered. There is also another way we might represent it with these rings of evenly spaced points, if we rotate the cube so points are at the north and south poles. Then the remaining six points are along two triangles, each triangle staggered with respect to the other.

The Metropolis-Hastings algorithm on eight free particles surprises us: it does not settle onto the cube. If we rotate the sphere we can find a way to represent it, as the vertices of two squares staggered with respect to one another. This shape, if we draw the edges between neighbors in, is the square anti-prism, and it is composed of triangular surfaces and two squares.

With twelve particles there are two obvious polyhedron configurations suggested by the work of Coxeter exploring simple polyhedrons and their duals [96]. The first is the icosahedron. The other configuration is the cuboctahedron.

The cuboctahedron is not a Platonic solid. This shape is not the lowest possible energy for the twelve-vortex problem. We can still view it, interestingly, as a set of three rings of four particles each, each ring staggered with respect to its neighbors.

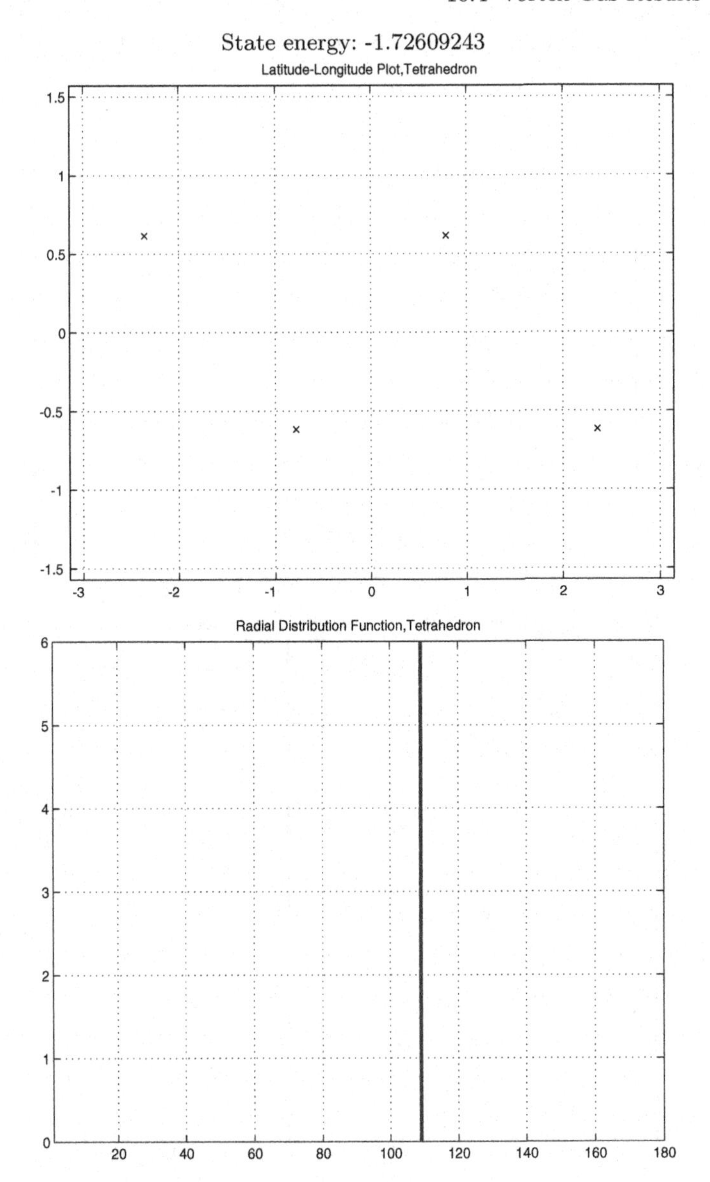

Fig. 10.1. The map and radial distribution function of the tetrahedron.

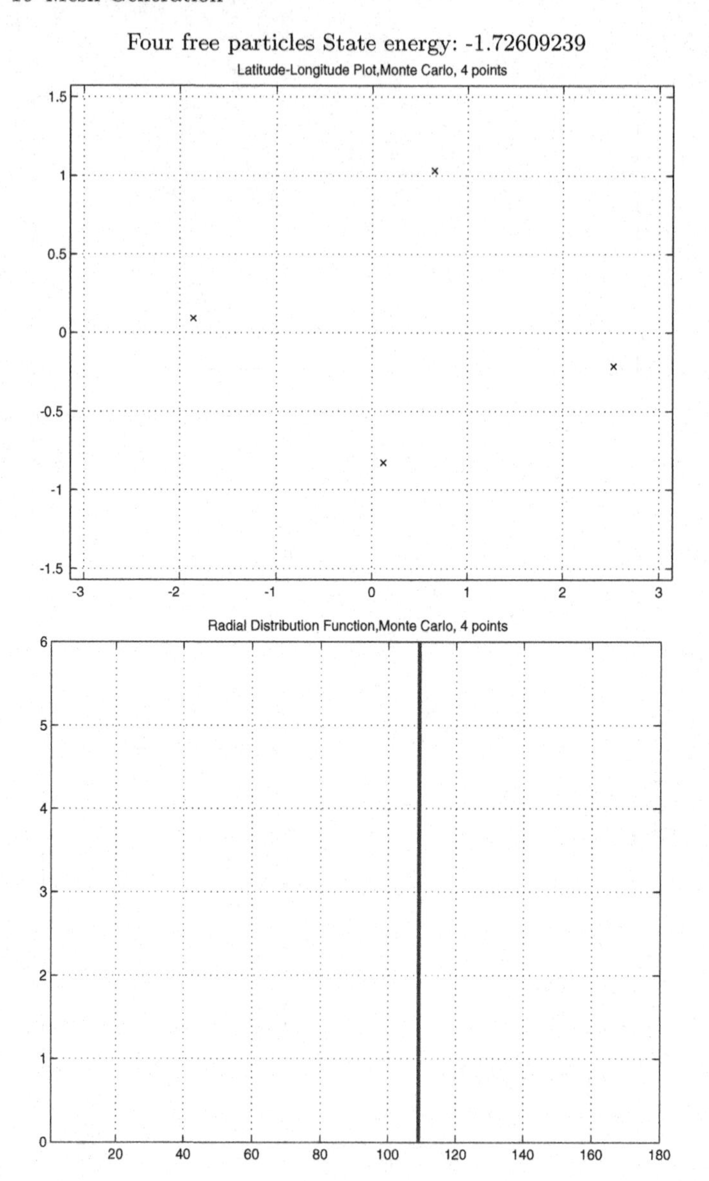

Fig. 10.2. The map and radial distribution function of four free particles.

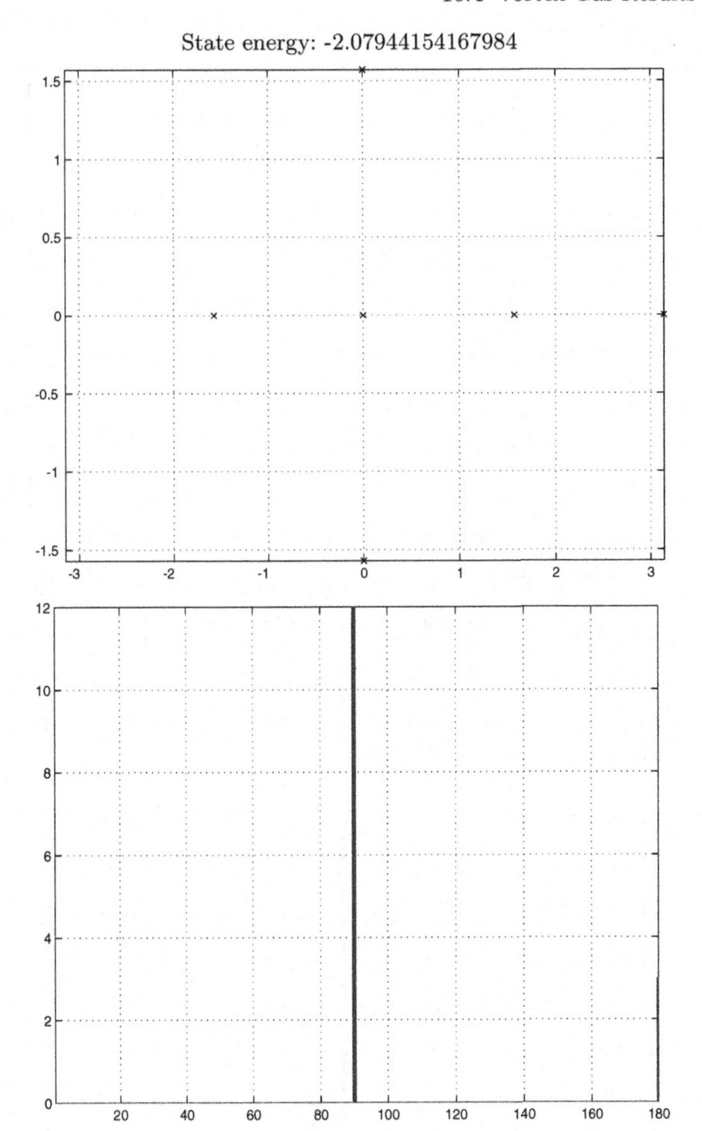

Fig. 10.3. The map and radial distribution function of the octahedron.

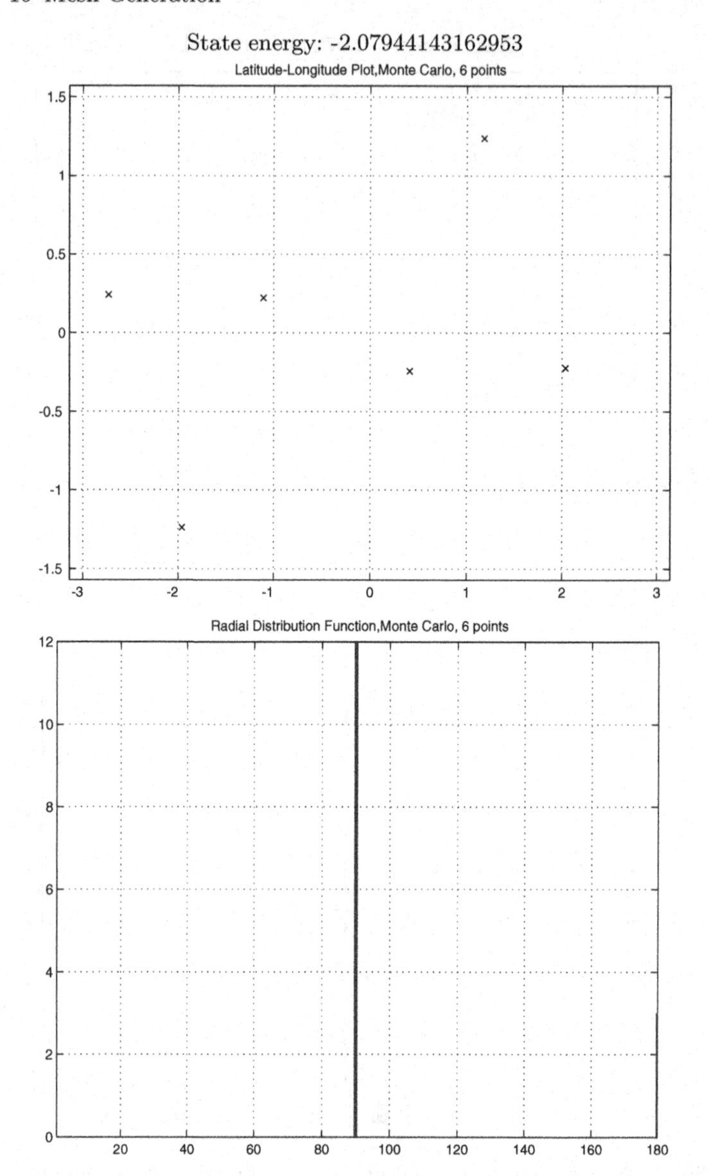

Fig. 10.4. The map and radial distribution function of six free particles.

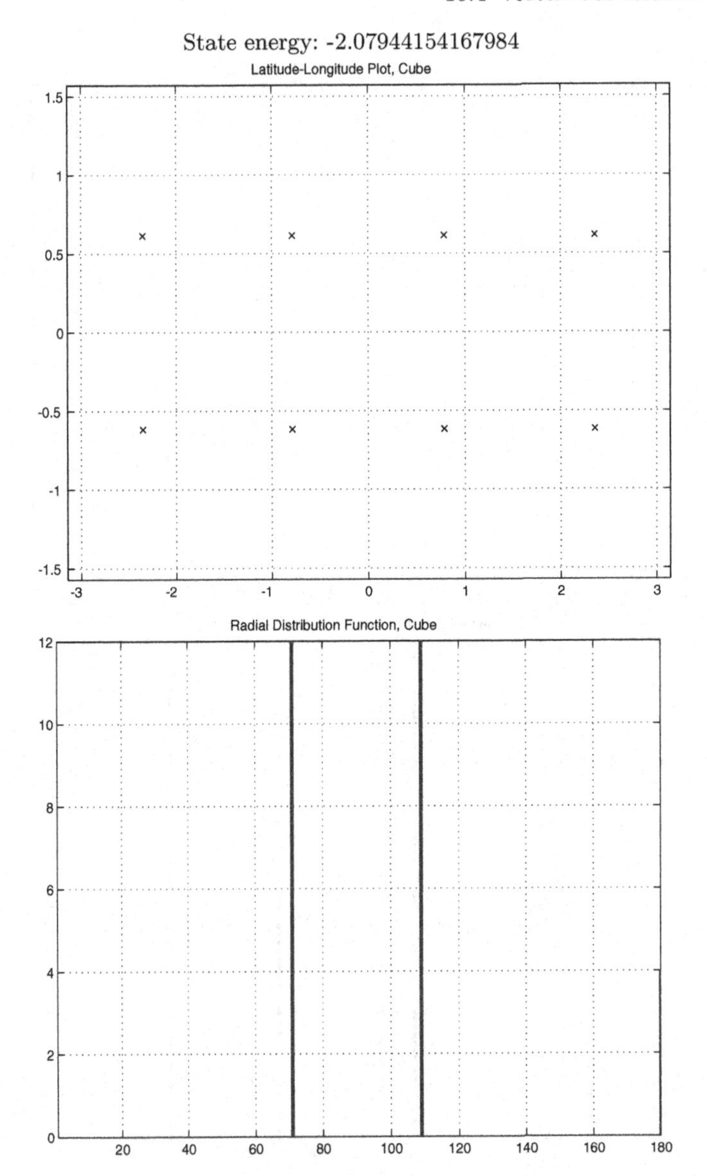

Fig. 10.5. The map and radial distribution function of the cube.

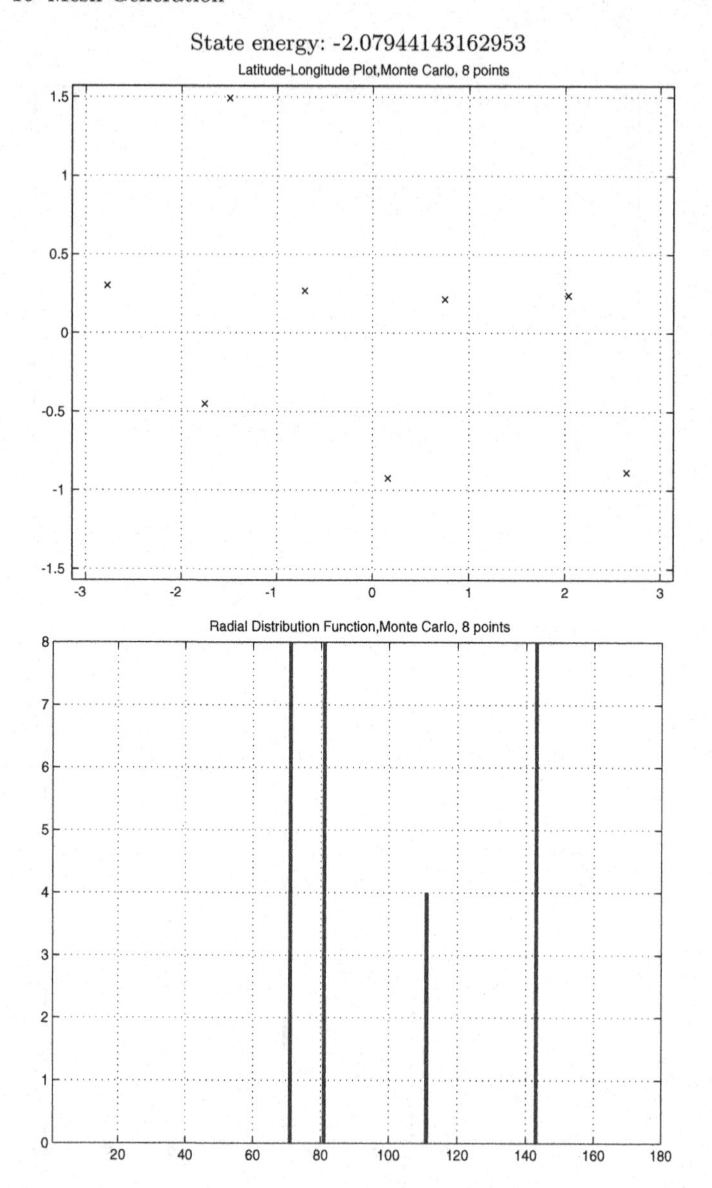

Fig. 10.6. The map and radial distribution function of eight free particles.

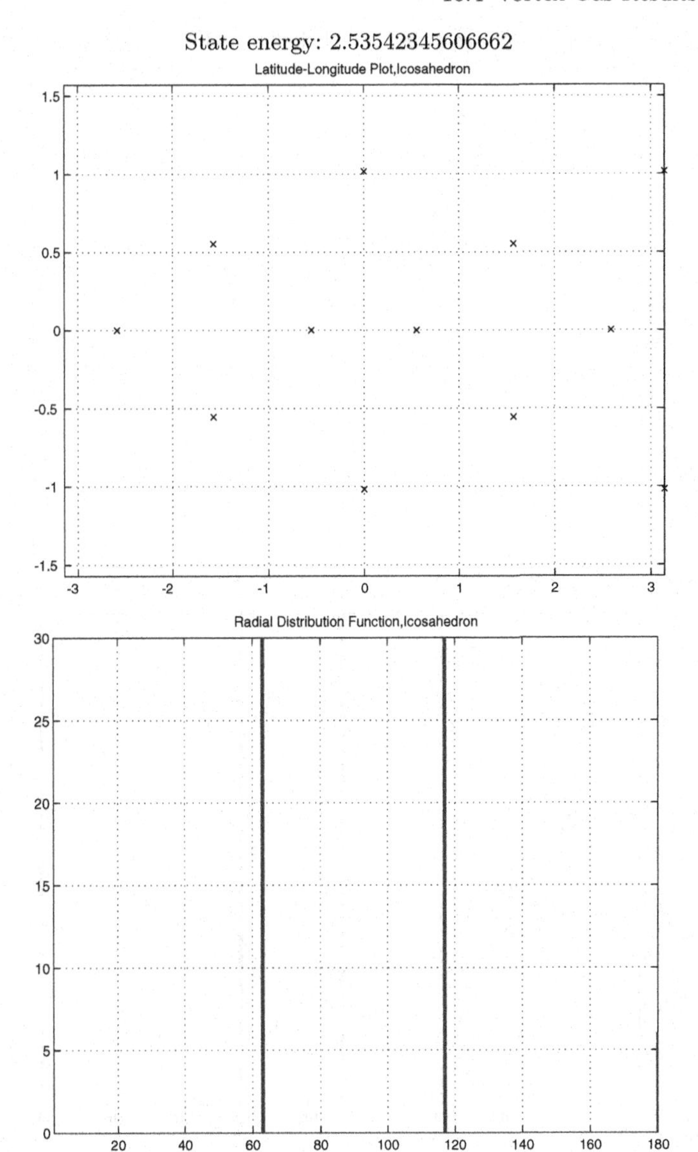

Fig. 10.7. The map and radial distribution function of the icosahedron.

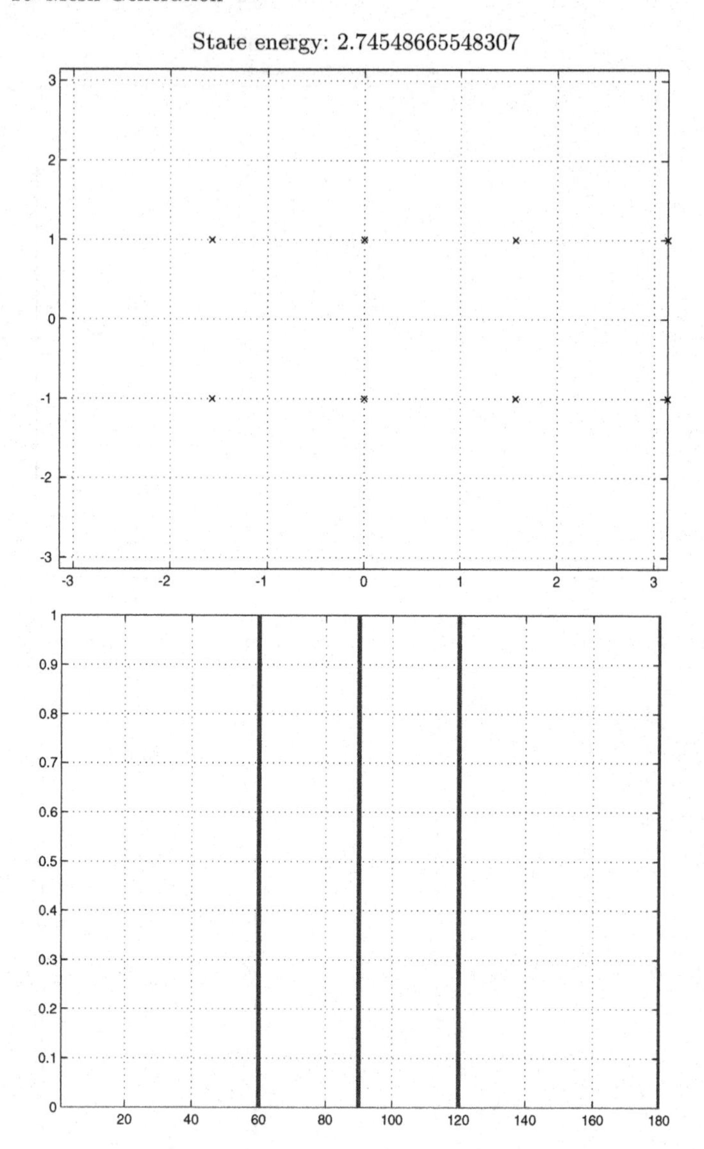

Fig. 10.8. The map and radial distribution function of six free particles.

With twelve free particles we may reasonably wonder whether we will find the icosahedron or the cuboctahedron as a result of the Metropolis-Hastings algorithm. Given that we expect our algorithm finds distributions that minimize energy it seems the icosahedron is more likely, and this is what we find numerically. That we know there is an interesting polyhedron configuration invites us to wonder if there are ways to get that shape from our algorithm. This will inspire our work in the next section.

We have another obviously interesting polyhedron to consider, the dodecahedron. This has twenty vertices and we can view its points as a set of four rings, each ring made of the vertices of a regular pentagon. The first and second rings are staggered longitudinally, and the third and fourth rings are the reflections of the second and first across the equator.

Twenty free particles do not settle to the dodecahedron. The points are more uniformly distributed around the sphere, and their configuration does not appear to be the vertices of any particular named polyhedron. We now have some interesting examples in which some of the Platonic solids were reached by our Metropolis-Hastings algorithm, but the others were not. There are other familiar solids we can explore – the Archimedean solids (polyhedra formed by two or more regular polygonal faces, as opposed to the single regular face which provides the Platonic solids) are the obvious set to continue researching – and we are drawn to wonder what polyhedrons can be found by Metropolis-Hastings and what ones cannot. We will continue to explore this line of thought in the chapter on polyhedra.

Putting aside polyhedra we know we can generate a mesh for any arbitrary number of mesh sites. As we take more and more points do we uniformly cover the surface of the sphere? We have two obvious ways to ask the question, by considering the energy and the radial distribution function.

Consider the energy first. If we have a uniform distribution of vorticity over the surface of the sphere we can calculate the energy this way. Letting ρ be the density of this uniform vorticity distribution, and taking dummy variables $\mathbf{x}$ and $\mathbf{y}$ the total energy is

$$E = -\frac{1}{2}\int_{S^2}\int_{S^2} \rho^2 \log|1 - \mathbf{x}\cdot\mathbf{y}| dA' dA \tag{10.16}$$

where dA' is the differential volume around $\mathbf{y}$ and dA is the differential of volume around $\mathbf{x}$. Integrating first with respect to $\mathbf{y}$ and dA', and taking without loss of generality the vector $\mathbf{x}$ to equal $\hat{k}$ then we have

$$E = -\frac{1}{2}\int_{S^2}\int_{S^2} \rho^2 \log|1 - \cos(\theta')| dA' dA \tag{10.17}$$

$$= -\frac{1}{2}\rho^2 \int_{S^2} dA \int_0^{2\pi} d\phi' \int_0^{\pi} \log|1 - \cos(\theta')| \sin(\theta') d\theta' \tag{10.18}$$

$$= -\frac{1}{2}\rho^2 \int_{S^2} dA (2\pi)\cdot(2\log 2 - 2) \tag{10.19}$$

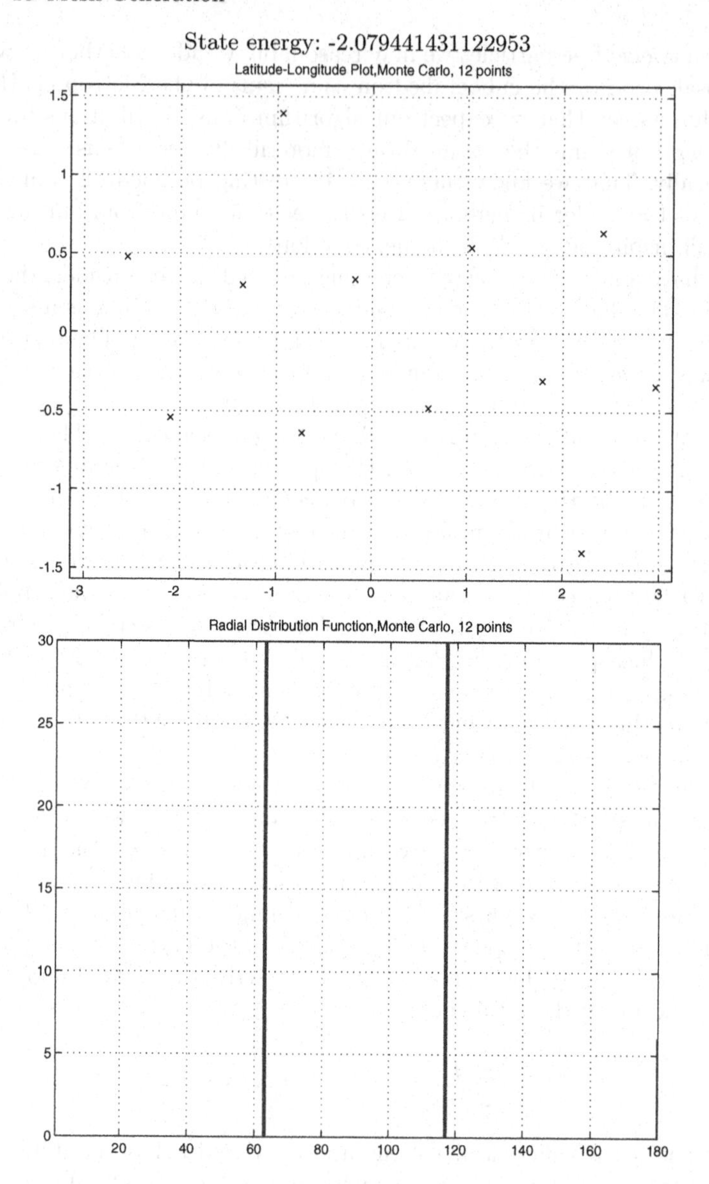

Fig. 10.9. The map and radial distribution function of twelve free particles.

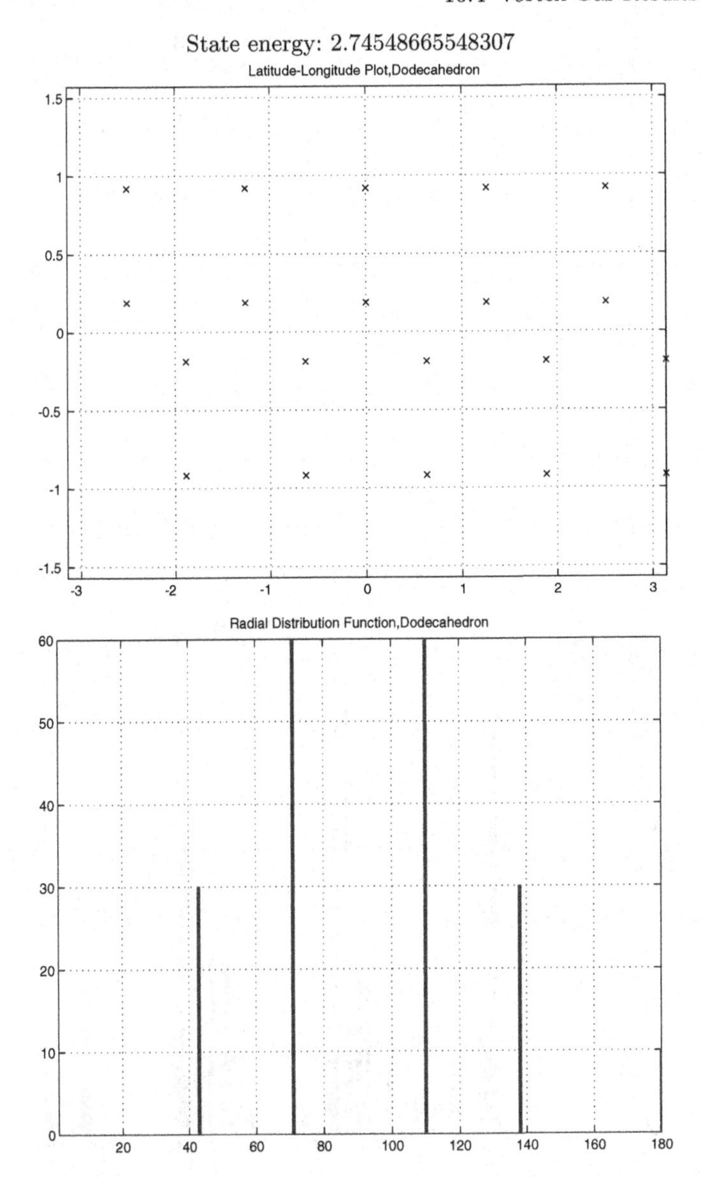

Fig. 10.10. The map and radial distribution function of the dodecahedron.

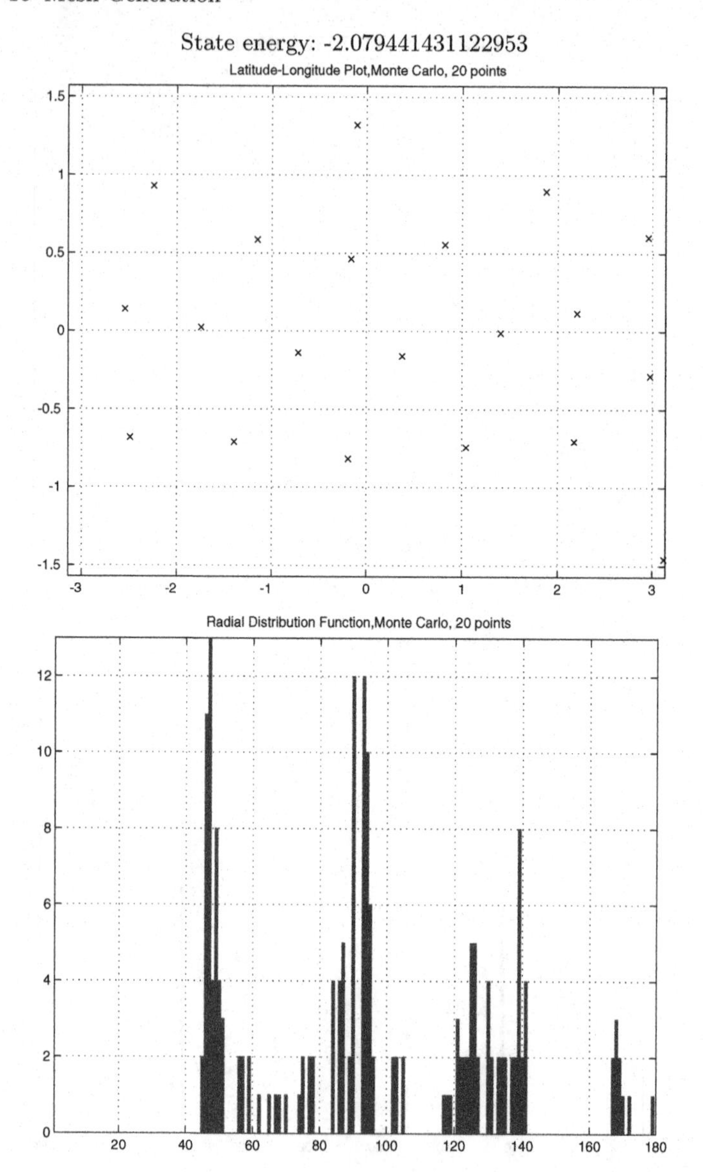

Fig. 10.11. The map and radial distribution function of twenty free particles.

$$= -\frac{1}{2}\rho^2 8\pi^2 (2\log 2 - 2) \tag{10.20}$$

where θ' and ϕ' are the longitude and latitude coordinates for $\mathbf{y}$.

If the N point vortices have strength 1 and are uniformly distributed across the sphere then the density of vorticity is $\rho = \frac{N}{4\pi}$. The energy of N perfectly uniformly distributed particles therefore is

$$E_N \approx -\frac{1}{2}\frac{N^2}{2}(2\log 2 - 2) \tag{10.21}$$

while we have, for the energy per particle pair,

$$\lim_{N\to\infty} \frac{E_N}{N^2} = -\frac{1}{2}(\log(2) - 1) \approx 0.153426 \tag{10.22}$$

which we term the mean field limit. It is the energy we get if we replace the actual vorticity field with its arithmetic mean.

Making such a mean field approximation – which is not always valid – often simplifies problems like evaluating the partition function. We will try to work without assuming the mean field, but will point out when it comes about or when we cannot do without it.

Now we measure the energy per particle pair for a variety of mesh sizes to determine whether it does approach $-\frac{1}{2}(\log(2) - 1)$. We also may compare the energy per particle pair against the number of sweeps completed to provide us with an estimate for how many Metropolis-Hastings sweeps are needed to reach an equilibrium.

From these energy observations it appears our meshes cover the sphere close to uniformly, and that we do not need much computational time to get these coverings. Several hundred sweeps brings us almost to our "final" positions. From this we are more confident that the Metropolis-Hastings approach is a good one to generating meshes satisfying our requirements.

Next we look at the radial distribution function for a large mesh – 1,024 points is convenient to examine. We want to compare this to the radial distribution function for a perfectly uniform mesh. We will take the north pole as reference point for this continuum radial distribution function. The "number of points" a distance θ from the north pole is the area of the slice of the sphere which is between $\theta - \frac{1}{2}d\theta$ and $\theta + \frac{1}{2}d\theta$ away from the north pole for a given differential angle $d\theta$. This strip is approximately a circular ribbon, of width $d\theta$ and of radius $\sin(\theta)$, so the area is $2\pi\sin(\theta)d\theta$ and therefore we expect as the number of points grows larger the radial distribution function should approximate a sine curve.

We do see this. Compare the radial distribution functions for 1,024 points and the sine curve; despite a few peaks reflecting how the mesh sites are close to forming lattices with their nearest neighbors.

It is worth investigating then whether or not our meshes have any rotational or mirror symmetries. The problem which prompted us to introduce the radial distribution function – the great number of possible permutations of points – makes it too challenging to prove or disprove symmetry directly. We will then try to create meshes that have a symmetry built into them, and compare those results to our free-particle equilibria.

10.5 Rigid Bodies

We have considered only particles allowed to freely move on the sphere. But we know from experiments that we get points configured on what look like the vertices of regular polygons, and we know some of these regular polygon vertex configurations are equilibria and energy minima. So let us consider particles that are not free, but that are rigidly locked to several other points spaced uniformly in longitude around a single line of latitude.

Each Metropolis-Hastings experiment attempts moving an entire **ring** – the set of points – by the same amount in the same (relative to the north-south axis) direction. An experiment changes the latitude and longitude of every point on the ring simultaneously. Right away we see this reduces the number of energy interactions that need to be calculated, so if we can reach energy minima by putting in rings we can expect potentially large savings in computational cost. Can we reach those minima?

A single ring of two points – both latitude p, one at longitude θ and the other at $\theta + \pi$ – will be pushed to the equator, where they become antipodal. We knew this to be a minimum. A single ring of three points – all at latitude p, and the three at longitudes θ, $\theta + \frac{2\pi}{3}$, and $\theta + \frac{4\pi}{3}$ – again go to the equator and form the equilateral triangle on a great circle we knew to be an energy minimum. We also realize a single ring of N point will always be drawn to the equator, and this will not be an energy minimum for any more than three points – though it will be an equilibrium.

We look to dividing the points of the problem into multiple rings then. The tetrahedron can be considered as a configuration of two rings of two points each [272], staggered with respect to one another (so that the line of longitude passing through one ring's points bisects the angle between two nearest of the opposite ring) and arranged symmetrically about the equator. It may also be regarded as a single ring of three evenly spaced points with one polar point.

Five free points form a shape called the bi-pyramid, and we cannot express it as a collection of several rings each with the same number of points. If we allow ourselves to write polar points as separate from the rings we can write it as a point at the north pole, an equatorial ring of three points, and the south pole.

With six points – the octahedron – we have two reasonable descriptions again. We can write it as the north pole, an equatorial ring of four points, and the south pole. Alternatively we can write it as a ring of three points and a staggered ring of three more points.

Seven points, like five points, cannot be divided into rings with a uniform number of points, but if we allow one point to be the north pole we can divide the others into two staggered rings of three points each.

Eight points – the Metropolis-Hastings configuration gives us the square anti-prism – we can write as one ring of four points and a staggered ring of the other four points. The vertices of the cube, meanwhile, we can write as one ring of four points and another ring of four points – with those rings aligned, so the points of both rings lie on the same line of longitude. (We can also write the vertices of the cube as one north and one south pole, and two rings of three points each staggered with respect to one another.)

The Metropolis-Hastings configuration for nine points distributes itself into three rings of three points each – one ring in either hemisphere and one on the equator. The equatorial ring is staggered with respect to the "temperate" rings, and the "temperate" rings are aligned with one another.

Obviously we can continue in this vein, but these are enough observations to begin forming hypotheses about these rings. First, since any number N is either composite or is one more than a composite number, we can form a configuration of N points using several rings of a uniform number of points, possibly with an extra polar point.

Second if there are an even number of rings they place as many in the northern as the southern hemisphere, and the latitudes of the rings are equal and opposite in sign. This follows from the mirror symmetry of the sphere across the equatorial plane. If there are an odd number – we start to suppose there will be one equatorial ring and then as many in the northern as the southern hemisphere, but the example of seven points is a counterexample. We can patch the hypothesis by taking it to be an odd number of rings all with the same number of points, or that both poles must have points. Then we have the equatorial ring and the mirror symmetry of the latitudes of the remaining rings.

We have found so far rings are either aligned with or staggered with respect to their neighbors. This suggests how to create a notation to describe meshes compactly. This notation is not more generally useful – though there are many ways to describe polyhedra with formulas – but it helps organize some thoughts about this problem, and so is a useful tool. Imagine a plane parallel to the equatorial plane which sweeps from the north to the south pole. Each time we pass a latitudinal ring we write the number of points in that ring, and note whether it is aligned with or staggered with respect to the previous ring. As we have seen the staggered case more than we have the aligned we will take the staggered case to be the default, and note with a superscript a when a ring is aligned to its predecessor.

Then we can write the triangle as (3); the tetrahedron as either $(2, 2)$ or as $(1, 3)$. The bi-pyramid is $(1, 3, 1)$ and the octahedron $(3, 3)$. Seven points gives us $(1, 3, 3)$. The square anti-prism is $(4, 4)$ while the cube is $(4, 4^a)$ or $(1, 3, 3, 1)$. Nine points are $(3, 3, 3)$.

Leaping to the dodecahedron, which we know is not an energy minimum, we realize we can write this shape as $(5, 5^a, 5, 5^a)$ – and now have a possibility to test. The polyhedra that were not energy minima had aligned rings, while those that were had staggered rings. We notice the icosahedron is the configuration $(3, 3, 3, 3)$ – and then the cuboctahedron is $(4, 4, 4)$. Obviously it cannot simply be having a staggered-only representation that makes a shape an energy minimum. If we rotate the cuboctahedron, though, we find it can be represented in a new way, as a ring of three, an equatorial ring of six, and another ring of three points, $(3, 6, 3)$. This equatorial ring is not aligned with either ring of three points, but neither is it staggered. It seems as though being representable with aligned or "other" rings may be the indicator that a shape is not an energy minimum.

So let us search for an energy minimum that has an aligned (or "other") alignment of rings. If we cannot find a counterexample easily we can start searching for ways to prove the hypothesis that aligned or other ring configurations indicate a shape cannot be an energy minimum.

When we try the Metropolis-Hastings algorithm for twelve points, we have several ways in which we may divide the rings – into six rings of two points each, four rings of three, and so on. Obviously a single ring of twelve points will be different from six rings of two, but do any of these ring set-ups give us the same pattern?

From experiment, measuring energy and radial distribution function, it appears that free particles, rings of two, and rings of three all give us the same shape, the icosahedron. Three rings of four points each gives the cuboctahedron and a higher energy. Two rings of six points each gives us a shape that might be called the hexagonal anti-prism – the two rings of six each are staggered around the equator, and its energy is considerably larger than that of the earlier cases. The highest-energy state we find is twelve rings on the equator. Apparently rings of few enough points are close to or indistinguishable from the free-particle case.

Let us then look at the abundant number 120 of points; there are many uniform rings we can consider. The energy for free particles, rings of two, three, and so on are plotted.

Then we repeat this experiment with 840 particles and again observe energy versus the number of particles in the rings.

The ultimate interesting question is, can our ring structures capture energy minima? If we compare the energy of the free particle case for 840 particles to that of rings of two, rings of three, and other small rings we find the rings are marginally higher in energy. In comparing the radial distribution functions we find some noticeable differences.

We are left to conclude these rings do not, in general, find energy minima. They can come quite close, and for few enough particles will even find the minima almost exactly. If we were concerned about quick computation of minima for a large number N of points we might use the Metropolis-Hastings algorithm on rings of $\sqrt{N}$ points each, and once we have found this use those points as the starting location for a second round with free particles.

Exploring rigid bodies has not helped us find the uniform meshes we want more easily, but it has given us reason to look more closely at polyhedra and symmetries. We will pick up this thought in another chapter.

10.6 Spherical Codes

There is a problem related to our process of finding equilibria for the vortex gas problem, called the **spherical coding problem**. A spherical code is a set of points on the unit sphere; each point corresponds to a message, but the translation to a message is not of interest. What we find interesting is a spherical code allows us to correct for errors.

We do not get from this the same arrangement of points we get from the vortex gas problem; but we do come close. Coxeter has explored the spherical coding problem further and found maximum minimum distances between points for a number of mesh sizes [96] and we compare them to the minimum distances between points from our vortex gas problem.

Coxeter points out that in the spherical coding problem the optimum arrangement of points, in general, is not symmetric.

Anyone can imagine a simple example of this, transmitting a sequence of bits 0 or 1 by sending, instead, 000 or 111. Then if any one bit is transmitted or received incorrectly it can be repaired by taking two of the three bits. This is an inefficient example – it requires tripling the length of a signal, and is forgiving of only a single error per transmitted bit – but it demonstrates the principle.

This is considered a spherical code because we can map each point to the surface of a sphere, 000 to the point with Cartesian coordinate $\left(-\frac{1}{\sqrt{3}}, -\frac{1}{\sqrt{3}}, -\frac{1}{\sqrt{3}}\right)$ and 111 to the point $\left(\frac{1}{\sqrt{3}}, \frac{1}{\sqrt{3}}, \frac{1}{\sqrt{3}}\right)$.

The question in general is how to distribute N points on the surface of the sphere so that the minimum distance between any two points – which corresponds to how much error may be made in any signal – is maximized? Each of these points is considered to be one coded message.

Only a few cases are analytically known to be optimum arrangements. Tóth[1] proved in 1943 that for N points there are invariably at least one pair of points separated by a distance D which satisfies

$$D \leq \sqrt{4 - \csc^2\left(\frac{\pi N}{6(N-2)}\right)} \qquad (10.23)$$

The optimal placement for two points is an antipodal pair, as one might expect. Similarly it is unsurprising that for three points we place them equally

[1] László Fejes Tóth. He also proved the regular hexagonal lattice provides the densest packing, on the plane, of circles of uniform size.

spaced along a great circle. And four points are optimally placed at the vertices of a regular tetrahedron, just as the optimum placement of four vortices of uniform strength are. Similarly the spherical code placement of six points are the vertices of the regular octahedron, exactly as the vortex gas problem finds.

The first surprising result comes about with eight points; it is natural to suppose the optimum placement would be the vertices of a cube. In fact – just as with the vortex gas problem – the best placement is the vertices of the square anti-prism. This figure has two planes consisting of squares, but the squares are staggered with respect to one another. This is again exactly the configuration found by the vortex gas problem, and one would naturally suppose we have found a method capable in general of approximating optimum placements of N points on the sphere for use as spherical code locations.

This optimism is a bit misplaced. The vortex gas problem finds a distribution of points $\mathbf{x}_1$, $\mathbf{x}_2$, $\mathbf{x}_3$, $\cdots$, $\mathbf{x}_N$ which minimizes the energy

$$H_N = -\frac{1}{2}\sum_{j=1}^{N}\sum_{k\neq j}^{N} \log|1 - \mathbf{x}_j \cdot \mathbf{x}_k| \tag{10.24}$$

which is incidentally equivalent to finding a minimum of

$$\exp(H_N) = \frac{1}{\sqrt{e}}\prod_{j=1}^{N}\prod_{k\neq j}^{N}(1 - \mathbf{x}_j \cdot \mathbf{x}_k) \tag{10.25}$$

The spherical coding problem meanwhile is to maximize the minimum distance

$$d = \min_{j,k}(1 - \mathbf{x}_j \cdot \mathbf{x}_k) \tag{10.26}$$

a different problem. That they correspond well in the examples above is a coincidence of the small number of vertices considered. With a greater number of points the difference between the maximum minimal separation for the spherical code and for the vortex gas problems becomes noticeable.

However we are justified in taking a Monte Carlo approach to finding spherical code sites. We write an algorithm for this much like we do for the vortex gas: given a set of points $\mathbf{x}_1$, $\mathbf{x}_2$, $\mathbf{x}_3$, $\cdots$, $\mathbf{x}_N$ and a statistical mechanics inverse temperature β conduct an experiment which consists of selecting a site j at random. (This may also be done in order.)

Take site $\mathbf{x}_j$ and consider a hypothetical move to a new position $\mathbf{x}_j'$. Calculate the change in the maximum minimum distance Δd which this move would cause. The proposed move is then accepted if it increases the maximum minimum distance (if $\Delta D > 0$); if the move would decrease the maximum minimum distance then a random number r is drawn from $(0, 1)$ and then the move is accepted if $r < \exp(\beta \Delta d)$ and rejected otherwise.

Repeat this sufficiently many times – which is not a great number, just as with the vortex gas problem – and we find the mesh sites converging quickly on sites near the best possible distributions.

There is an interesting note regarding spherical coding. There is no a priori reason to confine the points, or the codes, to the surface of the three-dimensional sphere. The problem is trivially easy on the two-dimensional circle (optimum placement being the placing of points at the vertices of a regular polygon); but interestingly it occasionally becomes considerably simpler in four or more dimensions.

Other problems closely related to the vortex gas or to the spherical coding problems which we may examine are the "kissing number" question – what is the maximum number of spheres of uniform size that may be placed tangent to a central sphere, and where should they touch the central sphere? – and the Thomson[2] problem of where to place N electrons constrained to the surface of the sphere so that they rest in a stable dynamic equilibrium. (In reality point charges cannot be held in a stable equilibrium by electric interactions alone, but the problem is still worth exploring, and statistical equilibria are a different sort of equilibrium.)

---

[2] Named for James J Thomson, 1856 - 1940, discoverer of the electron, who formed the "plum pudding" model of the atom. He also studied the dynamics of rings of vortices, and developed (independently of Herman Nernst, 1864-1941, who discovered that at absolute zero it was entropy, not energy, that was minimized) the Nernst-Thomson rule explaining the rate of ionization of compounds in water. [25]

11

Statistical Mechanics for a Vortex Gas

11.1 Introduction

We are not only interested in modelling fluid dynamics problems on the sphere. The unbounded plane is of obvious interest, and we want to generate meshes on it. We cannot hope to do more than cover a finite region of the plane. And yet if we run a Monte Carlo algorithm with N vortices all of the same strength initially placed randomly over any region of the plane and with a positive β we can get a reasonably uniform mesh – but it never settles to a statistical equilibrium. This is obvious in hindsight: whatever the arrangement of points $\mathbf{z}_1$, $\mathbf{z}_2$, $\mathbf{z}_3$, et cetera, one can always reduce the energy of the system by moving points farther away from the center of vorticity.

We need an algorithm including some natural limit against dispersing the points indefinitely far away. We are already familiar with the notion of Monte Carlo algorithms with two or more canonically conserved quantities and so need not introduce any new material to use them. But what constraint to use? Given that constraint what is the relationship between the constraining parameters – the Lagrange multipliers – and the radius over which the mesh of points will spread? Both the choice of quantity and its effect on the radius can be given definitive answers.

We have also the question of whether we can precisely define of the amount of space a set of mesh points covers. We have several alternatives; perhaps the most analytically satisfying is to say the area covered by the points is the area of the smallest circle which encloses all the mesh sites.

If we were considering a set of particles each with mass s_j rotating around the origin, the "moment of inertia" of the points $\mathbf{z}_1$, $\mathbf{z}_2$, $\mathbf{z}_3$, $\cdots$ $\mathbf{z}_N$ would be

$$\Omega\left(\mathbf{z}_1, \mathbf{z}_2, \cdots, \mathbf{z}_n\right) = \sum_{j=1}^{N} s_j \|\mathbf{z}_j\|^2 \tag{11.1}$$

(with $\|\mathbf{z}_j\|$ the norm of the vector from the origin to the point $\mathbf{z}_j$; if we represent points on the plane as complex coordinates this is just the absolute value of the point's coordinates). The moment of inertia grows as the points spread out from the origin. (We could also reasonably use the sum of squares of norms of the distance from $\mathbf{z}_c$, the center of vorticity; this would let our mesh spread out to center around points not at the origin – but we can place the center of vorticity at the origin by translation anyway.) Since we think of s_j as vorticity then we will call this sum the **moment of vorticity**.

With the moment of vorticity and with a chemical potential μ we then seek statistical equilibria. Choose at random a vortex, and consider moving it by a random amount within some region. The change in energy ΔH and the change in moment of vorticity $\Delta\Omega$ this displacement would cause is then calculated. Based on a randomly chosen number s uniformly distributed on $(0, 1)$, the change is accepted if $s < \exp(-\beta\Delta H - \mu\Delta\Omega)$ and rejected otherwise. The process of picking vortices repeated until an equilibrium is reached.

Do we know this will result in an equilibrium that covers some but not all of the plane? Experimentally we find it does, but we only know analytically that we will get such a covering when β equals 2 and μ equals 1. For other values we do not have exact results, but find unsurprisingly the radius of the disc covered by the N points will be proportional to $\sqrt{N}$ still, and will depend on the ratio of β and μ. But there is a difference in how uniform the covering of points is; a greater value of both constraints makes the disc more nearly circular and the average distance between neighboring points more nearly constant.

11.2 The Vortex Gas on the Plane

The interaction of N point vortices on the plane, each at the complex-valued coordinate z_j, and each with strength s_j, is

$$H_N(\mathbf{s}) = -\frac{1}{4\pi} \sum_{j=1}^{N} \sum_{k \neq j}^{N} s_j s_k \log |z_j - z_k| \tag{11.2}$$

where by the vector $\mathbf{s}$ we mean the set of vortex strengths and positions – that is, the state of the system.

We know, as outlined in the chapter on the dynamics of vortices, and equivalently by Noether's[1] theorem connecting symmetries to conserved quantities, of three conserved quantities, first integrals of motion. The first two are the real and imaginary components of the linear momentum:

[1] Emmy Amalie Noether, 1882 - 1935, studied mathematics despite university rules discouraging women attending classes. David Hilbert allowed her to teach courses in his name while she tried to get a university to allow her to do her habilitation. After her famous theorem on symmetries and conservation quantities she moved into ring theory and noncommutative algebras. [288]

$$\phi(\mathbf{z}) = \sum_{j=1}^{N} s_j z_j \tag{11.3}$$

from the translational invariance of the Hamiltonian (letting $\mathbf{z}$ represent point vortex positions $(z_1, z_2, z_3, \cdots, z_N)$); and the last is the moment of vorticity:

$$\Omega(\mathbf{z}) = \sum_{j=1}^{N} s_j |z_j|^2 \tag{11.4}$$

If the vorticities all have the same sign then no statistical equilibrium of the Hamiltonian alone will exist: for any configuration and any positive inverse temperature β the factor $\beta H_N(\mathbf{x})$ can invariably be increased by increasing the distances between points. But by placing canonical constraints on the energy and on the moment of vorticity simultaneously we generate a self-constraining system. Given β and the chemical potential μ there will be a partition function

$$Z(\beta, \mu) = \int_C dz_1 \int_C dz_2 \cdots \int_C dz_N \exp\left(-\beta H_N(\mathbf{s})\right) \exp\left(-\mu \Omega(\mathbf{z})\right) \tag{11.5}$$

bounded so long as μ is positive. The probability for any particular state $\mathbf{s}$ of

$$P(\mathbf{s}) = \frac{1}{Z} \exp\left(\beta H_N(\mathbf{s}) - \mu \Omega(\mathbf{s})\right) \tag{11.6}$$

As $\Omega(\mathbf{s})$ is practically positive-definite[2] and μ is positive this constraint effectively limits the points to be within a finite radius.

Generally the partition function cannot be precisely evaluated. J. Ginibre found – while examining random matrices – an exact value for the **one-particle reduced distribution function**, the probability of finding a function at a single point, which is derivable from the partition function, for the special case $\beta = 2$ and $\mu = 1$. In this particular case the probability of finding a particle at point z is

$$G_1^N(z) = \exp(-|z|^2) \sum_{k=0}^{N-1} \frac{|z|^{2k}}{k!} \tag{11.7}$$

which, as the number of points N grows infinitely large, tends toward 1.

Examining the function $G_1^N(z)$ for various N we find the density is very nearly one for z up to approximately $\sqrt{N}$, and decays exponentially to zero once z exceeds $\sqrt{N}$. Essentially every point at equilibrium will be within a circle centered at the origin and of radius $\sqrt{N}$, which we call the **containment radius.**

[2] A function is positive-definite if is always greater than or equal to zero, and can equal zero only when its independent variabes are always zero. $\Omega(\mathbf{s})$ is positive-definite if none of the z_j is at the origin.

Lim and Assad demonstrate a similar result for other β and μ, specifically that the most probable radius R enclosing the vortices at statistical equilibrium will be

$$R = \sqrt{\frac{\beta \Gamma}{2\mu}} \tag{11.8}$$

where Γ is the total vorticity, $\sum_{j=1}^{N} s_j$.

It is analytically taxing to explore a set of N discrete points. Our first derivation therefore uses a vortex density function $\rho(r,\theta)$ which we can take to be rotationally symmetric around the origin. The moment of vorticity is therefore

$$\Omega = \int \rho(r) z^2 d\mathbf{z} \tag{11.9}$$

$$= 2\pi \int_0^\infty \rho(r) r^3 dr \tag{11.10}$$

The Hamiltonian is

$$H = -\frac{1}{4\pi} \iint \rho(r_1)\rho(r_2) \log |\mathbf{r}_1 - \mathbf{r}_2| d\mathbf{r}_1 d\mathbf{r}_2 \tag{11.11}$$

$$= -\int_0^\infty dr_1 \int_0^\infty dr_2 \int_0^{2\pi} d\theta_1 \int_0^{2\pi} d\theta_2 r_1 r_2 \rho(r_1)\rho(r_2) \times \log \sqrt{r_1^2 + r_2^2 - 2r_1 r_2 cos(\theta_1 - \theta_2)} \tag{11.12}$$

$$= -\frac{1}{2} \int_0^\infty dr_1 \int_0^{r_1} dr_2 \int_0^{2\pi} d\theta_1 r_1 r_2 \rho(r_1)\rho(r_2) \times \log\left(r_1^2 + r_2^2 - 2r_1 r_2 cos(\theta_1)\right) \tag{11.13}$$

$$= -\frac{1}{2} \int_0^\infty dr_1 \int_0^{r_1} dr_2 r_1 r_2 \rho(r_1)\rho(r_2) \times 2\pi \log\left(\frac{r_1^2 + r_2^2 + \sqrt{(r_1^2 + r_2^2)^2 - 4r_1^2 r_2^2}}{2}\right) \tag{11.14}$$

$$= -\pi \int_0^\infty dr_1 \int_0^{r_1} dr_2 r_1 r_2 \rho(r_1)\rho(r_2) \log(r_1) \tag{11.15}$$

Our experience with vortices on the surface of the sphere suggests the equilibrium should be a nearly uniformly distributed set of points. Ginibre's result suggests we should expect these points to be uniformly distributed within a circle. If all the vortices are of the same strength s then the density is

$$\rho(r) = \left\{ \begin{matrix} \frac{Ns}{\pi R^2} & \text{if } z \leq R \\ 0 & \text{if } z > R \end{matrix} \right\} \tag{11.16}$$

We have not made any assumption about the radius R. We will find it by seeking the most probable radius, the radius which maximizes the Gibbs factor $\exp(-\beta H(\mathbf{s}) - \mu\Omega(\mathbf{s}))$.

Given the vortex density from equation 11.16, the moment of vorticity, which depends on the number of points N and the radius R, is

$$\Omega(R, N) = \int_0^R \frac{Ns}{\pi R^2}|r|^2 d\mathbf{r} \quad (11.17)$$

$$= \frac{Ns}{\pi R^2} 2\pi \int_0^R r^3 dr \quad (11.18)$$

$$= \frac{Ns}{\pi R^2} 2\pi \frac{R^4}{4} \quad (11.19)$$

$$= \frac{NsR^2}{2} \quad (11.20)$$

while the energy, similarly dependent on N and R, will be

$$H(R, N) = -4\pi^2 \frac{N^2 s^2}{\pi^2 R^4} \int_0^R dr_1 \int_0^{r_1} dr_2 r_1 r_2 \log(r_1) \quad (11.21)$$

$$= -\frac{4N^2 s^2}{R^4} \int_0^R dr_1 \frac{r_1^3}{2} \log(r_1) \quad (11.22)$$

$$= -\frac{N^2 s^2}{8} (4\log(R) - 1) \quad (11.23)$$

To maximize $\exp(-\beta H(\mathbf{s}) - \mu\Omega(\mathbf{s}))$ we must maximize $-\beta H(\mathbf{s}) - \mu\Omega(\mathbf{s})$. This requires finding the R for which

$$\frac{\partial}{\partial R}(-\beta H(\mathbf{s}) - \mu\Omega(\mathbf{s})) = 0 \quad (11.24)$$

$$-\beta \frac{N^2 s^2}{2R} + \mu N s R = 0 \quad (11.25)$$

$$R^2 = \frac{\beta N s}{2\mu} \quad (11.26)$$

$$R = \sqrt{\frac{\beta N s}{2\mu}} \quad (11.27)$$

So for a fixed total vorticity Ns the most probable radius is proportional to the ratio of β to μ. The ratio $\frac{\mu}{\beta}$ is known as the Larmor[3] frequency, a term borrowed from the physics of plasmas. The Larmor frequency describes the precession of an electron's magnetic moment around a magnetic field.

[3] Sir Joseph Larmor, 1857 - 1942, working with George Fitzgerald constructed the theory of the electron. He is also the first physicist known to have used the Lorenz transformations. [288]

The Monte Carlo simulation of this problem is straightforward. We expect from the above formula that the radius of a statistical equilibrium will depend on the number of points and the Larmor frequency, and can test at different sizes and different ratios for comparison.

But is there dependence on the size of β and μ, even if the ratio between is kept constant? Our physical interpretations of the quantities suggest there may be. The greater β the more unlikely an energy-increasing move is to be permitted. If β is small, however, energy-increasing moves are more likely to be permitted. Similarly a small μ will make more likely configurations with a high moment of vorticity.

So we predict that if the sizes of β and of μ are small, there will be many "exceptions," vortices scattered outside the most probable radius R. As β and μ increase, even as their ratio is kept the same, fewer vortices will fall outside this radius.

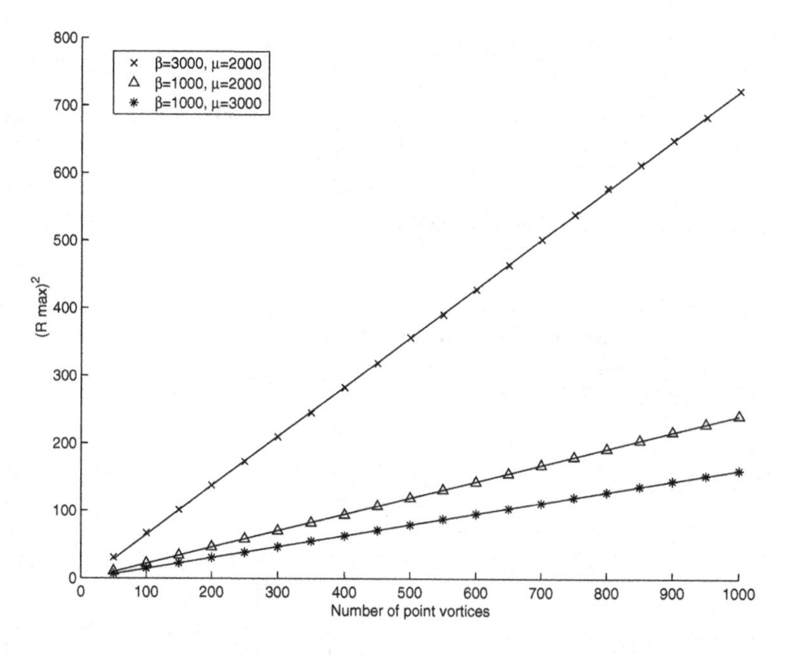

Fig. 11.1. How the radius of support for N vortices depends on the number of vortices and on the ratio of β to μ.

Monte Carlo experimentation confirms this prediction: with an 800-point vortex gas (with each vortex of strength $s = 1$) and $\beta = 0.001$ and $\mu = 0.002$ vortices are scattered into a cloud reaching out as far as 50 units from the origin, while the most probable radius is just under 14.

At $\beta = 1$ and $\mu = 2$, nearly all points fit within the given radius; no points even exceed 16 units from the origin. At $\beta = 1000$ and $\mu = 2000$

every point fits within the radius R, and similarly every point is within R at $\beta = 1,000,000$ and $\mu = 1,000,000$.

Besides the number of points fitting within the most probable radius R there is an obvious qualitative difference between the small β and the large β. Plots of the points show with larger β the points are more organized, more nearly crystallized in their placement. β of one and lower looks like the stars of a galaxy; β at 1000 and higher are rigid. We can attempt to quantify these differences.

Define the **radial density** function $rad(r)$ to be the number of vortices which fall in an annulus, centered at the origin, with interior radius r and exterior radius $r + 1$.

Define the **mean radial density** function $mrad(r)$ to be the mean of the number of vortices in the N distinct annuli of internal radius r and external radius $r + 1$, each annulus centered on a different vortex.

Define the **angular density** function $ang(\theta)$ to be the number of vortices falling inside the slice of the plane with a vertex at the origin, of angular width $d\theta$, centered along a ray at angle θ from the direction of the positive x-axis. This function is multiplied by a normalizing factor $\frac{2\pi}{N d\theta}$.

The **mean angular density** is the average of all angular densities for each of the points in the system.

Each of these various densities is demonstrated for a variety of β and μ – at the same ratio and with the same number of mesh sites, with the same long-duration Monte Carlo experiments equilibrating them – in the figures in this section.

Clearly the containment radius of the vortices on the plane depends on the ratio of the constraints applied to them; but we also find a dependence subtler to define, that the quality – the uniformity, how crystal-like the lattice is – depends on the magnitude of both constraints.

It is interesting to watch the positions of vortices in time as a system approaches equilibrium. Starting from the initial placement one sees a center as uniformly spaced as the final mesh will be accumulating in the center, with the remaining points as a rind surrounding it. This rind expands and, ultimately, dissolves as the full support is covered. Experimentally we can see the existence of this rind as evidence the system has not yet equilibrated.

11.2.1 Trapped Slender Vortex Filaments

There is another sort of simulation in Andersen's thesis, which can be done by a quantum Monte Carlo method, simulating trapped slender vortex filaments [5], in the Klein-Majda-Damodaran dynamical model [218] and the Lions-Majda statistical model [280]. A **vortex filament** is what the name suggests: a "whisker" in space with vorticity. One can picture it by imagining a small tube of fluid, with a fixed vorticity inside the tube. Now consider the limit as the radius of the tube shrinks to zero, but the circulation of the fluid on the

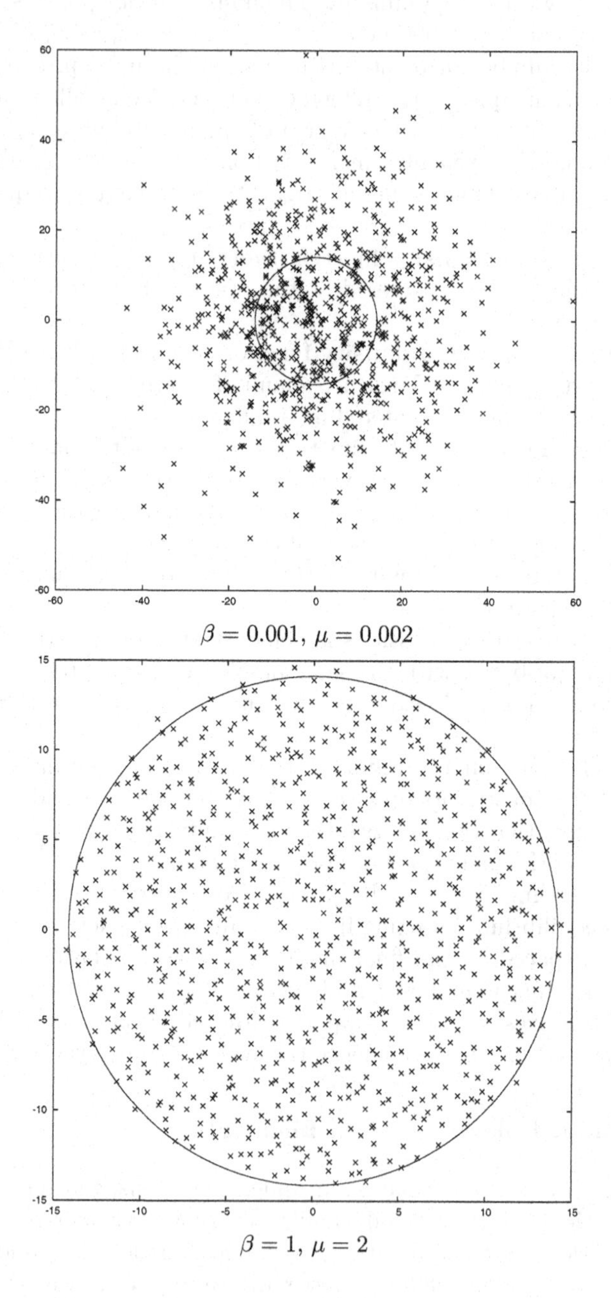

Fig. 11.2. vortex distribution on the unbounded plane following 100, 000 sweeps with 800 points, using different values of β and μ with the same ratio.

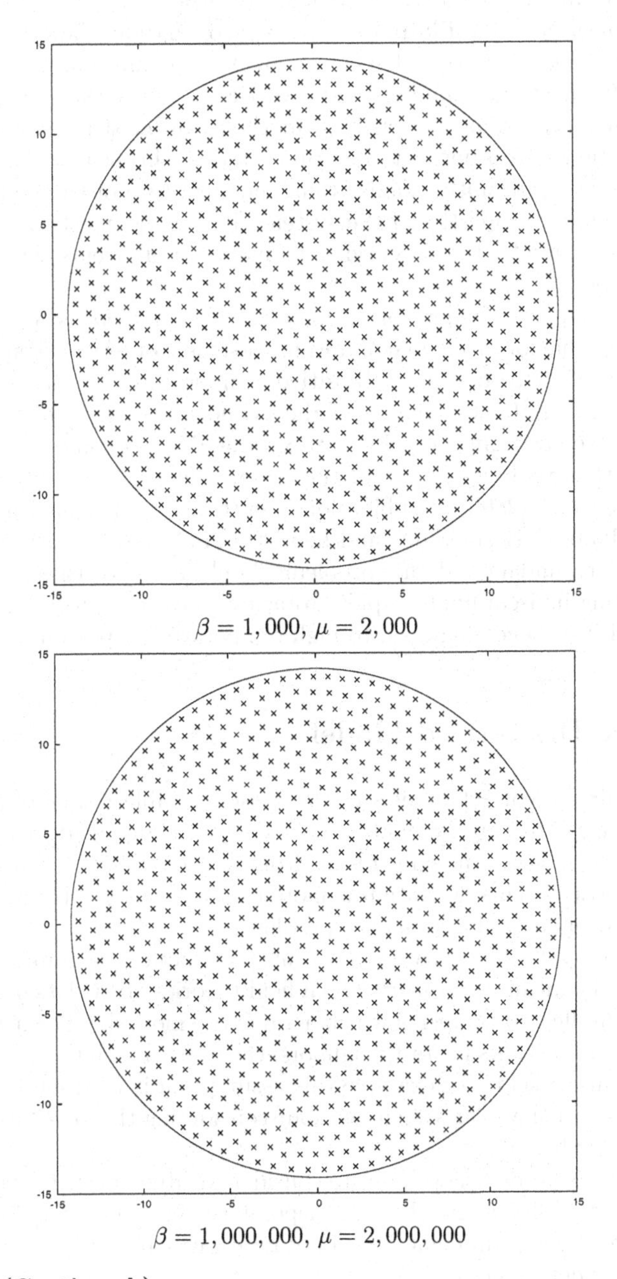

$\beta = 1,000,\ \mu = 2,000$

$\beta = 1,000,000,\ \mu = 2,000,000$

Fig. 11.2. (Continued.)

rim of the tube remains constant. This is a vortex filament, and it allows us to construct models of vorticity in three dimensions.

A **slender vortex filament** is an approximation made to resolve one of the problems of vortex filaments – by their definition, any point along a vortex filament has infinitely large vorticity, which leads to problems in treating them by ordinary mechanical methods. In the slender vortex filament approximation we do note let the "tube" radius shrink to zero, but instead to just a size significantly smaller than the curvature of the filament or any other interesting properties of the system. This works nicely, though it loses the ability to model tightly-curving filaments or filaments which grow too close together [425].

In this trapped slender vortex filament system there is a range of very interesting behavior, from an equilibrium ensemble of nearly straight and regular arrays of filaments at low positive temperatures to an entropy-driven regime where the bulk radius R of the cylindrical domain of the bundle of filaments diverges from the point vortex value discussed above.

Given that rigorous results of the mean field type are only available at low temperatures [280], a combination of cutting-edge simulation using path-integral Monte-Carlo methods and heuristic models for the free energy have to be used to understand this problem. Applications of these results range from rotating inviscid fluids, rapidly rotating convection, rotating superfluids and Bose-Einstein condensates, to high temperature superconductors.

11.3 The Discretized Model

While we derive correct models of the most probable radius of a statistical equilibrium by treating the vortex gas as a continuous field, it is fair to ask whether we can derive the same results from treating the system as a collection of point vortices. This derivation is more complex but yields insight into the treatment of many fluid mechanics problems.

Preliminary to the derivation we must introduce the Lundgren-Pointin scaling for the mean field theory. Mean field theories are a class of approximations to complicated problems which assume the most significant interactions a point has are with its nearest neighbors; the effect of points far away average out into a mean field. For example this assumption is used when one models the evolution of the solar system without considering the gravitational effects of the rest of the galaxy.

A scaling is needed as we want to consider the dynamics of a single fluid as discretized by different numbers N of point vortices. The equilibrium found at the same β across different N will not necessarily be equivalent; our scaling removes this dependence on N.

In representing a fluid with total vorticity Γ as a set of N point vortices each of strength s, the scaling we need – the **vortex method scaling** – is to set

$$\beta \to \tilde{\beta} N \tag{11.28}$$

$$\mu \to \tilde{\mu} N \tag{11.29}$$

$$s \to \frac{\Gamma}{N} \tag{11.30}$$

Even as the number of points grows infinitely large the total vorticity remains constant and the Gibbs factor $\beta H + \mu \Omega$ stays finite.

Let $q(z)$ be the distribution of vorticity on the complex plane. With Γ the fixed total vorticity (the net circulation) we define the energy functional $E[q]$

$$E[q] = H + \frac{\tilde{\mu}}{\tilde{\beta}} \Omega \tag{11.31}$$

The domain of the energy functional we limit the set $L^2(R^2)$ of functions q which are nonnegative square-integrable functions. The total circulation is itself a functional,

$$\Gamma[q] \equiv \int_{R^2} q(z) dz \tag{11.32}$$

This restricted energy functional we write as $E_\Gamma[q]$. We also have an entropy functional $S[q]$, which for any distribution of vorticity $q(z)$ in the set of non-negative square-integrable functions is

$$S_\Gamma[q] = -\int_{R^2} \frac{q(z)}{\Gamma} \log(q(z)) dz \tag{11.33}$$

The Helmholtz free energy, per particle, then is

$$F[q] = E_\Gamma[q] - \frac{1}{\tilde{\beta}} S_\Gamma[q] \tag{11.34}$$

We want the function which minimizes this. That distribution maximizes the Gibbs factor and so is the statistical equilibrium for given $\tilde{\beta}$ and $\tilde{\mu}$.

But a direct analytical approach, even after factoring in the simpler form of the free energy given by the mean field theory, is not tenable. However, Monte Carlo simulations in the temperature regime $0 < \frac{1}{\beta} \ll 1$ (which is a high temperature region) suggest some important properties are captured by the extremals of the internal energy E. We will use the approximate method of solving for extremals of the internal energy E in the following section.

To conclude this section we note something of the thermodynamics of the vortex gas at low temperatures. At $\frac{1}{\beta} = T = 0$, the free energy reduces to the internal energy E, and classical thermodynamics ceases to exist. In this case, the most probable vorticity distribution is that which extremizes the internal energy E. For low positive temperatures, the vorticity distributions q_e that minimize the free energy are close to the minimizers q_m of the internal energy E in any natural norms. Moreover, the Monte Carlo simulations (cf [26]) indicate that q_e tends to q_m as T tends to 0, in the same norms.

The heuristic picture we have is that the $T = 0$ solution is slightly "smeared" at the bounding circle by thermodynamic effects of the $-\frac{1}{\beta}S$ term. The rigorous proof of this scenario remains an open problem that awaits the exact $\beta > 0$ solution of the Onsager vortex gas in the unbounded plane. On the other hand, the vortex gas Problem in a given bounded domain has been solved, in the sense that the mean field free energy of the problem is known [67].

11.4 Extremizing E

Suppose the vorticity distribution $q(z)$ which minimizes the energy functional has radial symmetry; and suppose also that it has a **compact support**, that is, the set of points where the function is nonzero is a closed and bounded set[4]. Given the set D_R the disk of radius R centered on the origin define $L^2(D_R)$ to be all the nonnegative functions which are square-integrable on D_R, and let $V'(R)$ be the subset of $L^2(D_R)$ of functions which are bounded almost everywhere by a positive constant M.

We need a basis set for the functions $q(z)$ we will want to construct. Since the inverse Laplace-Beltrami operator $G(\mathbf{z}, \mathbf{z}')$ is a compact self-adjoint operator we are able to apply the Spectral Theorem: there exists an orthonormal basis of $L^2(D)$, consisting of the eigenfunctions ψ_j of the operator G, with associated eigenvalues s_j. Our orthonormal basis B is therefore the set of functions

$$B = \{I_D(\mathbf{x})\} \cup \{\psi_j\} \tag{11.35}$$

the combination of these eigenfunctions and the characteristic function for the disk D.

In trying to extremize the energy functional for given β and μ

$$E[q(\mathbf{z}), \beta, \mu] = H + \frac{\mu}{\beta}\Omega \tag{11.36}$$

we will find two cases. If $\log(R)$ is less than or equal to $\frac{1}{4}$ then E is bounded below by zero. When $\log(R)$ is greater than $\frac{1}{4}$ then the extremal $q_*(\mathbf{z})$ will be a saddle point, and there will be this relationship among circulation Γ, inverse temperature β, chemical potential μ, and radius R:

$$\frac{R^2}{|4\log R - 1|} = \frac{\Gamma\beta}{4\pi\mu} \tag{11.37}$$

Proof. We need to introduce first the integral operator

[4] Formally, a set S is compact if for any sequence of elements $\{x_j\}$ in S, the limit point $\lim_{j\to\infty} x_j$ is also in S. For sets in a Euclidean space this is equivalent to being closed and bounded.

$$G[q](\mathbf{z}) = -\frac{1}{2\pi} \int_{D_R} q(\mathbf{z}') \log |\mathbf{z} - \mathbf{z}'| d\mathbf{z}' \tag{11.38}$$

which one notices is the solution to the Laplace-Beltrami equation on the plane

$$-\nabla^2 w(\mathbf{z}) = q(\mathbf{z}) \tag{11.39}$$

as the Green's function for $-\nabla^2$ on the plane is $K(\mathbf{z}, \mathbf{z}') = \frac{1}{2\pi} \log |\mathbf{z} - \mathbf{z}'|$. Incidental to this is that the Hamiltonian functional for $q(\mathbf{z})$ is

$$H[q] = -\frac{1}{4\pi} \int_{D_R} d\mathbf{z} q(\mathbf{z}) \left(\int_{D_R} d\mathbf{z}' q(\mathbf{z}') \log |\mathbf{z} - \mathbf{z}'| \right) \tag{11.40}$$

$$= \frac{1}{2} \langle q, G[q] \rangle \tag{11.41}$$

The Euler-Lagrange equation for $q_*(\mathbf{x})$ is given by the equation

$$G[q_*](\mathbf{x}) = -\frac{\mu}{\beta} |\mathbf{x}|^2 \tag{11.42}$$

for functions $q_* \in V'(R)$. The only solution of this is the function

$$q_*(\mathbf{x}) = q_0 I_R(\mathbf{x}) \tag{11.43}$$

for some constant $q_0 > 0$, where $I_R(\mathbf{x})$ is the characteristic function for the set D_R. With this in equation 11.42 we have

$$0 = \left\langle q_*, G[q_*] + \frac{\mu}{\beta} |\mathbf{x}|^2 \right\rangle \tag{11.44}$$

$$= q_0 \int_{D_R} G[q_0 I_R] d\mathbf{x} + \frac{\mu q_0}{\beta} \int_{D_R} |\mathbf{x}|^2 d\mathbf{x} \tag{11.45}$$

Evaluating the integrals provides

$$0 = \frac{\pi R^2 q_0^2}{8} (1 - 4 \log (R)) + \frac{\pi R^4 \mu q_0}{2\beta} \tag{11.46}$$

making it obvious the solution depends on whether $\log (R)$ is less than or equal to $\frac{1}{4}$, or whether it is greater than $\frac{1}{4}$.

There are two solutions. The first is $q_0 = 0$ for all $R > 0$. However, the function $q_*(\mathbf{x}) = q_0 I_R(\mathbf{x})$ does not satisfy the Euler-Lagrange equation 11.42. The other solution, applicable when $\log (R) > \frac{1}{4}$, is

$$q_0 = \frac{4\mu}{\beta(4 \log (R) - 1)} \tag{11.47}$$

This is the only interesting stationary point.

Now consider an arbitrary other function $q(\mathbf{x})$, which can be written

$$q(\mathbf{x}) = q_0 I_R(\mathbf{x}) + q'(\mathbf{x}) \tag{11.48}$$

where the function $q'(\mathbf{x})$ is constrained by

$$\int_{D_R} q' d\mathbf{x} = 0 \tag{11.49}$$

From this we have

$$E[q] = H[q] + \frac{\mu}{\beta} \Gamma[q] \tag{11.50}$$

$$= E_0(q_0) + \langle q_0 I_R(\mathbf{x}), G[q'] \rangle + E_1[q'] \tag{11.51}$$

$$E_0(q_0) = \frac{\pi R^4 q_0^2}{16}(1 - 4\log(R)) + \frac{\mu \pi R^4 q_0}{2\beta} \tag{11.52}$$

$$E_1[q'] = \frac{1}{2} \langle q', G[q'] \rangle + \frac{\mu}{\beta} \langle q', |\mathbf{x}|^2 \rangle \tag{11.53}$$

Using the basis functions B then we can expand q' to get

$$\langle q_0 I_R(\mathbf{x}), G[q'] \rangle = q_0 \sum_m q'_m s_m \langle I_R, \psi_m \rangle = 0 \tag{11.54}$$

By the lemma 11.58 explained in the next section, $E_1[q']$ is strictly convex in q'. From that we know when $4\log(R) \leq 1$ that $E[q]$ is strictly convex in $V'(R)$, which implies

$$E[q] > E_0(0) = 0 \text{ for all } q \in V'(R) \tag{11.55}$$

If we close the set $V'(R)$ by including the trivial function $q' \equiv 0$ we find that is the unique minimizer of E.

If $4\log(R) > 1$ then the concavity of $E_0 q_0$ implies the extremal q_0 is a saddle point. The lower bound of $E[q]$ is on the boundary of $V'(R)$, that is, when $q(\mathbf{x}) = M I_R(\mathbf{x})$. And substituting the saddle point q_0 into the equation for total circulation gives us

$$\int_{D_R} q d\mathbf{x} = \Gamma \tag{11.56}$$

$$\frac{4\mu\pi R^2}{\beta(4\log(R) - 1)} = \Gamma \tag{11.57}$$

which is the relationship we hoped to prove.

11.5 Variational Problem on the Radius

The work we have above shows the most probable distribution of vorticity for a fixed total vorticity Γ and a given radius R. Now we let the radius vary to find the size of the disk which makes for the most probable distribution. We will eventually prove the most probable radius is the one matching the extrapolation of Ginibre's results and the Monte Carlo experiments.

Suppose we have a set of N vortices with a total circulation Γ. Define the set $V(\Gamma)$ to be the set of square-integrable functions $q(\mathbf{z})$ defined on R^2 for which the support is a disk of finite but nonzero radius R and such that $\int_{D_R} q d\mathbf{z} = \Gamma$.

Lemma 3. *The energy functional $E[w; \eta = \frac{\mu}{\beta}]$ is strictly convex for vorticity distributions w defined on the set*

$$V_R(\Gamma) = \{w \in V(\Gamma) \text{ with the support } D\} \tag{11.58}$$

Proof. Let $\lambda \in (0, 1)$; and let $w_1 \neq w_2$ be two distributions in $V_R(\Gamma)$. Then

$$w = \lambda w_1 + (1 - \lambda) w_2 \in V_R(\Gamma) \tag{11.59}$$

The strict convexity of functional $E[w; \eta]$ follows from that of $E'[w; \eta]$ as

$$E'[w; \eta] = E[w; \eta] - E_0[\Gamma, \eta] \tag{11.60}$$

and E_0 is a constant. The next objective is to show

$$E'[w; \eta] = E'[\lambda w_1 + (1 - \lambda) w_2; \eta] < \lambda E'[w_1; \eta] + (1 - \lambda) E'[w_2; \eta] \tag{11.61}$$

The left-hand side equals

$$\begin{aligned} &E[\lambda w_1 + (1 - \lambda) w_2; \eta] \\ &\quad = \int_{R^2} d\mathbf{x} \int_{R^2} d\mathbf{x}' \left(\lambda w_1 + (1 - \lambda) w_2\right)(\mathbf{x}) \left(\lambda w_1 + (1 - \lambda) w_2\right)(\mathbf{x}') \\ &\quad\quad \times g_{\eta,\Gamma}(\mathbf{x}, \mathbf{x}') - E_0[\Gamma; \eta] &(11.62) \\ &\quad = \lambda^2 E'[w_1; \eta] + (1 - \lambda)^2 E'[w_2; \eta] + 2\lambda(1 - \lambda) E'[w_1, w_2] &(11.63) \end{aligned}$$

where

$$E'[w_1, w_2] = \int_{R^2} d\mathbf{x} \int_{R^2} d\mathbf{x}' w_1(\mathbf{x}) w_2(\mathbf{x}') g_{\eta,\Gamma}(\mathbf{x}, \mathbf{x}') - E_0[\Gamma, \eta] \tag{11.64}$$

and finally for x and $\mathbf{x}'$ both on D

$$g_{\eta,\Gamma}(\mathbf{x}, \mathbf{x}') = -\frac{1}{4\pi} \log |\mathbf{x} - \mathbf{x}'| + \frac{\eta}{2\Gamma} \left(|\mathbf{x}|^2 + |\mathbf{x}'|^2\right) \geq c' > -\infty \tag{11.65}$$

The next step to prove is

$$\begin{aligned} &\lambda^2 E'[w_1; \eta] = (1 - \lambda)^2 E'[w_2; \eta] + 2\lambda(1 - \lambda) E'[w_1, w_2] \\ &< \lambda E'[w_1; \eta] = (1 - \lambda) E'[w_2; \eta] \end{aligned} \tag{11.66}$$

which is equivalent to showing

$$\lambda(1-\lambda)\left\{(E'[w_1;\eta]-E'[w_1,w_2])+(E'[w_2;\eta]-E'[w_1,w_2])\right\}>0 \tag{11.67}$$

The left-hand side equals

$$\lambda(1-\lambda)\left\{\begin{array}{l} \int_{R^2}d\mathbf{x}\int_{R^2}d\mathbf{x}'w_1(\mathbf{x})w_1(\mathbf{x}')g_{\eta,\Gamma}(\mathbf{x},\mathbf{x}') \\ -\int_{R^2}d\mathbf{x}\int_{R^2}d\mathbf{x}'w_1(\mathbf{x})w_2(\mathbf{x}')g_{\eta,\Gamma}(\mathbf{x},\mathbf{x}') \\ +\int_{R^2}d\mathbf{x}\int_{R^2}d\mathbf{x}'w_2(\mathbf{x})w_2(\mathbf{x}')g_{\eta,\Gamma}(\mathbf{x},\mathbf{x}') \\ -\int_{R^2}d\mathbf{x}\int_{R^2}d\mathbf{x}'w_1(\mathbf{x})w_2(\mathbf{x}')g_{\eta,\Gamma}(\mathbf{x},\mathbf{x}') \end{array}\right\} \tag{11.68}$$

$$=\lambda(1-\lambda)\left\{\begin{array}{l} \int d\mathbf{x}\int d\mathbf{x}'w_1(\mathbf{x})(w_1(\mathbf{x}')-w_2(\mathbf{x}'))g_{\eta,\Gamma}(\mathbf{x},\mathbf{x}') \\ -\int d\mathbf{x}\int d\mathbf{x}'w_2(\mathbf{x})(w_1(\mathbf{x}')-w_2(\mathbf{x}'))g_{\eta,\Gamma}(\mathbf{x},\mathbf{x}') \end{array}\right\} \tag{11.69}$$

$$=\lambda(1-\lambda)E[w_1-w_2;\eta] \tag{11.70}$$

Remaining to be proven is that if $w_1 \neq w_2$ when both are in $V_R(\Gamma)$ then

$$E[w_1-w_2;\eta]>0 \tag{11.71}$$

Since both w_1 and w_2 are in $V_R\Gamma$ then

$$\int_{R^2}(w_1-w_2)\,d\mathbf{x}=0 \tag{11.72}$$

and w_1-w_2 is not necessarily single-signed. This difference is therefore not necessarily in $V_R(\Gamma)$. But we have the properties

$$\int_{R^2}d\mathbf{x}\int_{R^2}d\mathbf{x}'\,(w_1-w_2)\,(\mathbf{x})\,(w_1-w_2)\,(\mathbf{x}')\frac{\eta}{2\Gamma}|\mathbf{x}|^2 \tag{11.73}$$

$$=\frac{\eta}{2\Gamma}\int_{R^2}d\mathbf{x}|\mathbf{x}|^2\,(w_1-w_2)\,(\mathbf{x})\int_{R^2}(w_1-w_2)\,(\mathbf{x}') \tag{11.74}$$

$$=0 \tag{11.75}$$

and also

$$\int_{R^2}d\mathbf{x}\int_{R^2}d\mathbf{x}'\,(w_1-w_2)\,(\mathbf{x})\,(w_1-w_2)\,(\mathbf{x}')\frac{\eta}{2\Gamma}|\mathbf{x}'|^2=0 \tag{11.76}$$

and therefore

$$E[w_1-w_2;\eta]=\int d\mathbf{x}\int d\mathbf{x}'\,(w_1-w_2)\,(\mathbf{x}) \times(w_1-w_2)\,(\mathbf{x}')G_{\eta,\Gamma}(\mathbf{x},\mathbf{x}') \tag{11.77}$$

$$=\int d\mathbf{x}\int d\mathbf{x}'\left[\begin{array}{l}(w_1-w_2)\,(\mathbf{x}) \\ \times(w_1-w_2)\,(\mathbf{x}') \\ \times(\frac{1}{4\pi}\log|\mathbf{x}-\mathbf{x}'|^{-1} \\ \times+\frac{\eta}{2\Gamma}\left(|\mathbf{x}|^2+|\mathbf{x}'|^2\right))\end{array}\right] \tag{11.78}$$

$$=\frac{1}{4\pi}\int_D d\mathbf{x}\int_D d\mathbf{x}'\,(w_1-w_2)\,(\mathbf{x}) \times(w_1-w_2)\,(\mathbf{x}')\log|\mathbf{x}-\mathbf{x}'|^{-1} \tag{11.79}$$

This last expression is equivalently written in the form $\frac{1}{2}\langle q, G(q)\rangle$, where G is a compact self-adjoint operator – for us, the kernel of the inverse Laplace-Beltrami operator.

The positive eigenvalues λ_j will tend to zero as j tends toward infinity, because $L^2(D)$ is an infinite-dimensional space. Therefore for any q in $L^2(D)$ we have

$$\langle q, G[q]\rangle = \left\langle \sum_{j=1}^{\infty} q_j \psi_j(\mathbf{x}), G\left[\sum_{k=1}^{\infty} q_k \psi_k(\mathbf{x}')\right]\right\rangle \tag{11.80}$$

$$= \sum_{j=1}^{\infty} q_j^2 \lambda_j \geq 0 \tag{11.81}$$

Since $q = w_1 - w_2 \neq 0$, then at least one of the coefficients $q_j(w_1 - w_2)$ is not zero; and this therefore implies

$$\langle w_1 - w_2, G(w_1 - w_2)\rangle = \sum_{j=1}^{\infty} q_j^2(w_1 - w_2)\lambda_j > 0 \tag{11.82}$$

This in turn proves the strict convexity of $E[w;\eta]$ on the space $V_R(\Gamma)$. □

Lemma 4. *If a functional $E[q]$ is strictly convex on a convex set V of functions q with support in a compact subset $D \subset R^2$, then the minimizer of E in V is unique and it is radially symmetric.*

Proof. By proposition 1.1 on page 35 in Ekeland and Teman, $min(E)$, the set of minimizers of E in V is a closed, convex set which is non-empty by the convexity of E. The strict convexity of E in V implies uniqueness of the minimizer of E in V:

Suppose there were two minimizers q_1 and q_2 of E in V. Then $\frac{1}{2}(q_1 + q_2)$ is also a minimizer of E in V. This contradicts the strict convexity of E in V:

$$\alpha = E\left(\frac{1}{2}(q_1 + q_2)\right) < \frac{1}{2}E(q_1) + \frac{1}{2}E(q_2) = \alpha \tag{11.83}$$

The uniqueness of the minimizer implies that the minimizer q is radial: Suppose q is not radial. Then a rotation q_θ of q is another element of the set of minimizers of E in V. This contradicts the uniqueness of the minimizer; and therefore, the lemma is proven. □

Theorem 13. *The variational problem of the minimization of the augmented energy functional*

$$E[q;\beta,\mu] = H[q] + \frac{\mu}{\beta}\Omega[q] \tag{11.84}$$

for $\beta > 0$, $\mu > 0$, in the set $V(\Gamma)$ of functions square-integrable, almost everywhere bounded and single-signed vorticity distributions $q(r,\theta)$ of fixed *total circulation $\Gamma > 0$ wish support equal to D_R of any finite radius $R \in (0, \infty)$ has a solution. That is, E takes its minimum in $V(R)$, at the unique radial minimizer*

$$q_m = \frac{4\mu}{\beta} I_R(r) \tag{11.85}$$

where the radius is given by

$$R = \sqrt{\frac{\Gamma\beta}{4\pi\mu}} \tag{11.86}$$

Proof. Consider the one-parameter family of orthonormal bases $\{\psi_j^{(R)}\}$, parametrized by $R \in (0, \infty)$, corresponding to the square-integrable classes $L^2(D_R)$. Each of these bases exists by the Spectral Theorem, applied separately to each of the $L^2(D_R)$ for finite R. Each vorticity distribution $q^{(R)}(\mathbf{x})$ in $L^2(D_R)$ can be written in the form

$$q^{(R)}(\mathbf{x}) = q_0^{(R)} I_R(\mathbf{x}) + q'^{(R)}(\mathbf{x}) \tag{11.87}$$

The fixed total circulation Γ completely determines the first term $q_0^{(R)} I_R(\mathbf{x})$, and the total circulation $q'^{(R)}(\mathbf{x})$ is zero.

Expanding on $E[q^{(R)}]$ yields

$$E[q^{(R)}] = E_0(R) + \left\langle q_0^{(R)} I_R(\mathbf{x}), G[q'] \right\rangle + E_1[q'^{(R)}] \tag{11.88}$$

$$E_0(R) = \frac{\pi R^4 \left[q_0^{(R)}\right]^2}{16} (1 - 4\log(R)) + \frac{\mu\pi R^4 q_0^{(R)}}{2\beta} \tag{11.89}$$

$$= \frac{\Gamma^2}{16\pi} (1 - 4\log(R)) + \frac{\mu\Gamma R^2}{2\beta} \tag{11.90}$$

$$E_1[q'^{(R)}] = \frac{1}{2} < q'^{(R)}, G[q'^{(R)}] > + \frac{\mu}{\beta} \left\langle q'^{(R)}, |\mathbf{x}|^2 \right\rangle \tag{11.91}$$

where on expanding $q_0^{(R)} I_R(\mathbf{x})$ and $q'^{(R)}$ in the orthonormal basis B for $L^2(D_R)$

$$\left\langle q_0^{(R)} I_R(\mathbf{x}), G[q'] \right\rangle = \left\langle q_0^{(R)} I_R(\mathbf{x}), G\left[\sum_m q'_m \psi_m\right] \right\rangle \tag{11.92}$$

$$= q_0^{(R)} \sum_m q'_m \lambda_m \left\langle I_R(\mathbf{x}), \psi_m \right\rangle = 0 \tag{11.93}$$

It follows from equation 11.93 that $E_0(R)$ is strictly convex in $R \in (0, \infty)$. For a fixed R, the remainder term $E_1[q'^{(R)}]$ is also strictly convex in the functions $q'^{(R)} \in L^2(D_R)$ such that $\int_{D_R} q'^{(R)} d\mathbf{x} = 0$, by the same proof as in lemma 11.58. For a fixed R, the term $\langle q_0 I_R(\mathbf{x}), G[q'] \rangle \geq 0$. Therefore for R varying over all finite positive values, it must be that $E[q^{(R)}]$ is the sum of two components, $E_0(R)$ and $\left(\langle q_0 I_R(\mathbf{x}), G[q'] \rangle + E_1[q'^{(R)}]\right)$. Each of these components is convex in its respective arguments, R and $q'^{(R)}$. This implies $E[q^{(R)}]$ is strictly convex in $V(\Gamma)$.

But to minimize $E[q^{(R)}]$ it follows from lemma 11.58 that the minimum of $\langle q_0 I_R(\mathbf{x}), G[q'] \rangle + E_1[q'^{(R)}]$ over $q'^{(R)} \in L^2(D_R)$ for a fixed R with the constraint $\int_{D_R} q'^{(R)} d\mathbf{x} = 0$ is achieved by the unique minimizer $q'^{(R)}(\mathbf{x}) \equiv 0$. Therefore $E[q^{(R)}]$ restricted to the one-parameter family

$$q_0^{(R)}(\mathbf{x}) = \frac{\Gamma}{\pi R^2} I_R(\mathbf{x}) \tag{11.94}$$

is given by

$$E_0(R) = \frac{\Gamma^2}{16\pi} \left(1 - 4 \log(R)\right) + \frac{\mu \Gamma R^2}{2\beta} \tag{11.95}$$

for fixed $\Gamma > 0$, $\beta > 0$, and $\mu > 0$.

So take the derivative with respect to R and set it equal to zero:

$$\frac{d}{dR} E_0(R) = -\frac{\Gamma^2}{4\pi} \frac{1}{R} + \frac{\mu \Gamma}{\beta} R = 0 \tag{11.96}$$

$$R = \sqrt{\frac{\Gamma \beta}{4 \pi \mu}} \tag{11.97}$$

just as in equation 11.86, and the form of the minimizer 11.85. The uniqueness of equation 11.85 follows, then, from the convexity of $E[q^{(R)}]$ on the convex set $V(\Gamma)$ which began this section and from lemma 11.58. This completes the proof. □

Remark 1. The form of equation 11.95 implies that relaxing the total circulation constraint in the variational problem leads to the uninteresting problem where the global minimizer is the trivial function $q = 0$, the case of zero total circulation.

12

Two-Layer Quasi-Geostrophic Models

12.1 Introduction

We develop a more realistic model of fluid flow based on the horizontal and restricted vertical motions of the atmosphere and oceans of a planet [442]. We will continue to study the vorticity field of an inviscid fluid, but will consider the atmosphere as having two (or, in principle, more) layers to it. The actual atmosphere is a complex of many layers, with interactions and behaviors on multiple time-scales. We also add to consideration the rotation of the earth; this is the "strophic" part of the term "geostrophic models." This model can be represented as a vorticity, dependent on latitude superimposed on the rest of the fluid motion. Several terms must be introduced before we can construct these models.

While this is only one example it demonstrates the techniques needed to add more complicated models to an atmosphere. The addition of further layers or of surface effects increases the analytic complexity, and marginally increases the numerical complexity, but practice in this two-layer model shows how to generalize our work here.

12.2 Two-Layer Quasi-Geostrophic Models

It is obvious the rotation of the planet should produce a Coriolis[1] acceleration on anything moving on its surface. Let Ω be the rate of the planet's rotation. We can then treat the effect of the planet's rotation as the addition at the point with latitude ϕ and longitude θ of a vortex of intensity $f = 2\Omega \sin(\phi)$. This f is known as the Coriolis parameter.

[1] Gaspard Gustave de Coriolis, 1792 - 1843, besides studying the dynamics of rotating coordinate systems from which we get Coriolis forces and accelerations, studied hydraulics, machine efficiency, and ergonomics. The modern meanings of "work" and of "kinetic energy" were defined by him. [288]

Now consider around the point (ϕ, θ)[2] the tangent plane from that point. Locally to that point there is little difference between the position on the sphere and on the tangent plane, and we can approximate the fluid flow on the sphere with the equations describing motion on the plane. The vorticity f we approximate with the Coriolis parameter f_0 for the point of tangency,

$$f_0 = 2\Omega \sin(\phi_0) \tag{12.1}$$

This tangent plane is known as the **f-plane**, and this approximation is therefore the **f-plane approximation** [144] [440].

The approximation can be improved by treating the Coriolis parameter not as constant, but as a linear term varying with the "vertical" direction – the y-coordinate direction in the plane, corresponding to latitude on the sphere. Letting the radius of the Earth be R we define the parameters

$$\beta = \frac{2\Omega \cos(\phi_0)}{R} \tag{12.2}$$

$$y = R(\phi - \phi_0) \tag{12.3}$$

we then construct the **β-plane** and the **β-plane approximation** by letting

$$f(y) \approx f_0 + \beta y \tag{12.4}$$

We examine a two-layer model for quasi-geostrophic flows on the f-plane. Such a model forms the basis of Nelson George Hogg and Henry Stommel's[3] derivation of the two-layer heton model [175]. We follow the discussion of Mark T DiBattista and Andrew J Majda [107]. Although a complete proof is still lacking, there is little doubt that the mean field theory can be rigorously shown to be asymptotically exact. The mean field partial differential equations are coupled non-linear elliptic equations in a pair of most probable "coarse-grained" stream functions $\bar{\psi}_1$ for the first layer and $\bar{\psi}_2$ for the second layer.

12.3 Governing Equations

The two-layer model for quasi-geostrophic flows on the f-plane is given by

$$\frac{\partial}{\partial t} q_1 + J(\psi_1, q_1) = 0$$

$$\frac{\partial}{\partial t} q_2 + J(\psi_2, q_2) = 0 \tag{12.5}$$

[2] In the section on the dynamics of point vortices on the sphere these coordinates were p and q respectively, or the z-coordinate and the longitude.

[3] Henry Stommel, 1920 - 1992, was one of the founders of modern oceonography, demonstrating early in his career how to explain the Gulf Stream from fluid dynamics. He would develop mathematically simple but physically potent models of ocean dynamics.

where the potential vorticities in layers one and two are respectively

$$\begin{aligned} q_1 &= \nabla^2 \psi_1 + F(\psi_2 - \psi_1) \\ q_2 &= \nabla^2 \psi_2 + F(\psi_1 - \psi_2) \end{aligned} \tag{12.6}$$

and F is Froude's[4] rotational number. This number is

$$F = \frac{1}{L_R^2} \tag{12.7}$$

where L_R is the quantity known as the **Rossby**[5] **radius of deformation**, the scale on which pressure variations horizontally create a vertical sheer, which depends on the strength of the interaction between the layers of fluid.

As we must conserve both potential vorticities q_1 and q_2 we have

$$\frac{\partial}{\partial t} q_1 + \left(\hat{k} \times \nabla \right) \psi_1 \cdot \nabla q_1 = 0 \tag{12.8}$$

$$\frac{\partial}{\partial t} q_2 + \left(\hat{k} \times \nabla \right) \psi_2 \cdot \nabla q_2 = 0 \tag{12.9}$$

The vorticity and the stream function can both be rewritten as having two components, the **barotropic** and the **baroclinic**. In a barotropic fluid the pressure of the fluid is a function of its altitude alone, so surfaces of a constant pressure keep a uniform distance from one another. In a baroclinic fluid the pressure is a function of more than the altitude alone. (The barotropic fluid is approached by the temperature and pressure of the deep ocean, while the turbulent surface is more baroclinic.) The barotropic flow is common to the entire atmosphere; the baroclinic is the difference between vertical layers. So define then

$$\psi_B = \frac{1}{2} (\psi_1 + \psi_2) \tag{12.10}$$

$$q_B = \frac{1}{2} (q_1 + q_2) \tag{12.11}$$

$$\psi_T = \frac{1}{2} (\psi_1 - \psi_2) \tag{12.12}$$

$$q_T = \frac{1}{2} (q_1 - q_2) \tag{12.13}$$

[4] William Froude, 1810 - 1879, was an assistant to railroad pioneer Isambard Kingdom Brunel, 1806 - 1859, on the Bristol and Exeter railway, and later a naval architect. He was the first to provide reliable laws describing the resistance of water against ships, and the stability of ships. [92]

[5] Carl-Gustaf Arvid Rossby, 1898 - 1957, was a pioneer in the large-scale movement of the atmosphere. Besides establishing the first civilian aviation weather service and the first Department of Meteorology (at Massachusetts Institute of Technology) in the United States, he identified and described the Rossby waves and the jet stream. [92]

Following from this

$$\nabla \psi_B = q_B \tag{12.14}$$
$$\nabla \psi_T - 2F\psi_T = q_T \tag{12.15}$$

which places the equations on familiar grounds, exactly akin to the point-vortex problems we have solved several times. To solve the inverse problems

$$\nabla^{-1} q_B = \psi_B \tag{12.16}$$
$$\nabla^{-1} q_T = \psi_T - \nabla^{-1}\left(2F\psi_T\right) \tag{12.17}$$

we use, on the plane, the two Green's functions

$$G_B\left(\mathbf{x}_1, \mathbf{x}_2\right) = \frac{1}{2\pi} \log |\mathbf{x}_1 - \mathbf{x}_2| \tag{12.18}$$
$$G_T\left(\mathbf{x}_1, \mathbf{x}_2\right) = -\frac{1}{2\pi} K_0\left(\sqrt{2F}|\mathbf{x}_1 - \mathbf{x}_2|\right) \tag{12.19}$$

where $K_0(r)$ is the modified Bessel[6] function of the first kind, defined by

$$K_n(r) = \frac{1}{2\pi \mathrm{i}} \oint e^{\frac{r}{2}\left(t+\frac{1}{t}\right)} t^{-n-1} dt \tag{12.20}$$

with the special case that

$$K_0(r) = \sum_{k=0}^{\infty} \frac{\left(\frac{1}{4}r^2\right)^k}{(k!)^2} \tag{12.21}$$

The Bessel's function of the first kind is a very short-range function, and is negligible at distances greater than $\frac{1}{\sqrt{F}}$.

Conserved by the symmetries of the system are three quantities, the pseudo-energy[7], the center of vorticity, and the angular momentum:

$$H = -\frac{1}{2}\int_{R^2} \psi_1 q_1 d\mathbf{x} - \frac{1}{2}\int_{R^2} \psi_2 q_2 d\mathbf{x} \tag{12.22}$$
$$\mathbf{L} = \int_{R^2} q_1 d\mathbf{x} + \int_{R^2} q_2 d\mathbf{x} \tag{12.23}$$
$$M = \int_{R^2} q_1 |\mathbf{x}|^2 d\mathbf{x} + \int_{R^2} q_2 |\mathbf{x}|^2 d\mathbf{x} \tag{12.24}$$

[6] Friedrich Wilhelm Bessel, 1784 - 1846, became interested in astronomy because of the need to better understand navigation for the import-export firm for which he was an accountant. He became the first person to measure the parallax of, thus the distance to, a star, 61 Cygni, and deduced the existence of Sirius's companion white dwarf. He pioneered the quantifying of observational and computational error. [288]

[7] The quantity given is actually a renormalized energy; if the circulations introduced later are nonzero – and we will have good reason for them to not be zero – then the interaction between layers will cause the actual energy to be infinitely large. This pseudo-energy treats only the energy in each layer separately.

Since quasi-geostrophic fluids preserve the potential vorticity, any functions of the potential vorticity have also to be constant. This offers potentially infinitely many further conserved quantities, but the only ones we find particularly useful are the individual layer circulations

$$\Gamma_1 = \int_{R^2} q_1 d\mathbf{x} \tag{12.25}$$

$$\Gamma_2 = \int_{R^2} q_2 d\mathbf{x} \tag{12.26}$$

from which we derive a **barotropic circulation**

$$\Gamma_B = \frac{1}{2}\left(\Gamma_1 + \Gamma_2\right) \tag{12.27}$$

and a **baroclinicity parameter**

$$\Gamma = \frac{\Gamma_1 - \Gamma_2}{\Gamma_1 + \Gamma_2} \tag{12.28}$$

from which $\Gamma_1 = \Gamma_B(1 + \Gamma)$ and $\Gamma_2 = \Gamma_B(1 - \Gamma)$. The model will be purely barotropic when $s_1 = s_2$; it is baroclinic when the vorticities are unequal.

Now to reach the mean field equations as DiBattista and Majda [107] do we will want to discretize the vorticity in each layer by the point-vortex form we have seen before. The vorticity in the upper layer we will represent with a set of N vortices all of uniform strength s_1 (so its total circulation is Ns_1); in the lower layer we represent it with vortices of uniform strength s_2 (so its total circulation is Ns_2). Since

$$q_1 \equiv \nabla^2 \psi_1 - F\left(\psi_1 - \psi_2\right) \tag{12.29}$$

$$= \Gamma_B\left(1 + \Gamma\right) \frac{\exp\left(\left(-\beta\psi_1(\mathbf{x}) - \mu|\mathbf{x}|^2\right) s_1\right)}{\int \exp\left(\left(-\beta\psi_1(\mathbf{x}) - \mu|\mathbf{x}|^2\right) s_1\right) d\mathbf{x}} \tag{12.30}$$

$$q_2 \equiv \nabla^2 \psi_2 + F\left(\psi_1 - \psi_2\right) \tag{12.31}$$

$$= \Gamma_B\left(1 - \Gamma\right) \frac{\exp\left(\left(-\beta\psi_2(\mathbf{x}) - \mu|\mathbf{x}|^2\right) s_2\right)}{\int \exp\left(\left(-\beta\psi_2(\mathbf{x}) - \mu|\mathbf{x}|^2\right) s_2\right) d\mathbf{x}} \tag{12.32}$$

In calculating the partition function for a set of N points $\mathbf{x}_{1,1}$, $\mathbf{x}_{1,2}$, $\mathbf{x}_{1,3}$, $\cdots$, $\mathbf{x}_{1,N}$ in the upper and N in the lower layer $\mathbf{x}_{2,1}$, $\mathbf{x}_{2,2}$, $\mathbf{x}_{2,3}$, $\cdots$, $\mathbf{x}_{2,N}$

$$Z_N = \iiint \exp\left(\begin{array}{c} -\beta H_N(\mathbf{x}) - \mu(\sum s_1 |\mathbf{x}_{1,k}|^2 \\ + \sum s_2 |\mathbf{x}_{2,k}|^2) - \gamma_1 s_1 N - \gamma_2 s_2 N \end{array}\right) ds d\mathbf{x} \tag{12.33}$$

where γ_1 and γ_2 are the Lagrange multipliers for a canonical constraint for the layer circulations.

The energy H is bounded by the interaction between point vortices being a logarithmic function; the terms it contributes to the energy are bounded by a product of terms like $|\mathbf{x}|^{\frac{\beta N}{4\pi}}$ while the contributions from the angular momentum constraint

$$\exp\left(-\mu s_j \sum_{k=1}^{N} |\mathbf{x}_{j,k}|^2\right) \tag{12.34}$$

for $j = 1$ and $j = 2$ decay rapidly provided μ, s_1, and s_2 are all the same sign. Between this and the bound above the partition function will be finite for all finite N provided

$$\frac{\beta N}{4\pi} > -2 \tag{12.35}$$

This suggests setting a scaled inverse temperature and chemical potential $\tilde{\beta} = \frac{\beta}{N}$ and $\tilde{\mu} = \frac{\mu}{N}$. We make the assumption that as the number of point vortices grows without bound they approximate the continuous vorticity distribution:

$$\frac{1}{N}\sum_{j=1}^{N} s_1 \delta(\mathbf{x} - \mathbf{x}_{1,j}) \to s_1 \rho_1(\mathbf{x}) \tag{12.36}$$

$$\frac{1}{N}\sum_{j=1}^{N} s_2 \delta(\mathbf{x} - \mathbf{x}_{2,j}) \to s_2 \rho_2(\mathbf{x}) \tag{12.37}$$

Any function defined on both layers $\mathbf{f} = (f_1, f_2)$ therefore satisfies

$$\sum_{j=1}^{N} \frac{s_1}{N} f_1(\mathbf{x}_{1,j}) + \sum_{k=1}^{N} \frac{s_2}{N} f_1(\mathbf{x}_{2,k}) \\ \to s_1 \int f_1(\mathbf{x})\rho_1(\mathbf{x})d\mathbf{x} + s_2 \int f_2(\mathbf{x})\rho_2(\mathbf{x})d\mathbf{x} \tag{12.38}$$

which principle can be extended to the energy functional H_N. For an arbitrary point $\mathbf{x}$ we can define

$$\frac{1}{N} H_N(\mathbf{x}, \mathbf{x}_{1,1}, \cdots, \mathbf{x}_{1,N-1}, \mathbf{x}_{2,1}, \cdots, \mathbf{x}_{2,N} \\ = \frac{1}{N} H^{(N-1,N)}(\mathbf{x}_{1,1}, \cdots, \mathbf{x}_{1,N-1}, \mathbf{x}_{2,1}, \cdots, \mathbf{x}_{2,N}) \\ - \frac{1}{N}\frac{s_1^2}{2} \sum_{j=1}^{N} (G_B(\mathbf{x}, \mathbf{x}_j) + G_T(\mathbf{x}, \mathbf{x}_j)) \\ - \frac{1}{N}\frac{s_1 s_2}{2} \sum_{j=1}^{N} (G_B(\mathbf{x}, \mathbf{x}_j) - G_T(\mathbf{x}, \mathbf{x}_j)) \tag{12.39}$$

$$\frac{1}{N} H_N(\mathbf{x}, \mathbf{x}_{1,1}, \cdots, \mathbf{x}_{1,N}, \mathbf{x}_{2,1}, \cdots, \mathbf{x}_{2,N-1}$$

$$= \frac{1}{N} H^{(N,N-1)}(\mathbf{x}_{1,1}, \cdots, \mathbf{x}_{1,N}, \mathbf{x}_{2,1}, \cdots, \mathbf{x}_{2,N-1})$$
$$-\frac{1}{N}\frac{s_1^2}{2}\sum_{j=1}^{N}(G_B(\mathbf{x},\mathbf{x}_j) + G_T(\mathbf{x},\mathbf{x}_j))$$
$$-\frac{1}{N}\frac{s_1 s_2}{2}\sum_{j=1}^{N}(G_B(\mathbf{x},\mathbf{x}_j) - G_T(\mathbf{x},\mathbf{x}_j)) \tag{12.40}$$

One should note the indices; the first two lines use $H^{(N-1,N)}$ in which the point $\mathbf{x}$ takes the place of the upper-level point $\mathbf{x}_{1,N}$. In the second $H^{(N,N-1)}$ represents the energy with $\mathbf{x}$ taking the place of the lower-level point $\mathbf{x}_{2,N}$.

As we assume our discrete approximations will approach the continuous vorticity field as the number of points grows infinitely large we have

$$-\frac{1}{N}\frac{s_1^2}{2}\sum_{j=1}^{N}(G_B(\mathbf{x},\mathbf{x}_j) + G_T(\mathbf{x},\mathbf{x}_j))$$
$$-\frac{1}{N}\frac{s_1 s_2}{2}(G_B(\mathbf{x},\mathbf{x}_j) + G_T(\mathbf{x},\mathbf{x}_j))$$
$$\to -\frac{s_1}{2}\int (G_B(\mathbf{x},\mathbf{y}) + G_T(\mathbf{x},\mathbf{y}))\,\rho_1(\mathbf{y})d\mathbf{y}$$
$$-\frac{s_1}{2}\int (G_B(\mathbf{x},\mathbf{y}) - G_T(\mathbf{x},\mathbf{y}))\,\rho_2(\mathbf{y})d\mathbf{y} \tag{12.41}$$

$$-\frac{1}{N}\frac{s_2^2}{2}\sum_{j=1}^{N}(G_B(\mathbf{x},\mathbf{x}_j) G_T(\mathbf{x},\mathbf{x}_j))$$
$$-\frac{1}{N}\frac{s_1 s_2}{2}(G_B(\mathbf{x},\mathbf{x}_j) + G_T(\mathbf{x},\mathbf{x}_j))$$
$$\to -\frac{s_2}{2}\int (G_B(\mathbf{x},\mathbf{y}) - G_T(\mathbf{x},\mathbf{y}))\,\rho_1(\mathbf{y})d\mathbf{y}$$
$$-\frac{s_2}{2}\int (G_B(\mathbf{x},\mathbf{y}) + G_T(\mathbf{x},\mathbf{y}))\,\rho_2(\mathbf{y})d\mathbf{y} \tag{12.42}$$

The one-point probability densities describing the likelihood of a vortex at any point $\mathbf{x}$ are then equal to

$$\rho_1(\mathbf{x}) \approx \rho_{1,N}(\mathbf{x}) \approx \exp\left(-\beta s_1 \psi_1(\mathbf{x}) - \mu s_1 |\mathbf{x}|^2\right)\frac{\tilde{Z}(N-1,N)}{Z(N)} \tag{12.43}$$

$$\rho_2(\mathbf{x}) \approx \rho_{2,N}(\mathbf{x}) \approx \exp\left(-\beta s_2 \psi_2(\mathbf{x}) - \mu s_2 |\mathbf{x}|^2\right)\frac{\tilde{Z}(N,N-1)}{Z(N)} \tag{12.44}$$

where the auxiliary functions are defined

$$\tilde{Z}(N-1,N) = \iiint \exp\begin{pmatrix} -\beta H^{(N-1,N)} - \mu\sum s_1|\mathbf{x}_{1,j}|^2 \\ -\mu\sum s_2|\mathbf{x}_{2,j}|^2 - \gamma_1 s_1 N - \gamma_2 s_2 N \end{pmatrix}$$
$$d\mathbf{x}_{1,1}\cdots d\mathbf{x}_{1,N-1}d\mathbf{x}_{2,1}\cdots d\mathbf{x}_{2,N} \tag{12.45}$$

$$\tilde{Z}(N, N-1) = \iiint \exp\begin{pmatrix} -\beta H^{(N,N-1)} - \mu \sum s_1 |\mathbf{x}_{1,j}|^2 \\ -\mu \sum s_2 |\mathbf{x}_{2,j}|^2 - \gamma_1 s_1 N - \gamma_2 s_2 N \end{pmatrix} d\mathbf{x}_{1,1} \cdots d\mathbf{x}_{1,N} d\mathbf{x}_{2,1} \cdots d\mathbf{x}_{2,N-1} \tag{12.46}$$

As $N \to \infty$ then we have the approximations

$$\frac{Z(N)}{\tilde{Z}(N-1, N)} \approx \int \exp\left(-\beta s_1 \psi_1(\mathbf{x}) - \mu s_1 |\mathbf{x}|^2\right) d\mathbf{x} \tag{12.47}$$

$$\frac{Z(N)}{\tilde{Z}(N, N-1)} \approx \int \exp\left(-\beta s_2 \psi_2(\mathbf{x}) - \mu s_2 |\mathbf{x}|^2\right) d\mathbf{x} \tag{12.48}$$

Substituting these into equations 12.43 and 12.44 gives at last the mean field equations

$$\begin{aligned} q_1 &\equiv \nabla^2 \psi_1 - F(\psi_1 - \psi_2) \\ &= \Gamma_1 \frac{\exp\left(-\beta \psi_1(\mathbf{x}) - \mu |\mathbf{x}|^2\right) s_1}{\int \exp\left(-\beta \psi_1(\mathbf{x}) - \mu |\mathbf{x}|^2\right) s_1 d\mathbf{x}} \end{aligned} \tag{12.49}$$

$$\begin{aligned} q_2 &\equiv \nabla^2 \psi_2 + F(\psi_1 - \psi_2) \\ &= \Gamma_2 \frac{\exp\left(-\beta \psi_2(\mathbf{x}) - \mu |\mathbf{x}|^2\right) s_2}{\int \exp\left(-\beta \psi_2(\mathbf{x}) - \mu |\mathbf{x}|^2\right) s_2 d\mathbf{x}} \end{aligned} \tag{12.50}$$

which thus descries the vorticities in the mean field approximation.

12.4 Numerical Models

In principle there are three main ways to approximate the partial differential equations 12.5. In a manner similar to that of the canonical Onsager vortex gas problem, these are:

a. The particle or point-vortex method
b. The lattice or spatial decomposition method
c. The wave or Fourier expansion method

Since equation 12.5 holds for functions q_1 and q_2 which are defined on the whole of R^2, the second method – lattice or spatial decomposition – is not suitable. Both methods **a** and **c**, however, should lead to tenable formations. We focus on the particle method **a** which was used to derive the mean field theory discussed above.

In this construction we represent the vorticity field of either layer as the combination of point vortices at locations $\mathbf{x}_{j,k}$, where j is either 1 or 2 and corresponds to the layer, and $k = 1, 2, 3, \cdots, N$. We treat each vortex as having a uniform strength, either s_1 in the upper layer or s_2 in the lower. Then we represent the vorticity distributions as

$$\rho_1(\mathbf{x}) = \frac{1}{N} \sum_{j=1}^{N} s_1 \delta(\mathbf{x} - \mathbf{x}_{1,j}) \tag{12.51}$$

$$\rho_2(\mathbf{x}) = \frac{1}{N}\sum_{j=1}^{N} s_2\delta(\mathbf{x} - \mathbf{x}_{2,j}) \tag{12.52}$$

As the number of points grows infinitely large these approximations grow sufficiently close to the actual distributions of vorticity; and the particles are relatively easy to manipulate analytically and extremely easy to treat numerically. This we do in the following section.

12.5 Numerical Vortex Statistics

The above approximation of q_1 and q_2 in terms of linear combinations of delta functions gives a heton model Hamiltonian H_N [26]. For finite N and using the conserved quantities

$$\Gamma_1 = \int_{R^2} q_1 d\mathbf{x} \tag{12.53}$$

$$\Gamma_2 = \int_{R^2} q_2 d\mathbf{x} \tag{12.54}$$

$$H = -\frac{1}{2}\left(\int_{R^2} q_1\psi_1 d\mathbf{x} + \int_{R^2} q_2\psi_2 d\mathbf{x}\right) \tag{12.55}$$

$$\Gamma = \int_{R^2} q_1|\mathbf{x}|^2 d\mathbf{x} + \int_{R^2} q_2|\mathbf{x}|^2 d\mathbf{x} \tag{12.56}$$

the Gibbs canonical partition function is

$$Z_N = \int_{R^2} dz_1^{(1)} \int_{R^2} dz_1^{(2)} \cdots \int_{R^2} dz_N^{(1)} \int_{R^2} dz_N^{(2)} \exp\left(-\beta H_N\right) \exp\left(-\mu\Gamma_N\right) \tag{12.57}$$

This partition function Z_N plays a role in the Monte Carlo simulation of the heton vortex gas with as before

$$Z_N = \int \exp\left(-\beta H(\mathbf{x}_N) - \mu\Gamma(\mathbf{x}_N)\right) d\mathbf{x}_N \tag{12.58}$$

taking here $d\mathbf{x}_N$ as shorthand for the entire state vector of vortex positions $\mathbf{x}_N \equiv (\mathbf{x}_{1,1}, \cdots, \mathbf{x}_{1,N}, \mathbf{x}_{2,1}, \cdots, \mathbf{x}_{2,N})$.

The Monte Carlo simulations are evaluated just as those with the vortex gas problem are: at each experiment select randomly one of the $2N$ vortices. Pick a small displacement also at random and find the change ΔH in enthalpy (the quantity $\beta H + \mu\Gamma$) which would result from moving that vortex by that position. If δH is negative we accept it; if δH is positive we draw a uniformly distributed random number r from between 0 and 1, and accept the move if $r < \exp(\Delta H)$. We reject it otherwise. We then pick another vortex at random and repeat the process. $2N$ attempted moves, whether accepted or rejected, is one sweep; and after sufficiently many sweeps the distribution settles to a statistical equilibrium. In the simulations done by Assad and Lim [26] the Rossby radius is set equal to 1, so the Froude's rotational number F is also 1.

The case of small Froude numbers in problems with a definite barotropic rim current or barotropic bias due for instance to pre-conditioning by open ocean gyres was simulated by Assad and Lim [26] using the Metropolis-Hastings algorithm. They find radially symmetric and concentric (with respect to the two layers) compactly supported coarse-grained vorticity distributions, that is indicative of a cold baroclinic core, consistent with previous statistical equilibria reported by DiBattista and Majda [108].

For moderate to large $\beta \gg 1$, we find that $\bar{q}_1$ and $\bar{q}_2$ are combinations of nearly a flat-top for the layer with the greater vortex strength ω_1, and a Gaussian-like profile for the other layer. As β grow infinitely large, both layers $\bar{q}_1$ and $\bar{q}_2$ tend to radially-symmetric flat-tops. We also find a power law for the radii R_1 and R_2 of the supports of $\bar{q}_1$ and $\bar{q}_2$.

These Monte Carlo simulations confirm that when the total circulation is nonzero, $\bar{q}_1$ and $\bar{q}_2$ have compact support and are radially symmetric. This simple fact was applied in [107] to explain the baroclinic stability of open-ocean convection that is preconditioned by a symmetry-breaking gyre.

As such, Monte Carlo simulation of the heton gas model is an efficient way to obtain physically relevant data on the two-layer model without solving the mean field equations 12.49 and 12.50.

When the inverse temperature β is zero the vortex model is solvable. In the barotropic case – when $s_1 = s_2$ – the probability of finding a vortex of strength s_1 at a distance from the origin r is

$$P(r) = \frac{\exp(-\mu s_1 r^2)}{\int_{R^2} \exp(-\mu s_1 r^2) d\mathbf{r}} \tag{12.59}$$

$$= \frac{\mu s_1}{\pi^2} \exp(-\mu s_1 r^2) \tag{12.60}$$

and so the vorticity of a single vortex is

$$q(r) = \frac{\mu s_1^2}{\pi} \exp(-\mu s_1 r^2) \tag{12.61}$$

while for N identical vortices it is

$$q^N(r) = \frac{N \mu s_1^2}{\pi} \exp(-\mu s_1 r^2) \tag{12.62}$$

The expectation value of the system of N vortices is then

$$\langle \Gamma(q^N) \rangle = \frac{N}{\mu} \tag{12.63}$$

In the baroclinic case, with two species of different strength, the most probable vortex profile is still independent Gaussian distributions:

$$q_1^N(R) = \frac{N\mu s_1^2}{\pi} \exp\left(-\mu s_1 r^2\right) \tag{12.64}$$

$$q_2^N(R) = \frac{N\mu s_2^2}{\pi} \exp\left(-\mu s_2 r^2\right) \tag{12.65}$$

Assad and Lim find in Monte Carlo experiments good fits to these expected exact distributions.

Negative β is potentially more problematic. The drive to find the greatest possible energy would encourage all the vortices to cluster together into as small a region as possible, ideally at the origin, but this configuration occupies such a small volume of phase space it is unlikely to be seen. More, analytically, the partition function diverges for sufficiently large-magnitude negative numbers. However, if

$$\beta > -\frac{8\pi}{N \max(s_1, s_2)} \tag{12.66}$$

the partition function is still finite and the probability distributions we derive from it therefore are still valid.

In the barotropic case we find numerically the vortices clustered more tightly toward the origin than they are in the positive temperature equivalents. In the baroclinic case there is a similar clustering of the vortices, with the amount by which each layer is clustered dependent on the relative strengths of the point vortices in each. As the magnitude of β increases in the negative numbers the clustering grows tighter and the space covered decreases until either we reach a configuration with too low an energy for the computer to simulate accurately.

So again we see the theme of numerical simulations picking up on and letting us expand on models which cannot be solved analytically or which simply are not meaningful in certain regions, and providing us a guide either to what the analytic solutions ought to be or simply providing us with a greater breadth of experience from which to build intuitions about the systems.

13

Coupled Barotropic Vorticity Dynamics on a Rotating Sphere

13.1 Introduction

In this chapter we apply the methods developed already to a simple geophysical flow with rotation with the aim of deriving and solving a crude statistical mechanics model for atmospheric super-rotation. Consider the system consisting of a rotating high density rigid sphere of radius R, enveloped by a thin shell of barotropic (non-divergent) fluid. The barotropic flow is assumed to be inviscid, apart from an ability to exchange angular momentum and energy with the heavy solid sphere. In addition we assume that the fluid is in radiation balance and there is no net energy gain or loss from insolation.

For a geophysical flow problem concerning super-rotation on a spherical surface there is little doubt that one of the key parameters is angular momentum of the fluid. In principle, the total angular momentum of the fluid and solid sphere is a conserved quantity. But by taking the sphere to have infinite mass, the active part of the model is just the fluid, which relaxes by exchanging angular momentum with an infinite reservoir.

It is also clear that a two-dimensional geophysical relaxation problem such as this one will involve energy and enstrophy. The total or rest frame energy of the fluid and sphere is conserved. Since we have assumed the mass of the solid sphere to be infinite, we need only keep track of the kinetic energy of the barotropic fluid – in the non-divergent case, there is no gravitational potential energy in the fluid since it has uniform thickness and density, and its upper surface is a rigid lid.

In particular, we will discuss an equilibrium statistical mechanics model for the coupled barotropic flow - rotating solid sphere system, known as the spherical model, where the energy is constrained canonically and the relative enstrophy microcanonically. This approach differs from previous work on the statistical mechanics of barotropic flows by allowing the angular momentum in the fluid to change modulo the fixed relative enstrophy. In other words the fluid exchanges energy explicitly with an infinite reservoir that is held at fixed statistical temperatures; at the same time it also exchanges angular momentum implicitly with the spinning sphere. Freeing the angular momentum in the barotropic system is designed to produce a crude model for the complex torque driven interactions between a vertically averaged atmosphere and a spinning planet.

The classical energy-enstrophy theory [227] suffers from a serious technical defect: it is equivalent to a Gaussian model, where the partition function is not defined for low temperatures. We will derive the spin-lattice Hamiltonian in the Gaussian energy-enstrophy model, and review the differences between the Gaussian model and the spherical model.

The classical energy-enstrophy theory of Kraichnan and its usually spectral counterparts in geophysical flows have been modified and extended by the first author, turning them into powerful spin-lattice models which allow us to use the time-tested methods of equilibrium statistical mechanics to study their critical phenomenology. Contrary to expectations from many previous works, this body of work, combining theoretical and simulation techniques, have shown that correct and careful applications of Planck's theorem on extremals of the free energy functional can produce physically relevant results on phase transitions in large-scale geophysical flows in both positive and negative temperatures. This body of results is remarkable for its explicit dependence on only a handful of easily measured key parameters in atmospheric and oceanic flows such as total kinetic energy, potential energy, relative enstrophy, potential enstrophy, rates of spin and the density of the fluid.

This includes new results on the transition to super-rotating almost-solid-body flows in modified barotropic models where the angular momentum constraint is relaxed. Dependence of this method on the extremization of the full free energy functional, instead of maximizing the entropy alone, is key to its success. For example, in the case of the transitions to organized solid-body flows, which occur only when the kinetic energy to relative enstrophy ratio is high enough (or negative temperature is near zero), these new results cannot be obtained using maximum entropy methods. The reason is simple: in these cases, the critical temperatures have very small numerical values and the nearly solid-body macrostates have low entropy, thence, the free energy is dominated by the internal energy term whose extremals are the corresponding pure ground states. Minimizing the full free energy when temperature is positive (as prescribed by the usual Planck's theorem) and maximixing the full free energy (as prescribed by the modified Planck's theorem for negative temperatures) led directly to the discovery of these new critical phenomena.

13.2 The Coupled Barotropic Vorticity – Rotating Sphere System

We begin with the emphasis that what we present in this chapter is not a study of the standard Barotropic Vorticity Equation (BVE) but rather a model of the coupled barotropic flow - rotating sphere system which does not conserve angular momentum of the fluid. In taking this approach, we adopt the age-old tradition in equilibrium statistical mechanics of not dealing with the dynamical partial differential equations or ordinary differential equations that governs the time evolution of the system, but rather its energy and some of its constraints. It will be clear below that there are advantages to not constrain the angular momentum. As useful background however, we give next a short review of the Barotropic Vorticity Equation.

In the canonical spherical coordinates ($\cos(\theta)$, where θ is the colatitude; and longitude ϕ), the Barotropic Vorticity Equation is

$$\omega_t + J(\psi, q) = 0 \tag{13.1}$$
$$\omega(t; \cos(\theta), \phi) = \Delta\psi \tag{13.2}$$

where $\omega = \Delta\psi$ is the relative vorticity given in terms of a stream function ψ, and where the total vorticity q is the sum of the relative vorticity w and planetary vorticity due to the planetary spin $\Omega > 0$, i.e.,

$$q = w + 2\Omega\cos(\theta) \tag{13.3}$$

It is important to notice here that we now use Ω to represent the planetary spin around the z-axis, which agrees with the frequent use in mechanical problems of ω and Ω as symbols for the rotation of particles and solid bodies, and for atmospheric rotations. When we will need to refer to the enstrophy of the vortices again we will use Q. Mathematics is unfortunately always short of mnemonic and easily written symbols.

Steady states of the Barotropic Vorticity Equation are easily characterized. Let $\omega = \omega(\psi)$ be any smooth function[1] of the stream function ψ. Then any solutions of the following non-linear elliptic partial differential equation

$$\Delta\psi = \omega(\psi) \tag{13.4}$$
$$\int_{S^2} d\mathbf{x}\Delta\psi = 0 \tag{13.5}$$

are steady state stream functions. This, one can see by setting the Jacobian in (13.1) to zero. In particular any zonal stream function $\psi = \psi(\theta)$ represents a steady-state solution of the barotropic vorticity equation.

13.2.1 Physical Quantities of the Coupled Flow – Rotating Sphere Model

The most important quantity in the coupled barotropic vorticity - rotating sphere model is the **rest frame pseudo kinetic energy** given in a frame that is rotating at the fixed angular velocity of the sphere,

[1] A smooth function has continuous derivatives on the interior of the domain up to some arbitrary order.

$$H[w] = -\frac{1}{2}\int_{S^2} d\mathbf{x}\psi q. \tag{13.6}$$

We emphasize the fact that, because the total vorticity q is the sum of relative vorticity ω and the vorticity of the rotating frame, and angular momentum of the fluid is not fixed by the angular velocity of this frame, this expression for the rest frame kinetic energy of the barotropic fluid contains a non-standard second term that has the form of varying net angular momentum. Energy in the relative vorticity is continuously exchanged between the first term with zero angular momentum (relative to the frame rotating at fixed angular velocity) and the second term with nonzero net angular momentum.

The dynamics of this exchange is reviewed in the section on a non-standard variational approach for the Coupled Barotropic Flow – Rotating Sphere Model. We shall find that on a fixed relative enstrophy manifold - such a constraint is justified by the requirement for a well-defined problem and by the Principle of Selective Decay - the steady states with extremal pseudo kinetic energy are precisely the prograde (respectively, retrograde) solid-body flows that have the most (respectively, the least) fluid angular momentum compared to the flow rotating rigidly at the angular velocity of the rotating frame.

More surprising is the related statistical mechanics results that follow from exact solutions of the vector spherical models obtained by Lim as well as the Monte-Carlo simulations in Ding's thesis.

The **vector spherical model** is similar to the Ising model in putting a lattice over the domain and assigning spins to each site, and having the energy of the system depend on pairwise interactions – generally just between nearest or the very near neighbors – and interaction with an external field. Where it differs from the Ising model is that the spin in the vector model is taken to be a three-dimensional normal vector instead of being a scalar. There are similar models treating the spin as a two-dimensional vector (this is known as the **XY model**), or to vectors of any number of dimensions, known in general as **n-vector models** or **$O(n)$ models**.

We find that no matter where we place the initial state on the corresponding pseudo energy manifold for a given negative statistical temperature (with sufficiently small absolute value), the Markov chain generated by the Metropolis-Hastings algorithm equilibrates around a state with some of the fluid angular momentum allowed by the relative enstrophy constraint. For instance, we could choose an initial state on this manifold with all its energy in the zero angular momentum part and find that the statistical equilibrium is the state with most of its energy in the second part with positive angular momentum relative to the rotating frame.

As a consequence of Stokes' theorem on the sphere, the total circulation is fixed at zero,

$$\Gamma[\omega] \equiv \int_{S^2} \omega d\mathbf{x} = 0. \tag{13.7}$$

Unlike the BVE, the next vorticity moment, the **relative enstrophy**

$$Q[w] = \int_{S^2} d\mathbf{x}\omega^2, \tag{13.8}$$

or the square of the L^2 norm of the relative vorticity ω, is not conserved in the coupled fluid-rotating sphere model. However, since the coupled fluid-sphere model is not well-defined without an additional constraint on the size of the argument of the energy functional – such as a bound on the relative enstrophy – and Selective Decay states that only the ratio of energy to enstrophy should play a role in the asymptotic and statistical equilibrium states, we will fix the relative enstrophy in the following discussion.

All other **moments of vorticity** $\int d\mathbf{x}\omega^n$ are not considered here and in many physical theories of fluid motions, including some equilibrium statistical mechanics models [85] [227] [369], and the variational problems below.

13.3 Rotating versus Non-rotating Sphere – Variational Results

In recent years, a substantial mathematical framework has been built around the problem of vorticity dynamics and statistics on a rotating sphere (cf Shepherd [393], Cho and Polvani [84], Yoden and Yamada [449] and the references therein). A recent paper of one of the authors [257] points out that in the case of a non-rotating sphere where $\Omega = 0$, equilibrium statistical mechanics is allowed because, unlike the situation of a rapidly rotating sphere, there is no evidence of the dynamics of the system not being ergodic.

Negative temperatures in a new equilibrium statistical mechanics model for the Coupled Barotropic Vorticity - Rotating Sphere System (where the enstrophy is microcanonically constrained) we will show are associated with global energy maximal distributions, in the form of solid-body rotation vorticity states $\omega_M^{\pm} = \alpha_{1,0}\psi_{1,0}$. We also will show these extremals are unique modulo the subspace $span\{\psi_{10}, \psi_{1,1}, \psi_{1,-1}\}$, that is:

Theorem 14. *Let $H[\omega_M^{\pm}]$ be the energy (13.6) of the solid-body rotation vorticity states $\omega_M^{\pm}$, and let $Q[\omega_M^{\pm}]$ be their enstrophy (13.8). Suppose that vorticity ω satisfies*

$$Q[\omega] = Q[\omega_M^{\pm}] \tag{13.9}$$

and the zero total circulation condition

$$\Gamma[\omega] \equiv \int_{S^2} \omega d\mathbf{x} = 0. \tag{13.10}$$

Then

$$H[\omega] = H[\omega_M^{\pm}] \tag{13.11}$$

implies $\omega \in span\{\psi_{1,0}, \psi_{1,1}, \psi_{1,-1}\}$.

The converse of theorem 14 is also true, that is, with $\psi_{l,m}$ denoting the spherical harmonics, we have

Theorem 15. *If* $\omega \in span\{\psi_{1,0}, \psi_{1,\pm 1}\}$ *satisfies (13.9), then* $H[\omega] = H[\omega_M^{\pm}]$.

The fact that

$$H[\omega] \leq H[\omega_M^{\pm}] \tag{13.12}$$

holds for any vorticity ω that satisfies (13.9) and (13.10) was shown in Lim [257]. We stress that the variational principle in this and the following sections of the book do not have the standard Barotropic Vorticity partial differential equation as its Euler-Lagrange equation. Instead we are stating a new variational principle for the enstrophy-constrained optimization of the total kinetic energy of barotropic flows which models the complex torque mechanism for the transfer of angular momentum between a barotropic flow and a spinning sphere.

13.4 Energy-Enstrophy Theory for Barotropic Flows

The only constraint in this theory, besides the zero total circulation condition, is the fixed enstrophy

$$\int_{S^2} d\mathbf{x}\omega^2 = Q > 0. \tag{13.13}$$

We apply the constraint $\|\omega\|_2 = Q > 0$ in the definition of the constraint manifold[2],

$$V = \left\{ \omega \in L^2(S^2) | \|\omega\|_2^2 = Q > 0, \Gamma[\omega] = \int_{S^2} \omega d\mathbf{x} = 0 \right\}, \tag{13.14}$$

By the method of Lagrange multipliers (see [122], [401]), the energy-enstrophy functional

$$E[\omega] = H[\omega] + \lambda \|\omega\|_2^2 \tag{13.15}$$

is the **augmented objective functional** for the unconstrained optimization problem which is associated with the constrained optimization problem, to extremize $H[\omega]$ on the set

$$V = \left\{ \omega \in L^2(S^2 | \|\omega\|_2^2 = Q > 0, \int_{S^2} d\mathbf{x}\omega = 0 \right\} \tag{13.16}$$

The unconstrained optimization of the augmented energy-enstrophy functional which we get from the Lagrange multiplier method yields the necessary

[2] A manifold is a topological space for which, around every point, is a ball topologically equivalent to an open unit ball in $\Re^n$. That is, it is any surface that, locally, looks flat.

conditions for extremals of the constrained optimization problem. It turns out that extremals of the augmented objective functional in this case are also extremals of the original objective functional.

We will focus then on sufficient conditions for the existence of extremals of the constrained optimization problem in the non-rotating case.

13.4.1 Extremals of the Augmented Energy Functional: Non-rotating Sphere

Using the eigenfunction expansion

$$\omega = \sum_{l \geq 1,m} \alpha_{l,m} \psi_{l,m}, \tag{13.17}$$

of vorticity ω satisfying $\int_{S^2} \omega d\mathbf{x} = 0$, the energy and enstrophy functionals are given in terms of the orthonormal spherical harmonics, that is,

$$H[\omega] = -\frac{1}{2} \int \psi w = -\frac{1}{2} \sum_{l \geq 1,m} \frac{\alpha_{l,m}^2}{\lambda_{l,m}} \tag{13.18}$$

$$Q[\omega] = \sum_{l \geq 1,m} \alpha_{l,m}^2. \tag{13.19}$$

where the sums are taken for l the positive integers to ∞, and m the integers from $-l$ to l inclusive. We will show the extremals of the constrained optimization problem in (13.16) consist of elements in $span\{\psi_{1,0}, \psi_{1,\pm 1}\}$, the vector space of functions formed by linear combinations of these functions.

A direct application of (13.18) and (13.19) yields

Lemma 5. *The energy and enstrophy of the vorticity projection in the space* $M \equiv span\{\psi_{1,0}, \psi_{1,\pm 1}\}$, *that is,*

$$\omega_M = \alpha_{1,0}\psi_{1,0} + \alpha_{1,1}\psi_{1,1} + \alpha_{1,-1}\psi_{1,-1} \tag{13.20}$$

are given, respectively, by

$$H[\omega_M] = \frac{1}{4}(\alpha_{1,0}^2 + \alpha_{1,1}^2 + \alpha_{1,-1}^2) = \frac{1}{4}Q, \tag{13.21}$$

$$Q[\omega_M] = \alpha_{1,0}^2 + \alpha_{11}^2 + \alpha_{1,-1}^2. \tag{13.22}$$

The Euler-Lagrange method gives the necessary conditions for extremals of a constrained problem. *Sufficient* conditions for the existence of extremals can nonetheless be found from the geometrical basis of the method itself.

An extremal must be contained in the set of common tangents of the level surfaces of the energy H and enstrophy Q, but only those common tangency points P where one level surface remains on the same side of the other level surface in a full neighborhood of P correspond to constrained energy maximizers or minimizers. From those points, moreover, if the level surface of H is on the outside of the surface of Q in a full neighborhood of a point $P_{\max}$, measuring "outside" with respect to the origin, then $P_{\max}$ is a constrained energy maximum.

If, on the other hand, the level surface of H is on the inside of the surface of Q in a full neighborhood of $P_{\min}$, with respect to the origin, then $P_{\min}$ is a constrained energy minimum.

Theorem 16. *The extremals of the constrained optimization problem presented in equation (13.16) are degenerate global energy maximizers, consisting of*

$$\mathit{span}\,(\psi_{1,0}, \psi_{1,\pm 1})\,. \tag{13.23}$$

Proof: By (13.18) and (13.19) respectively, the energy surface is an infinite-dimensional ellipsoid in the space $L^2(S^2)$ of square-integrable functions defined on the unit sphere. The enstrophy surface is itself a sphere in $L^2(S^2)$. Since the eigenvalues $\lambda_{l,m} = -l(l+1)$, we deduce (i) the three shortest semi-major axes of the ellipsoid have equal length, and these correspond to the Fourier components of $\psi_{1,0}$, $\psi_{1,1}$, and $\psi_{1,-1}$; and (ii) their lengths $L(l)$ increase quadratically with the azimuthal wave-number l.

Thus, the extremals for a fixed enstrophy Q consist of any point

$$P = (\alpha_{10}, \alpha_{11}, \alpha_{1,-1}) \tag{13.24}$$

in the three-dimensional subspace $\mathit{span}\,(\psi_{1,0}, \psi_{1,\pm 1})$ which satisfies the spherical condition

$$\alpha_{1,0}^2 + \alpha_{1,1}^2 + \alpha_{1,-1}^2 = Q. \tag{13.25}$$

By deduction (ii), the length $L(l)$ of a semi-major axis for $l \geq 2$ exceeds the corresponding radius of the enstrophy sphere

$$\sum_{l \geq 1, m} \alpha_{l,m}^2 = Q, \tag{13.26}$$

and therefore, apart from $\mathit{span}\,(\psi_{1,0}, \psi_{1,\pm 1})$, the ellipsoid lies on the outside of this enstrophy sphere.

Thus, any of the above extremals is a degenerate global energy maximizer. This is confirmed also by the simple calculation: for any $P' = \{\alpha'_{l,m}\}$ on the sphere (13.26)

$$H[P'] = -\frac{1}{2} \sum_{l \geq 1, m} \frac{(\alpha'_{l,m})^2}{\lambda_{l,m}} \tag{13.27}$$

$$< \frac{1}{4} p^2 + \frac{1}{4} \left(Q - p^2\right) \cdots \tag{13.28}$$

$$= \frac{1}{4} (\alpha_{1,0}^2 + \alpha_{1,1}^2 + \alpha_{1,-1}^2). \tag{13.29}$$

□

The proof of the **converse theorem 15** follows from an application of theorem 16.

13.4.2 Rotating Sphere

The most important way the rotating case differs from the non-rotating case is that the nonzero spin $\Omega > 0$ breaks the $SO(3)$ rotational symmetry. This removes the energy degeneracy stated in theorem 16, and allows the pair of solid-body rotation steady states, $w_M^{\pm} = \pm\sqrt{Q}\psi_{1,0}$ to achieve non-linear stability under certain parameter regimes.

The main results for the rotating case is summarized in the following statement, where $C = \|\cos\theta\|_2$ and the relative enstrophy is given by the square of the L^2 norm. Clearly, the symmetry breaking removes a corresponding degeneracy in the non-rotating case, resulting in much richer physical consequences. The proof of the counterpart of Theorem 16 for a rotating sphere, the **Rotating Sphere Theorem**, is lengthy and will not be given here [256].

Theorem 17. ***Rotating Sphere Theorem:*** *For fixed spin $\Omega > 0$, there is a critical value Ω^2C^2 of the relative enstrophy Q, below which both the prograde and retrograde solid-body flows exist as Lyapunov*[3] *stable steady states of the Coupled Barotropic Vorticity - Rotating Sphere System, with the retrograde solution having smaller energy.*

For $\Omega^2C^2 < Q < 4\Omega^2C^2$, only the prograde solution exists as a Lyapunov stable steady state; it corresponds to the global energy maximizer of the constrained variational problem (these are the rotational counterparts of optimizing $H[\omega]$ subject to (13.16)). The retrograde solution is unstable in the context of the coupled barotropic flow - rotating solid sphere system, because it is now a ***saddle point.***

If $Q > 4\Omega^2C^2$, then the prograde state is a global energy maxima and Lyapunov stable. The retrograde state is now a restricted local energy maxima, in the sense that it is a local minima in the span$\{\psi_{1,\pm1}\}$. In particular, the retrograde steady-state of the Coupled Barotropic Vorticity - Rotating Sphere System is non-linearly stable to zonally symmetric perturbations when $Q > 4\Omega^2C^2$.

An equilibrium is **Lyapunov stable** if all perturbations of the equilibrium remain bounded for all time. A linear system is Lyapunov stable if and only if all its eigenvalues have non-positive real parts, and if there are no repeated imaginary eigenvalues.

Because the additional stabilizing factor of angular momentum conservation can be invoked for the standard Barotropic Vorticity Model, a Lyapunov

[3] Aleksandr Mikhailovich Lyapunov, 1857 - 1918, was a friend in school to Markov, and a student of Chebyshev, who challenged him to describe equilibria of rotating fluids. [288]

Stability result in the Genaralized Model implies Lyapunov Stability in the standard one. The symmetry-breaking retrograde motions in the Generalized Model can lose stability at a saddle point, but has no counterpart in the standard model with fixed angular momentum [395].

13.5 Statistical Mechanics

Under a necessary non-extensive continuum limit, as the number of mesh points N grows infinitely large, the classical energy-enstrophy theories for inviscid two-dimensional geophysical flows are ill-defined for negative temperatures with small absolute value. This family of theories is equivalent to the Gaussian spin-lattice model which has a **low temperature defect.**

Resolution of this foundational problem is obtained by a microcanonical constraint on the enstrophy. This leads to an exactly solvable model known historically as Kac's spherical model. This new energy-enstrophy theory is well-defined for all temperatures.

13.5.1 Derivation of Spin-Lattice Hamiltonians

A family of spin-lattice models reminiscent of the Ising model with external fields are derived for the Coupled Barotropic Vorticity - Rotating Sphere System under a decomposition of the sphere S^2 into nearly uniform Voronoi cells. Using a uniform mesh M consisting of N points $\{\mathbf{x}_1, \mathbf{x}_2, \mathbf{x}_3, \cdots \mathbf{x}_N\}$ on S^2, and the Voronoi cells based on this mesh, we approximate

$$\omega(x) \simeq \sum_{j=1}^{N} s_j H_j(\mathbf{x}) \tag{13.30}$$

where $s_j = \omega(x_j)$ and H_j is the characteristic or indicator function on the Voronoi cell D_j centered at $\mathbf{x}_j$ [112]. The truncated energy (13.6) takes the standard form of a spin-lattice Hamiltonian,

$$H_N = -\frac{1}{2}\sum_{j=1}^{N}\sum_{k=1}^{N} J_{jk} s_j s_k - \sum_{j=1}^{N} F_j s_j \tag{13.31}$$

where for this problem

$$J_{j,k} = \int_{S^2} d\mathbf{x} H_j(x) \int_{S^2} d\mathbf{x}' \log|1 - \mathbf{x}\cdot\mathbf{x}'| H_k(\mathbf{x}') \tag{13.32}$$

$$\to \frac{16\pi^2}{N^2} \log|1 - \mathbf{x}_j \cdot \mathbf{x}_k| \text{ as } N \to \infty; \tag{13.33}$$

and the external fields

$$F_j = \Omega\|\cos\theta\|_2 \int_{S^2} d\mathbf{x}' H_j(\mathbf{x}') \int_{S^2} d\mathbf{x}\psi_{10}(\mathbf{x}) \log|1 - \mathbf{x}\cdot\mathbf{x}'| \quad (13.34)$$

$$\to \frac{2\pi}{N}\Omega\|\cos\theta\|_2\psi_{10}(\mathbf{x}_j) \text{ as } N \to \infty, \quad (13.35)$$

where $C = \|\cos\theta\|_2$ is the L^2 norm of the zonal function $\cos(\theta)$ and the spherical harmonic $\psi_{1,0}$ represents the relative vorticity of solid-body rotation. Again we note that the external field term in the kinetic energy is essentially the changing global angular momentum contained in the relative flow. This quantity is clearly fixed in the standard Barotropic Vorticity Model because it conserves the angular momentum of the fluid.

The truncated relative enstrophy is given by

$$Q_N = \int_{S^2} d\mathbf{x}\omega^2 \simeq \int_{S^2} d\mathbf{x}\left(\sum_{j=1}^{N} s_j H_j(\mathbf{x})\right)^2 = \frac{4\pi}{N}\sum_{j=1}^{N} s_j^2. \quad (13.36)$$

Lastly, the truncated total circulation is given by

$$\Gamma_N = \int_{S^2} d\mathbf{x}\omega \simeq \int_{S^2} d\mathbf{x}\sum_{j=1}^{N} s_j H_j(\mathbf{x}) = \frac{4\pi}{N}\sum_{j=1}^{N} s_j. \quad (13.37)$$

13.5.2 Gaussian Energy-Enstrophy Model

The classical energy-enstrophy theory is now given in terms of the truncated energy H_N and relative enstrophy Q_N by the Gaussian partition function,

$$Z_N = \left(\frac{b}{2\pi}\right)^{N/2} \int \left[\prod_{j=1}^{N} ds_j\right] \exp\left(-b\sum_{j=1}^{N} s_j^2\right) \exp\left(-\beta H_N[\mathbf{s};\Omega]\right) (13.38)$$

for the spin vector $\mathbf{s} = \{s_1, \cdots, s_N\}$, with each $s_j \in (-\infty, \infty)$ and $b = \frac{4\pi\mu}{N}$, where μ is the chemical potential associated with the relative enstrophy constraint. This Gaussian model can be solved exactly:

$$Z_N[h;\beta,\Omega] = \left(\frac{2\pi\mu}{N}\right)^{N/2} \frac{1}{\sqrt{\det K}} \exp\left[\frac{1}{2}\langle h|K^{-1}|h\rangle\right] \quad (13.39)$$

in terms of

$$K_{j,k} = -\beta J_{j,k} + \frac{4\pi\mu}{N}\delta_{j,k} \quad (13.40)$$

and its determinant $\det(K)$, and the external fields $h = (h_1, h_2, \cdots, h_N)^T$, with

$$h_k = \beta F_k = \frac{2\pi\beta}{N}\Omega\|\cos\theta\|_2\psi_{1,0}(\mathbf{x}_k). \quad (13.41)$$

For Z_N to be well-defined, it is clear that the matrix K must not have any negative eigenvalues. In that case, the exact solution is given by

$$Z_N[h;\beta,\Omega] = \left(\frac{2\pi\mu}{N}\right)^{N/2} \frac{1}{\sqrt{\det K}} \exp\left[\frac{1}{2N}\sum_{l,m}^{N} \frac{|\tilde{h}(l,m)|^2}{\tilde{K}(l,m)}\right] \tag{13.42}$$

in terms of the discrete Fourier transform,

$$\frac{1}{2}\langle h|K^{-1}|h\rangle = \frac{1}{2N}\sum_{l,m}^{N} \frac{|\tilde{h}(l,m)|^2}{\tilde{K}(l,m)}, \tag{13.43}$$

where $\tilde{K}(l,m)$ is the eigenvalue of the matrix K corresponding to the l,m mode.

13.6 Spherical Model

The resolution of this serious foundational difficulty in many of the classical energy-enstrophy statistical mechanics theories for two- and 2.5-dimensional geophysical flows lies in replacing the classical canonical enstrophy constraint by a microcanonical one [264]. Usually, this leads to great analytical difficulties. In this case, however, the resulting spin-lattice models fall into a well-known exactly-solvable category called Kac's spherical models [41]. There is however an important difference between the spherical models for non-rotating flows and for barotropic flows on a rotating sphere.

We use the first author's approach, based on forming vortex statistics as Ising-like models. In this view the effects of rotation are treated as an external magnetic field.

In the rotating sphere case, the spherical model partition function is

$$Z_N = \int \prod_{j=1}^{N} ds_j \delta\left(NQ_N - 4\pi\sum_{j=1}^{N} s_j^2\right) \exp\left[-\beta H_N\right] \tag{13.44}$$

where the spin-lattice Hamiltonian H_N and the external fields F_j are given in [111] and [271]. The microcanonical relative enstrophy constraint takes the Laplace integral form and gives

$$Z_N = \frac{1}{4\pi \mathrm{i}} \int \prod_{j=1}^{N} ds_j \int_{a-i\infty}^{a+i\infty} d\eta \exp\left[\frac{1}{2}\eta N Q_r - \frac{1}{2}\langle \mathbf{s}|K|\mathbf{s}\rangle\right] \exp\left[\beta\sum_{j=1}^{N} F_j s_j\right] \tag{13.45}$$

in terms of the matrix

$$K_{j,k} = \eta - \beta J_{j,k} \tag{13.46}$$

and the spin vector $\mathbf{s} = \{s_1, \cdots, s_N\}$.

The approach based on Kac's spherical model gives rise to exactly-solvable statistical models for vortex dynamics on a sphere that do not have the foundational difficulty of the classical energy-enstrophy models. This spherical model can be efficiently simulated by Monte Carlo methods [166]. We present some results of this simulation in the figures below [110].

13.7 Monte Carlo Simulations of the Spherical Model

Some Monte Carlo results are presented in the figures below, which show the most probable relative vorticity distributions under different values of the three key parameters in the spherical model for the coupled Barotropic Fluid - Rotating Solid Sphere Model. These parameters are (i) the rate of spin $\Omega > 0$, (ii) the relative enstrophy $Q_{rel} > 0$ and (iii) the inverse temperature β, which is allowed to be negative. Indeed, to achieve the highest mean kinetic energies $\langle H \rangle$ for given spin rate Ω and relative enstrophy Q_{rel}, negative values of β must be used.

The most probable relative vorticity in the spherical model for negative β (as in figure 13.1) and the global energy maximizer in the variational theory is the prograde solid-body flow. In figures 13.1, the color convention for positive vorticity is red and that for negative vorticity is blue. Thus, with the predominance of red in the northern hemisphere (where $\Omega > 0$ denotes west-to-east planetary rotation), we deduce that the vorticity in figure 13.1 is a prograde solid-body flow state.

Other significant relative vorticity patterns will be presented in a forthcoming paper by Ding and Lim [110], along with an exhaustive discussion of the Monte Carlo exploration of key parameter space.

13.8 The Vector Spherical Model for Barotropic Statistics

We give a more natural vectorial reformulation of the spin-lattice models than those mentioned earlier. Instead of representing the local relative vorticity $\omega(\mathbf{x}_j)$ at lattice site $\mathbf{x}_j$ by a scalar s_j, it is natural to represent it by the vector

$$\mathbf{s}_j = s_j \mathbf{n}_j \tag{13.47}$$

where $\mathbf{n}_j$ denotes the outward unit normal to the sphere S^2 at the point $\mathbf{x}_j$. Similarly, represent the spin $\Omega > 0$ of the rotating frame – the spinning planet – by the vector

$$\mathbf{h} = \frac{2\pi}{N} \Omega \mathbf{n} \tag{13.48}$$

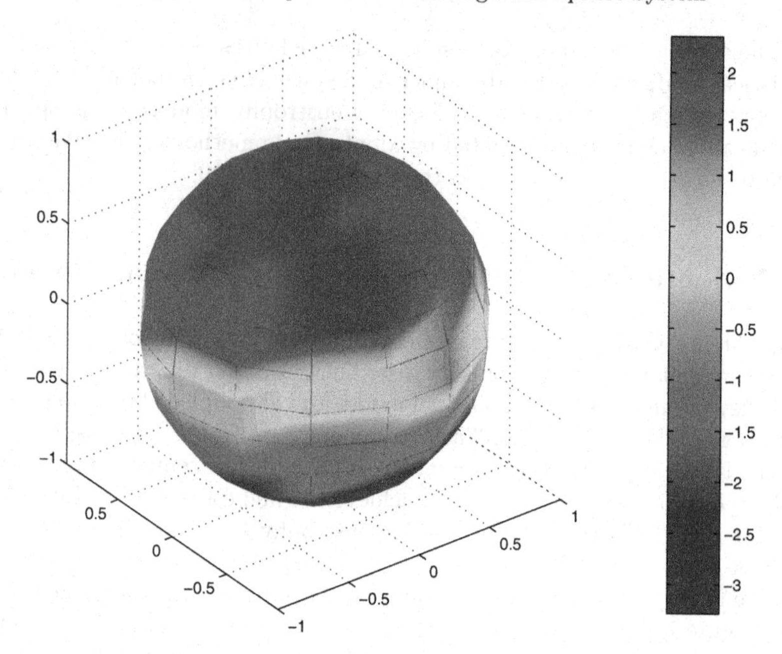

Fig. 13.1. Most probable vorticity in the spherical model for the barotropic vorticity equation on a rotating sphere ($\beta = -10$, $\Omega = 100$, $Q_{rel} = 100$).

where $\mathbf{n}$ is the outward unit normal at the north pole of S^2. Denoting by $\gamma_{j,k}$ the angle subtended at the center of S^2 by the lattice sites $\mathbf{x}_j$ and $\mathbf{x}_k$, we obtain the following vectorial Ising model for the total (fixed frame) kinetic energy of a barotropic flow in terms of a rotating frame at spin rate Ω,

$$H_H^N = -\frac{1}{2} \sum_{j \neq k}^{N} J_{jk} \mathbf{s}_j \cdot \mathbf{s}_k - \mathbf{h} \cdot \sum_{j=1}^{N} \mathbf{s}_j \tag{13.49}$$

where the interaction matrix is now given by the infinite range

$$J_{j,k} = \frac{16\pi^2}{N^2} \frac{\log\left(1 - \cos\left(\gamma_{j,k}\right)\right)}{\cos\left(\gamma_{j,k}\right)} \tag{13.50}$$

with the dot denoting the usual inner product in $\Re^3$ and $\mathbf{h}$ denoting a fixed external field.

The Kac-Berlin method we can then modify [259] to treat the vector spherical model, which consists of H_H^N and the spherical or relative enstrophy constraint,

$$\frac{4\pi}{N} \sum_{j=1}^{N} \mathbf{s}_j \cdot \mathbf{s}_j = Q \tag{13.51}$$

In addition, the Stokes theorem implies that it is natural to treat only the case of zero circulation,

$$\frac{4\pi}{N}\sum_{j=1}^{N}\mathbf{s}_j\cdot\mathbf{n}_j=0. \tag{13.52}$$

Looking ahead, we notice the important fact that the following vectorial sum or magnetization

$$\Gamma=\frac{4\pi}{N}\sum_{j=1}^{N}\mathbf{s}_j \tag{13.53}$$

will turn out to be a natural order parameter for the statistics of barotropic flows on a rotating sphere.

Can the spin-lattice model H_H^N on S^2 support phase transitions? More precisely, we check what the Mermin-Wagner theorem has to say about H_H^N. This theorem, also known as the Mermin-Wagner-Hohenberg or as the Coleman theorem, shows that phase transitions cannot happen in two- or lower-dimensional systems that have continuous symmetries and finite interactions.

What we know about H_H^N is

1. It has spatial dimension $d=2$,
2. It has a continuous global symmetry group. Namely, for each element $g\in SO(3)$,

$$H_H^N(g\mathbf{s})=H_H^N(\mathbf{s}) \tag{13.54}$$

3. It has an infinite-range interaction. That is, for any sequence of uniform lattices of N sites on S^2,

$$\lim_{N\to\infty}\frac{4\pi}{N}\sum_{j=1}^{N}J_{j,k}=-\infty \tag{13.55}$$

Properties 1 and 2 by themselves would imply via the Mermin-Wagner theorem that H_H^N does not support phase transitions. Since all $d\leq 2$, finite range models with a continuous symmetry group cannot have phase transitions. However, property 3 violates the finite range condition of this theorem. Hence, H_H^N on S^2 can in principle have phase transitions in the thermodynamic limit.

It is interesting to compare this vector model H_H^N on S^2 with the Ising-type model H_N for the same barotropic flow in the last section. The Mermin-Wagner theorem there allows H_N to have phase transitions in the thermodynamic limit for a different reason. There the Ising type interaction $J_{j,k}=\frac{16\pi^2}{N^2}\log\left(1-\cos\left(\gamma_{j,k}\right)\right)$ has finite range instead of infinite range, but H_N does not have a continuous symmetry group, only the discrete symmetry

Z_2. The Ising type of the spherical model H_N in the last section is emphasized in the one-step renormalization derivation of the mean field theory for rotating barotropic flows by the Bragg method [269].

The extension of the Kac-Berlin method to this vector spherical model was used to calculate a closed form expression for its partition function in the non-extensive continuum limit, which share many properties of the exact solution for the spherical model that was formulated in the inertial frame [258] and discussed in chapter 7. Phase transitions from disordered states without global angular momentum to organized large-scale flows with net angular momentum occurs at finite negative and positive values of the scaled statistical temperature and was found to persist for all values of relative enstrophy, independent of the angular velocity of the rotating frame used in the formulation of the models [259].

13.9 Remarks

This shows the relatively simple Coupled Barotropic Vorticity - Rotating Sphere model will support non-degenerate super and sub-rotating steady state solutions, provided the rate of spin Ω is positive. These are completely characterized by the prograde and retrograde solid-body vorticity distributions $w_M^{\pm} = \pm\sqrt{Q}\psi_{1,0}$, which are, moreover, non-linearly stable in certain well-defined and verifiable parameter regimes. We derived a complete picture of possible bifurcations as either (a) the rate of spin Ω is increased with relative enstrophy fixed at Q, or (b) the relative enstrophy is increased with spin fixed.

One of the significant physical consequence of this work is that the vertically averaged barotropic component of a more realistic damped driven atmospheric system is expected to have asymptotic states close to solid-body rotational flows. These states furthermore have the defining characteristics that, (i) only when the kinetic energy H exceeds $H_c(\Omega, Q)$ are super-rotating asymptotic states observable; and (ii) the sub- rotating asymptotic state is allowed or observable for arbitrarily small kinetic energy H provided the relative enstrophy Q does not exceed the threshold $Q_c(\Omega) = \Omega^2 C^2$.

In a heuristic sense, the rotating sphere theorem gives verifiable necessary and sufficient conditions relating three key bulk properties of the barotropic component of a damped driven stratified rotating flow in order for super and sub-rotational quasi-steady states to be realized by the long time asymptotics of atmospheric balance.

At this point, it is not clear that a rigorous theory can be constructed for super-rotating asymptotic states which involves the full Initial Value Problem of a damped, driven, stratified, rotating atmospheric system. Nonetheless, there is much room for improvement of this variational model without compromising the rigor of the subsequent analysis [395]. One of this directions is to construct a similar variational theory for the steady states of the Coupled

Shallow Water equations on a rotating sphere [111]. Another promising direction has been taken by Newton and Shokraneh [326] who derived the correct dynamical point vortex model on a rotating sphere.

References

1. Ralph Abraham, Jerrold E Marsden, *Foundations of Mechanics: A mathematical exposition of classical mechanics with an introduction to the qualitative theory of dynamical systems and applications to the three-body problem.* W. A. Benjamin, Inc, New York, 1967.
2. L A Ahlfors, *Complex analysis,* McGraw-Hill, New York,1979.
3. M P Allen, D J Tildesley, *Computer simulation of liquids,* Oxford University Press, 1987.
4. Eric Lewin Altschuler, Timothy J Williams, Edward Ratner, Robert Tipton, Richard Stong, Farid Dowla, Frederic Wooten. *Possible minimum lattice configurations for Thompson's problem of charges on a sphere,* Physical Review Letters Volume 78 Number 14 (7 April 1997) 2681-2685.
5. T Andersen and C C Lim, *Trapped slender vortex filaments - statistical mechanics,* preprint 2006, presented at the GAMM Berlin 2006.
6. David G Andrews, *An Introduction to Atmospheric Physics,* Cambridge University Press, 2000.
7. H Aref, *Integrable, chaotic and turbulent vortex motion in two-dimensional flows,* Annu Rev Fluid Mech 15 (1983) 345-389
8. H Aref, *Motion of three vortices revisited,* Phys Fluids Vol 31 1988 No 6 1392-1409.
9. H Aref, *On the equilibrium and stability of a row of point vortices,* J Fluid Mech 290 (1995) 167-191.
10. H Aref, *Point vortex motions with a center of symmetry,* Phys Fluids 25 (12) (1982) 2183-2187.
11. H Aref, *Three vortex motion with zero total circulation – addendum,* J App Math Phys (ZAMP) 40 (1989) 495-500.
12. H Aref, M Bröns, *On stagnation points and streamline topology in vortex flows,* J Fluid Mech 370 (1998) 1-27.
13. Hassan Aref, James B Kadtke, Ireneusz Zawadzki, Laurence J Campbell, Bruno Eckhardt, *Point vortex dynamics: recent results and open problems,* Fluid Dynamics Research 3 (1988) 63-74.
14. H Aref, N Pomphrey, *Integrable and chaotic motions of four vortices I. The case of identical vortices,* Proc R Soc Lond, A380 (1982) 359-387.
15. H Aref, M Stremler, *On the motion of three point vortices in a periodic strip,* J Fluid Mech, 314 (1996) 1-25

16. H Aref, D L Vainchtein, *Point vortices exhibit asymmetric equilibria,* Nature 392 (23 April 1998) 767-770.
17. V I Arnold, V V Kozlov, A I Neishtadt, *Mathematical aspects of the classical and celestial mechanics,* Moscow: VINITI. Results of a science and engineering. Contemporary problems of mathematics. Fundamental directions. Vol 3 (1985) 304-
18. V I Arnold, V V Kozlov, A I Neishtadt, *Mathematical aspects of classical and celestial mechanics,* 2nd Ed, Encyclopedia of mathematical Sciences, v 3 (Dynamical Systems III), New York, Springer, 1996.
19. V I Arnold, *Conditions for non-linear stability of stationary plane curvilinear flows of an ideal fluid,* Sov. Math Dokl., 6, 773-777, 1965.
20. V I Arnold, *Geometrical methods in the theory of ordinary differential equations,* 2nd Ed, Springer-Verlag, New York, 1988.
21. V I Arnold, *Mathematical methods of classical mechanics,* New York, Springer v.60, 1984.
22. V I Arnold, *On an a priori estimate in the theory of hydrodynamical stability,* AMS Translations, Ser. 2, 79, 267-269, 1969.
23. V I Arnold, A B Givental, *Symplectic Geometry,* M., VINITI, v 4, (1985), 5-140.
24. George Arfken, *Mathematical Methods for Physicists,* 3rd Edition. Academic Press, San Diego, California, 1985.
25. Isaac Asimov, *Asimov's Chronology of Science and Discovery.* Harper and Row Publishers, New York City, 1989.
26. Syed M Assad, Chjan C Lim, *Monte Carlo simulation of baroclinic vortices in the Heton model,* Proc. of the 3rd MIT Conf on Comp Fluid and Solid Mech., Elsevier, Cambridge, June 2005.
27. Syed M Assad, Chjan C Lim, *Statistical equilibrium of the Coulomb/vortex gas on the unbounded two-dimensional plane,* Discrete and Continuous Dynamical Systems B, 5(1), 1-14, 2005.
28. Syed M Assad and Chjan C Lim, *Statistical Equilibrium Distributions of Baroclinic Vortices in the Rotating Two-layer model at low Froude numbers,* Geo and Astro Fluid Dyn accepted for publication 2006.
29. A Babiano, G Boffetta, G Provenzale, A Vulpiani, *Chaotic advection in point vortex models and two-dimensional turbulence,* PHys Fluids 6 (1994) 2465-2474.
30. A A Bagrets, D A Bagrets, *Non-integrability of two problems in vortex dynamics,* Chaos Vol 7 Num 3 (1997) 368-375.
31. Radu Balescu, *Equilibrium and non-equilibrium statistical mechanics,* John Wiley and Sons, New York City, 1975.
32. A Barut, R Raczka, *Theory of Group Representations and Applications,* PWN, Polish Scientific Publishers (1977).
33. G K Batchelor, *Computation of the energy spectrum in homogeneous two-dimensional turbulence,* Phys Fluid Suppl 12 (1969) 233-
34. G K Batchelor, *The theory of homogeneous turbulence,* Cambridge University Press, Cambridge, 1953.
35. G K Batchelor, *An introduction to fluid mechanics,* Cambridge University Press, Cambridge, 1967.
36. R J Baxter, *Exactly Solved Models in Statistical Mechanics,* Academic Press, London, 1982.

37. J T Beale, A J Majda, *Vortex Methods I*, Math Comp 39 (1982) 1-
38. T B Benjamin, *Theory of the vortex breakdown phenomenon,* J Fluid Mech 14 (1962) 593-
39. B Bergersen, D Boal, and P Palffy-Muhoray, *Equilibrium configurations of particles on a sphere: the case of logarithmic interactions.* J Phys A: Mah Gen 27 (1994) 2579-2586.
40. José M Bernardo, Adrian F M Smith, *Bayesian Theory.* John Wiley and Sons, Chichester, England, 1994.
41. T H Berlin, M Kac, *The Spherical Model of a Ferromagnet*, Physical Review 86, (1952) 821 - 835.
42. Jeff Biggus, *Sketching the History of Statistical Mechanics and Thermodynamics.* http:// history.hyperjeff.net/ statmech.html, visited May 2003.
43. Arthur Beiser, *Concepts of Modern Physics.* 4th Edition, McGraw-Hill, New York City, 1987.
44. J J Binney, N J Dowrick, A J Fisher, M E J Newman, *The Theory of Critical Phenomena,* Oxford University Press, Oxford, 1995.
45. Denis Blackmore, Jyoti Champanerkar, Chengwen Wang. *A generalized Poincaré-Birkhoff theorem with applications to coaxial vortex ring motion,* Discrete and Continuous Dynamical Systems B 5 (2005) 15-48.
46. D Blackmore, O Knio, *KAM theory analysis of the dynamics of three coaxial vortex rings,* Physica D 140 (2000) 321-348
47. S Boatto, R T Pierrehumbert, *Dynamics of a passive tracer in a velocity field of four identical point vortices,* J Fluid Mech 394 (1999) 137-174.
48. G Boer, T Shepherd, *Large scale two-dimensional turbulence in the atmosphere,* J Atmos Sci 40 (1983) 164-184.
49. A V Bolsinov, *Compatible Poisson brackets on Lie algebras and the completeness of families of involutive functions,* Izv AN USSR, ser math, v 55 (1991) No 1, 68-92.
50. A V Bolsinov, A V Borisov,*Lax representation and compatible Poisson brackets on Lie algebras,* (to be published) 1999.
51. A V Bolsinov, I S Mamaev, *Lie Algebras in Vortex Dynamics and Celestial Mechanics IV.*, Regular and Chaotic Dynamics Vol 4 Num 1 (1999) 23-50.
52. A V Borisov, K V Emelyanov, *Non-integrability and stochastisity in rigid-body dynamics,* Izhevsk, Izd Udm Univ, 1995.
53. A V Borisov, A E Pavlov, *Dynamics and Statics of Vortices on a Plane and a Sphere I,* Regular and Chaotic Dynamics Vol 3 Num 1 (1998) 28-38.
54. A V Borisov, V G Lebedev, *Dynamics of Three Vortices on a Plane and a Sphere II. General compact case,* Regular and Chaotic Dynamics Vol 3 Num 2 (1998) 99-114.
55. A V Borisov, V G Lebedev, *Dynamics of Three Vortices on a Plane and a Sphere III. Non-compact Case,* Regular and Chaotic Dynamics Vol 3 Num 4 (1998) 76-90.
56. M Bowick, A Cacciuto, D R Nelson, A Traesset. *Crystalline order on a sphere and the generalized Thompson problem,* http:// arxiv.org/ PS_cache/ cond-mat/ pdf/ 0206/ 0206144.pdf
57. H Brands, P H Chavanis, R Pasmanter, J Sommeria, *Maximum entropy versus minimum enstrophy vortices,* PHys Fluids 11(11) (1999) 3465-3477.
58. Pierre Brémaud, *Markov Chains: Gibbs Fields, Monte Carlo Simulations, and Queues.* Springer-Verlag, New York City, 1999.

59. V A Bogomolov, *Dynamics of vorticity at a sphere,* Fluid Dyn 6 (1977) 863-
60. V A Bogomolov, *Two dimensional fluid dynamics on a rotating sphere,* Izv Atmos Ocean Phys 15 (1979) 18-
61. V A Bogomolov, *On the motion of a vortex on a rotating sphere,* Izv Atmos Ocean Phys 21 (1985) 298-
62. L N Brillouin, *Science and information theory,* Academic Press, New York, 1962.
63. William J Broad, *M N Rosenbluth, 76, an H-Bomb Developer, Is Dead.* The New York Times, 30 September 2003.
64. Stephen G Brush, *The Kind of Motion We Call Heat: A History of the Kinetic Theory of Gases in the 19th Century.* North-Holland Pub Company, Amsterdam, 1976.
65. Richard L Burden, J Douglas Faires, *Numerical Analysis.* 5th Edition, Prindle, Weber and Schmidt, Boston, 1993.
66. R Caflisch, *Mathematical analysis of vortex dynamics,* Proc Workshop on Math Aspects of Vortex Dynamics, 1-24, Society for Industrial and Applied Math, Philadelphia, 1988.
67. E Caglioti, P L Lions, C Marchioro, M Pulvirenti, *A special class of stationary flows for two-dimensional Euler equations: a statistical mechanics description,* Commun Math Phys 143 (1992) 501-525.
68. E Caglioti, P L Lions, C Marchioro, M Pulvirenti, *A special class of stationary flows for two-dimensional Euler equations: a statistical mechanics description: Part II,* Commun Math Phys 174 (1995) 229-260.
69. Florian Cajori, *A History of Mathematical Notations.* Open Court Publishing Company, 1928.
70. E F Cladin, *Chemical thermodynamics,* Clarendon Press, Oxford, 1961.
71. F Calogero, *Integrable many-body problems,* Lectures given at NATO Advanced Study Institute on Non-linear Equations in Physics and Mathematics, Istanbul, August 1977, 3-53.
72. L J Campbell, K O'Neil, *Statistics of two-dimensional point vortices and high energy vortex states,* J Stat Phys 269 (1993)
73. L J Campbell, R M Ziff, *A catalogue of two-dimensional vortex patterns,* Los Alamos Sci Lab Rep Num LA-7384-MS P 40.
74. L J Campbell, R M Ziff, *Vortex patterns and energies in a rotation super-fluid,* Phys Rev B Vol 20 Num 5 (1979) 1886-1901.
75. B Cantwell, *Organized motion in turbulent flow,* An Rev Fluid Mech 13 (1981) 457-515.
76. Toby N Carlson, *Mid-Latitude Weather Systems.* Harper-Collins Academic, London, 1991.
77. S Chandrasekhar, *Hydrodynamic and Hydromagnetic Stability,* Oxford University Press, New York City, 1961.
78. C L Charlier, *Die Mechanik des Himmels,* Walter de Gruyter and Company, 1927.
79. D Chandler, *Introduction to statistical mechanics,* Oxford University Press, 1987.
80. J G Charney, *Geostrophic turbulence,* J Atmos Sci 28 (1971) 1087-1985.
81. P H Chavanis, *From Jupiter's great red spot to the structure of galaxies: statistical mechanics of two-dimensional vortices and stellar systems,* in *Non-linear dynamics and chaos in astrophysics: A festschriff in honor of George Contopoulas,* Annals of the New York Academy of Sciences 867 (1998) 120-140.

82. P H Chavanis, *Systematic drift experienced by a point vortex in two-dimensional turbulence,* Phys Rev E 58(2) (1998) R1199-R1202.
83. S J Chern, *Math Theory of the Barotropic Model in GFD,* PhD thesis, Cornell University, 1991.
84. J Cho and L Polvani, *The emergence of jets and vortices in freely evolving, shallow-water turbulence on a sphere,* Phys Fluids 8(6), 1531 - 1552, 1995.
85. A J Chorin. *Vorticity and Turbulence*, Springer-Verlag, New York City, 1994.
86. A J Chorin, *Partition functions and equilibrium measures in two dimensional and quasi three dimensional turbulence,* Phys Fluids 8 (1996) 2656-
87. A J Chorin, Ole Hald, *Vortex renormalization in three space dimensions,* Phys Rev B 51 (1995) 11969-11972.
88. A J Chorin, J E Marsden, *A Mathematical Introduction to Fluid Mechanics.* Springer-Verlag, New York City, 1979.
89. A T Conlisk, Y G Guezennec, G S Elliot, *Chaotic motion of an array of vortices above a flat wall,* Phys Fluids A1(4) (1989) 704-717.
90. A T Conlisk, D Rockwell, *Modelling of vortex-corner interactions using point vortices,* Phys Fluids 24 (1981) 2133-2142.
91. P Constantin, *Geometric and analytical studies in turbulence,* in *Trends and perspectives in applied mathematics,* Edited by L Sirovich, Springer-Verlag, New York City, 1993.
92. *Concise Dictionary of Scientific Biography,* 2nd Edition, Timothy J DeWerf, Managing Editor. Charles Scribner's Sons, New York City, 2000.
93. P Constantin, Ch Fefferman, *Scaling exponents in fluid turbulence: some analytic results,* Nonlinearity 7 (1994) 41-57.
94. P Contantin, C Foias, I Kukavica, A Majda, *Dirichlet quotients and periodic two-dimensional Navier-Stokes equations,* J Math Pure Appl 76 (1997) 125-
95. I P Cornfeld, S V Fomin, Ya G Sinai. *Ergodic Theory,* Springer-Verlag, New York City, 1982.
96. H S M Coxeter, LL D, FRS, *Regular Polytopes 2nd Edition.* MacMillan Company, New York, 1963.
97. R Courant and D Hilbert, *Methods of Mathematical Physics*, John Wiley and Sons, New York City, 1953.
98. R Courant and D Hilbert, *Methods of Mathematical Physics*, John Wiley and Sons, New York City, 1962.
99. S Crow, F Champagne, *Orderly structure in jet turbulence,* J Fluid Mech 48 (1971) 547-
100. U Dallman, *Three-dimensional vortex structures and vorticity topology,* Fluid Dyn Res 3 (1988) 183-189.
101. Charles Darwin, Ralph Howard Fowler. *On the partition of energy,* Philos Mag 44, 450479 (1922).
102. Robert L Devaney, *Singularities in classical mechanical systems*, Ergodic Theory and Dynamical Systems I Proceedings, Special Year, Maryland, 1979-80, A Katok, Editor. Birkhäuser, Boston, 1981.
103. D A Dawson, *Stochastic models for complex systems,* Entropy and Ergodic Theory, University Press of Canada-Toronto, 1975.
104. P Deift, *Orthogonal polynomials and random matrices: a Riemann-Hilbert approach,* Courant Lecture Notes 3, AMS Press, 1999.
105. K G Denbigh, *The principles of chemical equilibrium,* Cambridge University Press, 1971.

106. Mark T DiBattista, A Majda, *An equilibrium statistical model for the spreading phase of open-ocean convection,* Proc Natl Acad Sci 96 (1999) 6009-6013.
107. Mark T DiBattista, Andrew J Majda. *Equilibrium statistical predictions for baroclinic vortices: the role of angular momentum.* Theoretical and Computational Fluid Dynamics 14 (2001) 293-322.
108. Mark T DiBattista, Andrew J Majda, *An Equilibrium Statistical Theory for Large-Scale Features of Open-Ocean Convection,* J Phys Oceanography, 30 (2000), 1325 - 1353.
109. Mark T DiBattista, L Polvani, *Barotropic vortex pairs on a rotating sphere,* J Fluid Mech 358 (1998) 107-
110. X Ding, *Statistical equilibria of barotropic vorticity and the shallow water model on a rotating sphere,* PhD thesis Rensselaer Polytechnic Institute, 2007.
111. X Ding and Chjan C Lim, *A Variational theory of the steady states of the SWE on a rotating sphere,* preprint, 2006.
112. X Ding and Chjan C Lim, *Phase transitions in a Spherical energy-relative enstrophy model for barotropic flows on a rotating sphere,* accepted for publication in Physica A, 2006.
113. R DiPerna, A Majda, *Concentrations and regularizations for 2D incompressible flow,* Comm Pure App Math 40 (1987) 301-345.
114. Malcolm Dole, *Introduction to Statistical Thermodynamics.* Prentice-Hall, New York City, 1954.
115. D G Dritschel, *A fast contour dynamics method for many-vortex calculations in two-dimensional flows,* Phys Fluids A 5 (1993) 173-
116. D G Dritschel, L M Polvani, *The roll-up of vorticity strips on the surface of a sphere,* J Fluid Mech, 234 (1992) 47-
117. F J Dyson, *Statistical theory of the energy levels of complex systems I, II, II,* J Math Phys 3 (1962) 140-156, 157-165, 166-175.
118. B Eckhardt, *Irregular scattering of vortex pairs,* Europhys Lett 5(2) (1988) 107-111.
119. B Eckhardt, *Integrable four vortex motion,* Phys Fluids Vol 31(10) 1988, 2796-2801.
120. B Eckhardt, H Aref, *Integrable and chaotic motion of four vortices II: Collision dynamics of vortex pairs,* Philos Trans R Soc London Vol 31(10), (1988) 2796-2801.
121. I Ekeland, R Temam, *Convex analysis and variational problems,* North-Holland, 1976.
122. I Ekeland and R Temam, *Convex Analysis and Variational Problems,* North-Holland, 1976.
123. A R Elcrat, C Hu, K G Miller, *Equilibrium configurations of point vortices for channel flow past interior obstacles,* Eur J Mech B/Fluids 16 (1997) 277-
124. A R Elcrat, K G Miller, *Rearrangements in steady multiple vortex flows,* Comm in PDE 20 (1995) 1481-1490.
125. G J Erikson, C K Smith (Eds), *Maximum Entropy and Bayesian models in science and engineering,* Kluwar, Dordrecht, 1988.
126. G L Eyink, H Spohn, *Negative-temperature states and large-scale, long-lived vortices in two-dimensional turbulence,* J Stat Phys 70 (1993) 833-886.
127. Marie Farge, *Vortex motion in a rotating barotropic fluid layer,* Fluid Dynamics Research 3 (1988) 282-288.
128. R P Feynman, *Statistical Mechanics,* Addison-Wesley, Massachusetts, 1998.

129. K S Fine, A C Cass, W G Flynn, C F Driscoll, *Relaxation of 2D turbulence to vortex crystals,* Phys Rev Lett 75 (18) (1995), 3277-3280.
130. Faculty web page of Michael E Fisher. http:// www.ipst.umd.edu/ Faculty/ fisher.htm visited February 2004.
131. M Flucher, *Variational problems with concentration, Progress in non-linear differential equations and their application* 36 (1999) Birkhäuser, Basel.
132. A T Fomenko, *Symplectic Geometry,* GU, 1998.
133. R H Fowler, E A Guggenheim, *Statistical thermodynamics,* Cambridge University Press, 1949.
134. D G Fox, S A Orszag, *Inviscid dynamics of two-dimensional turbulence,* Phys Fluids 16 (1971) 169-
135. S Friedlander, *Interaction of vortices in a fluid on the surface of a rotating sphere,* Tellus XXVII (1975) 15-
136. J Frohlich, D Ruelle, *Statistical mechanics of vortices in an inviscid two-dimensional fluid,* Comm Math Phys 87 (1982) 1-36.
137. Robert P H Gasser, W Graham Richards, *An introduction to statistical thermodynamic,* World Scientific, Singapore, 1995.
138. I M Gel'fand, *Generalized Functions,* Academic Press, New York, 1964.
139. I M Gel'fand, S V Fomin, *Calculus of Variations,* Prentice-Hall, Englewood Cliffs, NJ, 1963.
140. S Geman, *The spectral radius of large random matrices,* Ann Prob 14 (1986) 1318-1328.
141. A Del Genio, W Zhuo and T Eichler, *Equatorial Super-rotation in a slowly rotating GCM: implications for Titan and Venus,* Icarus 101, 1-17, 1993.
142. James E Gentle, *Random Number Generation and Monte Carlo Methods.* 1st edition, Springer-Verlag, New York, 1998.
143. M Gharib, E Rambod, K Shariff, *A universal time scale for vortex ring formation,* J Fluid Mech 360 (1998) 121-
144. A E Gill, *Atmosphere – Ocean Dynamics,* International Geophysics Series Volume 30, Academic Press, New York City, 1982.
145. J Ginibre, *Statistical Ensembles of Complex, Quaternion, and Real Matrices,* J Math Phys 6(3), 440 - 449.
146. Herman Goldstein, *Classical Mechanics.* Addison-Wesley, 1950.
147. G M Goluzin, *Geometric theory of functions of a complex variable,* American Mathematical Society, Providence, Rhode Island, 1969.
148. J Goodman, *Convergence of the random vortex method,* CPAM 40 (1987) 189-
149. J Goodman, T Y Hou, J Lowengrub, *Convergence of the point vortex method for 2-D Euler equations,* Comm Pure App Math 43 (199) 415-430.
150. D N Goryachev, *On some cases of the motion of straight line vortices,* Moskwa, 1898.
151. Ya L Granovski, A S Zhedanov, I M Lutsenko, *Quadratic algebras and dynamics in a curved space II. Kepler's problem,* Teor i mat fiz Vol 91 Num 3 (1992) 396-410.
152. G H Grant, W G Richards, *Computational Chemistry,* Oxford University Press, 1995.
153. A G Greenhill, *Plane vortex motion,* Quat J of Pure Appl Math (1878) Vol 15, 10-29.
154. H P Greenspan, *The Theory of Rotating Fluids,* Cambridge University Press, 1968.

155. W Gröbli, *Specialle probleme über die Bewegung Geredliniger paralleler Wirbelfäden,* Vierteljahrsch. D. Naturforsch. Geselsch. Zürich. V. 22 (1887) p 37-81, 129-165.
156. I S Gromeka, *On vortex motions of a liquid on a sphere,* Collected papers, Moscow, AN USSR 1952.
157. Richard Grotjahn, *Global Atmospheric Circulations: Observations and Theories.* Oxford University Press, Oxford, 1993.
158. V M Gryanik, *Dynamics of singular geostrophic vortices in a two-level model of the atmosphere (or ocean),* Bull Izv Acad Sci USSR, Atmospheric and Oceanic Physics 19(3) (1983) 171-179.
159. V M Gryanik, textitDynamics of localized vortex perturbations – "vortex changes" in a baroclinic fluid, Isvestiya, Atmospheric and Oceanic Physics, 19(5) (1983) 347-352.
160. B Gutkin, U Smilansky, E Gutkin, *Hyperbolic billiards on surfaces of constant curvature,* Comm Math Phys 208 (1999) 65-90.
161. D Ter Haar, *Elements of Thermostatistics.* 2nd Edition, Holt, Reinhart and Winston, New York City, 1966.
162. O Hald, *Convergence of vortex methods for Euler's equations III,* SIAM J Num Anal 24 (1987) 538-
163. O H Hald, *Convergence of vortex methods,* in *Vortex methods and vortex motion,* eds K E Gustafson, J A Sethian, SIAM, Philadelphia, 1991.
164. D Hally, *Stability of streets of vortices on surfaces of revolution with a reflection symmetry,* J Math Phys 21 (1980) 211-
165. J P Hansen and I R McDonald, *Theory of simple liquids, 2nd ed.*, Academic Press, London, 1986.
166. J M Hammersley and D C Handscomb, *Monte Carlo Methods.* Methuen & Co, London; John Wiley & Sons, New York City, 1964.
167. J K Harvey, F J Perry, *Flow field produced by trailing vortices in the vicinity of the ground,* AIAA Journal 9 (1971) 1659-1660.
168. L G Hazin, *Regular polygons of point vortices and resonance instability of stationary states,* DAN USSR Vol 230 Num 4 (1976) 799-802.
169. Hidenori Hasimoto, *Elementary aspects of vortex motion,* Fluid Dynamics Research 3 (1988) 1-12.
170. T H Havelock, *The stability of motion of rectilinear vortices in ring formation,* Phil Mag 7(11) (1931) 617-633.
171. D Hilbert, *S Cohn-Vossen,* Anschauliche Geometrie. Berlin, 1932.
172. T L Hill, *An introduction to statistical thermodynamics,* Dover, New York, 1986.
173. C Ho, P Huerre, *Perturbed free shear flows,* Annu Rev Fluid Mech 16 (1984) 365-
174. G Holloway, M Hendershott, *Stochastic closure for non-linear Rossby waves,* J Fluid Mech 82 (1977) 747-765.
175. N G Hogg, H M Stommel, *The Heton: an elementary interaction between discrete baroclinic vortices, and its implication concerning eddy heat flow,* Proc Roy Soc Lond A397 (1985) 1-20.
176. N G Hogg, H M Stommel, *Hetonic explosions: the break-up and spread of warm pools as explained by baroclinic point vortices,* J Atmos Sci 42 (1985) 1465-1476.
177. D D Holm, J E Marsden, T S Ratiu, *Euler-Poincaré models of ideal fluids with non-linear dispersion,* Phys Rev Lett 349 (1998) 4173-4177.

178. D D Holm, J E Marsden, T S Ratiu, *Euler-Poincaré equations and semi-direct products with applications to continuum theories,* Adv in Math 137 (1998) 1-81.
179. D Holm, J Marsden, T Ratiu and A Weinstein, *Non-linear stability of fluid and plasma equilibria,* Physics Report 123, 1-116, 1985.
180. E Hopf, *Statistical hydromechanics and functional calculus,* J Rat Match Anal 1 (1952) 87-123.
181. T Y Hou, J Lowengrub, *Convergence of the point vortex method for three-dimensional Euler equations,* CPAM 43 (1990) 965.
182. R Iacono, *On the existence of Arnold-stable barotropic flows on a rotating sphere*, Phys Fluids, 15(12), 3879-3882, 2003.
183. Ernst Ising Obituary, web page, http://www.bradley.edu/ las/ phy/ personnel/ isingobit.html, visited June 2005.
184. Claude Itzykson, Jean-Michel Drouffe, *Statistical Field Theory,* Cambridge University Press, 1989.
185. C G J Jacobi, *Vorlesungen über dynamic,* Aufl 2 Berlin: G Reimer, 1884, 300 S.
186. E Atlee Jackson, *Equilibrium Statistical Mechanics.* Prentice-Hall, Englewood Cliffs, New Jersey, 1968.
187. J D Jackson, *Classical Electrodynamics,* Wiley, New York City, 1963.
188. M Jamaloodeen, *Hamiltonian methods for some geophysical vortex dynamics models,* PhD thesis, Department of mathematics, University of Southern California, 2000.
189. R Jastrow, R I Rasool Eds, *The Venus atmosphere,* Gordon and Breach, 1969.
190. E T Jaynes, in *Maximum entropy and Bayesian methods in science and engineering,* Volume 1, Edited by G J Erikson and C R Smith, Kluwer, Dordrecht, 1988.
191. E T Jaynes, *Where do we stand on maximum entropy,* in R D Levine, M Tribus, Eds, *The maximum entropy formalism,* MIT Press, Cambridge, 1979.
192. J H Jeans, *The mathematical theory of electricity and magnetism,* 5th edition, Cambridge University Press, Cambridge, 1933.
193. R Jordan, B Turkington, *Ideal magneto-fluid turbulence in two dimensions,* J Stat Phys 87 (1997) 661-
194. G Joyce, D Montgomery, *Negative temperature states for the two-dimensional guiding center plasma,* J Plasma Phys 10 (1973) 107-121.
195. E R van Kampen, A Wintner, *On a symmetrical canonical reduction of the problem of three bodies,* Amer J Math Vol 59 Num 1 (1937) 153-166.
196. M Karweit, *Motion of a vortex pair approaching an opening in a boundary,* Phys Fluids 18 (1975) 1604-1606.
197. T Kato, J T Beale, A Majda, *Remarks on the breakdown of smooth solutions for the three-dimensional Euler equations,* Comm Math Phys 94 (1989) 61-
198. Yitzhak Katznelson, Benjamin Weiss. *When all points are recurrent/ generic,* Ergodic Theory and Dynamical Systems I Proceedings, Special Year, Maryland, 1979-80, A Katok, Editor. Birkhäuser, Boston, 1981.
199. Joseph H Keenan, *Thermodynamics,* MIT Press, Cambridge, Massachusetts, 1970.
200. Joseph B Keller, *The Scope of the Image Method,* Communications on Pure and Applied Mathematics (VI) 1953, 505-512.
201. Lord Kelvin, *On vortex atoms,* Proc R Soc Edinburg 6 (1867) 94-105.
202. Lord Kelvin, *On vortex motion,* Trans Royal Soc Edinburgh 25 (1868) 217-260.

203. K M Khanin, *Quasi-periodic motions of vortex systems,* Physica 4 D Vol 3 (1982) 261-269.
204. K M Khazin, *Regular polygons of point vortices,* Sov Phys Dokl 21 (1976) 567-569.
205. S Kida, *Statistics of the system of line vortices,* J Phys Soc Japan 39 (1975) 1395-3404.
206. Rangachari Kidambi, Paul K Newton, *Point vortex motion on a sphere with solid boundaries,* Physics of Fluids 12 (2000) 581-588.
207. Rangachari Kidambi, Paul K Newton, *Motion of three point vortices on a sphere*, Physica D 116 (1998) 143 - 175.
208. Rangachari Kidambi, Paul K Newton, *Collision of three vortices on a sphere,* to appear in Nuovo Cimento, 2000.
209. Rangachari Kidambi, Paul K Newton, *Collapse of three vortices on a sphere,* Proc at Int Workshop on Vortex Dynamics in Geophysics Flow, Castro Marina, Italy, June 1998.
210. Rangachari Kidambi, Paul K Newton, *Streamline topologies for integrable vortex motion on a sphere*, Physica D 140 (2000) 95 - 125.
211. M Kiessling, *Statistical mechanics of classical particles with logarithmic interactions,* Commun Pure Appl Math 46 (1993) 27-56.
212. J Kim, P Moin, R K Moser, *Turbulence statistics in fully developed channel flow at low Reynolds numbers,* J Fluid Mech 177 (1987) 133-
213. Yoshifumi Kimura, *Chaos and collapse of a system of point vortices,* Fluid Dynamics Research 3 (1988) 98-104.
214. Yoshifumi Kimura, *Vortex motion on surfaces with constant curvature,* Proc R Soc London, Seris A 455 (1999) 245-259.
215. Y Kimura and H Okamoto, *Vortex motion on a sphere.* J Phys Soc Japan **56** 1987, 4203-4206
216. G Kirchoff, *Vorlesungen über mathematische Physik,* Leipzig (1891) 272 S.
217. F Kirwan, *The topology of reduced phase spaces of the motion of vortices on a sphere,* PHysica D 30 (1988) 99-
218. R Klein, A Majda and K Damodaran, *Simplified equation for the interaction for nearly parallel vortex filaments,* J. Fluid Mech 288 (1995) 201-
219. K V Klyatskin, G M Reznik, *Point vortices on a rotating sphere,* Oceanology 29 (1989) 12-
220. H Kober, *Dictionary of Conformal Representations,* Dover, New York City, 1952.
221. J Koiller, S Carvalho, R da Silva Rodriques, L C Concalves de Olivera, *On Aref's vortex motions with a symmetry center,* Physica D 16 (1985) 27-61.
222. Andrey Nikolaevich Kolmogorov, *The local structure of turbulence in incompressible viscous fluid for very large Reynolds numbers,* C R Acad Sci USSR 30 (1941) 301-
223. Andrey Nikolaevich Kolmogorov, *On the logarithmically normal law of distribution of the size of particles under pulverization,* Dokl Akad Nauk SSSR 31 (1941) 99-101.
224. Andrey Nikolaevich Kolmogorov, *A refinement of previous hypotheses concerning the local structure of turbulence in a viscous incompressible fluid at high Reynolds number,* J Fluid Mech 13 (1962) 82-85.
225. T P Konovalyuk, *Classification of interaction of a vortex pair with a point vortex in an ideal fluid,* Hydromechanics, resp meshved sbornik, AN USSR Vol 62 No 2(8) (1990) 588-597.

226. V V Kozlov, *Dynamical Systems X: General Theory of Vortices*, Encyclopedia of Mathematical Sciences volume 67, Springer-Verlag, 2003.
227. R H Kraichnan, *Statistical dynamics of two-dimensional flows.* J Fluid Mech **67** 1975, 155-175.
228. R H Kraichnan, *Inertial ranges in two-dimensional turbulence,* Phys Fluids 10 (1967) 1417-
229. R H Kraichnan, Y S Chen, *Is there a statistical mechanics of turbulence,* Physica D 37 Num 1-3 (1989) 160-172.
230. R H Krachnan, D Montgomery, *Two dimensional turbulence,* Rep Prog Phys 43(5) (1980) 547-619.
231. I Kunin, F Hussain, X Zhou, *Dynamics of a pair of vortices in a rectangle,* Int J Eng Sci 32 (1994) 1835-
232. R Kupferman, A Chorin, *One-sided polarization and renormalization flow in the Kosterlitz-Thouless phase transition,* SIAM J Applied Math 59 (1000).
233. L Kuznetsov, G M Zavlavsky, *Regular and chaotic advection in the flow field of a three-vortex system,* Phys Rev E 58(6) (1998) 7330-7349.
234. L Kuznetsov, G M Zavlavsky, *Passive particle transport in three-vortex flow,* Phys Rev E 61(4) 3777-3792.
235. M Lagally, *Über ein verfahren zur transformation ebener wirbelprobleme,* Math Z 10 (1921) 231-239.
236. H Lamb, *Hydrodynamic,* Dover, New York, 1932.
237. D Landau, K Binder, *Monte Carlo simulations,* Cambridge University Press.
238. L D Landau, E M Lifschitz, *Fluid Mechanics,* Pergamon Press, Oxford, 1959.
239. L D Landau, E M Lifschitz, *Mechanics,* Pergamon Press, Oxford, 1976.
240. L D Landau, E M Lifschitz, *Statistical Physics,* Pergamon Press, Oxford, 1980.
241. E Laura, *Sul moto parallelo ad un plano un fluido in cul vi sono N votioi elementari,* Atti della Reale Accad Torino 37 (1902) 369-476. bibitemlauraSulle E Laura, *Sulle equazioni differenziali canoniche del moto di un sistema di vortici elementari, rettilinei e paralleli in un fluido imcompressibile idefinito,* Atti della Reale Accad Torino 40 (1905) 296-312.
242. David A Lavis, George M Bell, *Statistical Mechanics of Lattice Systems 1: Closed-Form and Exact Solutions*, Springer-Verlag, 1999.
243. P D Lax, *Functional Analysis,* Wiley-Interscience, 2002.
244. J L Lebowitz, H Rose, E Sper, *Statistical mechanics of the non-linear Schrödinger equation,* J Stat PHys 50 (1988) 657-687.
245. T D Lee, *On some statistical properties of hydrodynamical and magnetohydrodynamical fields,* Q Appl Math 10 (1952) 69-
246. S Legg, J Marshall, *A Heton model of the spreading phase of open ocean deep convection,* j Phys Oceanography 23(6) (1993) 1040-1056.
247. S Legg, H Jones, M Visbeck, *A Heton perspective of baroclinic eddy transfer in localized ocean deep convection,* J Phys Oceanography 26 (1996) 2251-2266.
248. S Legg, J Marshall, *The influence of the ambient flow on the spreading of convective water masses,* J Mar Res 56 (1998) 107-139.
249. C Leith, *Diffusion approximation for two-dimensional turbulence,* Phys Fluids 11 (1968) 671-
250. R D Levine, M Tribus (Eds), *The maximum entropy formalism,* MIT Press, Cambridge, 1979.
251. D Lewis, T Ratiu, *Rotating n-gon/kn-gon vortex configurations,* J Non-linear Sci 6 (1996) 385-414.

252. E Lieb, M Loss, *Analysis,* 2nd edition, Graduate Studies in Mathematics vol 14, AMS.
253. M J Lighthill, *An introduction to Fourier analysis and generalized functions,* Cambridge Press, Cambridge, 1958.
254. Chjan C Lim, *Barotropic Vorticity Dynamics and Statistical Mechanics – Application to super-rotation,* AIAA 2005-2005-5060.
255. Chjan C Lim, *Properties of energy-enstrophy theories for two-dimensional vortex statistics,* in *Proceedings of the fourth IUTAM Conference,* Shanghai, China, August 2002. Edited by W-Z Chien, Shanghai University Press, Shanghai, 2002, 814-819.
256. Chjan C Lim, *Energy extremals and non-linear stability in an Energy-relative enstrophy theory for barotropic flows on a rotating sphere,* preprint, 2006.
257. Chjan C Lim, *Energy maximizers and robust symmetry breaking in vortex dynamics on a nn-rotating sphere,* SIAM J Applied Math, 65(6), 2005, 2093-2106.
258. Chjan C Lim, *Exact solution of the spherical model for Barotropic Vortex statistics and negative critical temperature,* preprint 2006.
259. Chjan C Lim, *A Heisenberg model for Barotropic Vortex statistics on a rotating sphere and condensation of energy into super-rotating ground states,* preprint 2006, presented at the IUTAM Symp Moscow, August 2006.
260. Chjan C Lim, *A combinatorial perturbation method and Arnold whiskered tori in vortex dynamics,* Physica 64D (1993) 163-184.
261. Chjan C Lim, *A long range spherical model and exact solutions of an energy enstrophy theory for two-dimensional turbulence,* Physics of Fluids 13 (2001) 1961-1973
262. Chjan C Lim, *Mean field theory and coherent structures for vortex dynamics on the plane,* Physics of Fluids 11 (1999) 1201-1207, 3191.
263. Chjan C Lim, *Non-existence of Lyapunov functions and the instability of the von Kármán vortex streets,* Phys Fluids A 5 (9), (1993) 2229 - 2233.
264. Chjan C Lim, *Coherent Structures in an Energy-Enstrophy Theory for Axisymmetric Flows.* Phys Fluids, 15 (2003) 478-487.
265. Chjan C Lim, *Relative equilibria of symmetric n-body problems on the sphere: Inverse and Direct results.* Comm Pure and Applied Math **LI** 1998, 341-371.
266. Chjan C Lim, *Statistical equilibrium in a simple Mean Field Theory of Barotropic Vortex Dynamics on a Rotation Sphere,* Submitted for publication, 2006.
267. Chjan C Lim, Syed M Assad. *Self-containment radius for low temperature single-signed vortex gas and electronic plasma,* Regular and Chaotic Dynamics, 10(3), (2005), 239-255.
268. Chjan C Lim, Andrew Majda. *Point vortex dynamics for coupled surface/ interior quasi-geostrophic and propagating Heton clusters in models for ocean convection,* Geophys Astrophys Fluid Dyn 94(3-4) (2001) 177-220.
269. Chjan C Lim and R Singh Mavi, *Phase transitions of Barotropic Flow on the Sphere by the Bragg Method,* preprint 2006, presented at the AMS meeting Atlanta Feb 2006.
270. Chjan C Lim, James Montaldi, Mark Roberts, *Relative equilibria of point vortices on the sphere,* Physica D 148 (2001) 97-135.
271. Chjan C Lim and J Nebus, *The Spherical Model of Logarithmic Potentials As Examined by Monte Carlo Methods,* Phys Fluids, 16(10), 4020 - 4027, 2004.

272. Chjan C Lim, Joseph Nebus, Syed M Assad. *A Monte Carlo Algorithm for Free and Coaxial Ring Extremal States of the Vortex N-Body Problem on a Sphere.* Physica A 328 (2003) 53-96.
273. Chjan C Lim, Joseph Nebus, Syed M Assad. *Monte Carlo and Polyhedron based Simulations I: Extremal States of the Logarithmic N-Body Problem on a Sphere.* Discrete and Continuous Dynamical Systems B 3 (2003) 313-342.
274. Chjan C Lim, Lawrence Sirovich. *Non-linear vortex trail dynamics,* Physics of Fluids 31 (1988) 991-998.
275. C C Lin, *On the motion of vortices in 2D I. Existence of the Kirchoff-Routh Function,* Proc Natl Acad Sci 27 (1941) 570-575.
276. C C Lin, *On the motion of vortices in 2D II. Some further investigations on the Kirchoff-Routh function,* Proc Natl Acad Sci 27 (1941) 575-577.
277. C C Lin, *Theory of spiral structures,* Theoretical and Applied Mechanics, W T Koiter Ed., 57-69, North-Holland, Amsterdam, 1976.
278. P L Lions, *Mathematical topics in fluid mechanics Vol 1: Incompressible models,* Oxford Lecture Series in Mathematics and its Applications 3, Oxford, 1996.
279. P L Lions, *On Euler equations and statistical physics,* Galileo Chair 1: Scuola Normale Superiore, Classe di Scienze, Pisa, 1998.
280. P L Lions, A Majda, *Equilibrium statistical theory for nearly parallel vortex filaments,* Comm Pure Appl Math Vol LIII(1) (2000) 76-142.
281. Richmond W Longley, *Elements of Meteorology.* John Wiley and Sons, New York, 1976.
282. T S Lundgren, Y B Pointin, *Non-Gaussian probability distributions for a vortex fluid,* Phys Fluids 20(3) (1977) 356-363.
283. T S Lundgren, Y B Pointin, *Statistical mechanics of two-dimensional vortices,* J Stat Phys 17 (1977) 323-355
284. H J Lugt, *Wirbelströmung in Natur und Technik.* Braun, 1979.
285. D Lynden-Bell, *Statistical mechanics of violent relaxation in stellar systems,* Mon Not R Astr Soc 136 (1967) 101-121.
286. Lawrence Livermore National Laboratory: Edward Teller, web page http:// www.llnl.gov/llnl/06news/NewsMedia/teller_edward/teller_index.html visited January 2004.
287. Lars Onsager – Bibliography web page http:// nobelprize.org/ chemistry/ laureates/ 1968/ onsager-bio.html, visited October 2004.
288. The Mactutor History of Mathematics Archive, http:// www-gap.dcs.st-and.ac.uk/ ~history/ index.html visited May 2003-October 2005.
289. A J Majda, *Simplified asymptotic equations for slender vortex filaments,* in *Recent advances in partial differential equations, Venice 1996,* Proceedings of symposia in applied math, 54, R Spigerl, S Venakides Eds, 237, American Mathematical Society, Providence, 1998.
290. A J Majda, A Bertozzi, *Vorticity and incompressible fluid flow,* Cambridge Texts in Applied Mathematics 27, Cambridge Press, Cambridge, 2001.
291. A J Majda, M Holen, *Dissipation, topography, and statistical theories of large scale coherent structure,* Commun Pure Applied Math L (1997) 1183-1234.
292. M E Maltrud, G K Vallis, *Energy spectra and coherent structures in forced two-dimensional and beta-plane turbulence,* J Fluid Mech 228 (1991) 321-
293. Carlo Marchioro, Mario Pulvirenti, *Mathematical Theory of Incompressible Nonviscous Fluids*, Springer-Verlag, New York City, 1994.
294. H Marmanis, *The kinetic theory of point vortices,* Proc R Soc Lond A 454 (1998) 587-606.

295. Jerrold E Marsden, Tudor S Raitu, *Introduction to Mechanics and Symmetry.* 2nd Edition. Springer-Verlag, New York City, 1999.
296. J Marsden, A Weinstein, *Reduction of symplectic manifolds with symmetry,* Rep on Math Phys, Vol 5 Num 5 (1974) 121-130.
297. J Marshall, F Schott, *Open-ocean convection: observations, theory and models,* Rev Geophysics 37 (1999) 1-64.
298. A Masotti, *Atti Pont Accad Sci Nuovi Lincei* 84 (1931) 209-216, 235-245, 464-467, 468-473, 623-631.
299. J E Mayer, M G Mayer, *Statistical mechanics,* Wiley, New York, 1940.
300. Barry M McCoy and Tai Tsun Wu, *The two-dimensional Ising model,* Harvard University Press, Cambridge, Massachusetts, 1973.
301. P B Medawar, *The art of the soluble,* Methuen, London, 1967.
302. M L Mehta, *Random Matrices,* Academic Press, New York City, 1990.
303. V V Meleshko, M Yu Konstantinov, *Dynamics of vortex structures,* Kiev, Naukova Dumka, 1993.
304. G J Mertz, *Stability of body-centered polygonal configurations of ideal vortices,* Phys Fluids 21 (1978) 1092-1095.
305. I Mezić, I Min, *On the equilibrium distribution of like-signed vortices in two dimensions,* in *Proceedings of the Workshop on Chaos, Kinetics, and Non-linear Dynamics in Fluids and Plasmas,* Marseille, France, 1997.
306. J Michel, R Robert, *Large deviations for Young measures and statistical mechanics of infinite dimensional dynamical systems with conservation law,* Comm Math Phys 159 (1994) 195-215.
307. Reinhold Meise, Deitmar Vogt, *Introduction to Functional Analysis*, Oxford University Press, 1997.
308. L M Milne-Thompson, *Theoretical Hydrodynamics,* MacMillen, 1968.
309. J Miller, *Statistical mechanics of Euler equations in two dimensions,* Phys Rev Lett 65 (1990) 2137-2140.
310. J Miller, P Weichman, M C Cross, *Statistical mechanics, Euler's equations, and Jupiter's red spot,* Phys Rev A 45 (1992) 2328-
311. T Miloh, D J Shlein, *Passage of a vortex ring through a circular aperture in an infinite plane,* Phys Fluids 20 (1977) 1219-1227.
312. I A Min, I Mezić, A Leonard, *Lévy stable distributions for velocity and velocity difference in systems of vortex elements,* Phys Fluids 8(5) (1996) 1169-1180.
313. D Minnhagen, *The two-dimensional Coulomb gas, vortex unbinding, and superfluid-superconducting films,* Rev Mod Phys 59(4) (1987) 1001-1066.
314. D Montgomery and G Joyce, *Statistical mechanics of "negative temperature" states,* Phys Fluids 17 (1974) 1139-1145.
315. K Mohseni, *Statistical equilibrium theory for axisymmetric flows: Kelvin's variational principle and an explanation of the vortex ring pinch-off process,* Phys Fluids 13 (2001) 1924-
316. K Mohseni, H Ran, T Colonius, *Numerical experiments on vortex ring formation,* J Fluid Mech 430 (2001) 267-
317. J Montaldi, R M Roberts, *Relative equilibria of molecules,* J Non-linear Sci 9 (1999) 53-88.
318. Giuseppe Morandi, *Statistical Mechanics: An Intermediate Course,*World Scientific, Singapore, 1996.
319. G K Morikawa, E V Swenson, *Interacting motion of rectilinear geostrophic vortices,* Phys Fluids 14 (1971) 1058-1073.

320. P M Morse, H Feshbach, *Methods of Theoretical Physics,* McGraw-Hill, New York City, 1953.
321. Mu Mu and T G Shepherd, *On Arnol'd's second non-linear stability theorem for two-dimensional quasi-geostrophic flow,* Geophys Astrophys Fluid Dyn. 75 21-37.
322. Frederic Nebeker, *Calculating the weather: Meteorology in the 20th century.* Academic Press, San Diego, 1994.
323. Joseph Nebus, *The Dirichlet Quotient of point vortex interactions on the surface of the sphere examined by Monte Carlo experiments,* Discrete and Continuous Dynamical Systems B 5 (2005) 125-136.
324. Paul K Newton, *The N-Vortex Problem: Analytical Techniques.* Springer-Verlag, New York City, 2001.
325. Paul K Newton, R. Kidambi, *Streamline topologies for integrable vortex motion on a sphere,* Physica D, 140, 95 - 125, (2000)
326. Paul K Newton and H Shokraneh, *The N-Vortex Problem on a Rotating Sphere,* Proc R. Soc London Series A, 462, 149 - 169, (2006).
327. E A Novikov, *Dynamics and statistics of a system of vortices,* Sov Phys JETP 41(5) (1975) 937-943.
328. E A Novikov, Yu B Sedov, *Collapse of vortices,* Zh Eksp Teor Fiz Vol 77 no 2(8), 1979, 588-597.
329. T Nozawa, S Yoden, *Formation of zonal band structure in forced two-dimensional turbulence on a rotating sphere,* Phys Fluids 9(7) (1997) 2081.
330. A M Oboukhov, *Some specific features of atmospheric turbulence,* J Fluid Mech 13 (1961) 77-81.
331. Akira Ogawa, *Vortex Flow,* CRC Press, Boca Raton, 1992.
332. H Okamoto, Y Kimura, *Chaotic advection by a point vortex in a semidisk,* Topological Fluid Mechanics, Proc IUTAM Symp Ed. H K Moffat, Tsinobar, 105-113, Cambridge Press, Cambridge, 1989.
333. K A O'Neil, *Stationary configurations of point vortices,* Trans AMS 302(2) (1987) 383-425.
334. T M O'Neil, *Trapped plasmas with a single sign of charge,* Phys Today (1999) 24-30.
335. Manfred Opper, David Saad, *Advanced Mean Field Methods: Theory and Practice,* MIT Press, Cambridge, Massachusetts, 2001.
336. Edward W Packel, *Functional Analysis: A Short Course,* Robert E Krieger Publishing Company, New York City, 1974, 1980.
337. L Onsager, *Statistical hydrodynamics,* Nuovo Cimento Suppl 6 (1949) 279-287.
338. S A Orzsag, *Representation of isotropic turbulence by scalar functions,* Stud in Appl Math 48 (1969) 275-279.
339. S A Orszag, *Analytical theories of turbulence,* J Fluid Mech 41 (1970) 363-386.
340. *Oxford Dictionary of Scientists,* John Daintith, Derek Gjertsen Editors, Oxford University Press, 1993, 1999.
341. J I Palmore, *Relative equilibria of vortices in two dimensions,* Proc Natl Acad Sci USA 79 (1982) 716-718.
342. R K Pathria, *Statistical Mechanics.* Butterworth-Heinemann, Oxford, 1996.
343. V E Paul, *Bewegung eines wirbels in geradlining begrenzten gebieten,* Z Angew Math Mech 14 (1932) 105-116.
344. J Pedlosky, *Geophysical Fluid Dynamics,* 2nd Edition, Springer-Verlag, 1987.
345. J Pedlosky, *The instability of continuous heton clouds,* J Atmos Sci 42 (1985) 1477-1486.

346. S Pekarsky, J E Marsden, *Point vortices on a sphere: Stability of relative equilibria,* J Math Phys Vol 39 Num 11 (1998) 5894-5907.
347. A Péntek, T Tél, Z Toroczkai, *Chaotic advection in the velocity field of leapfrogging vortex pairs,* J Phys A: Math Gen 28 (1995) 2191-2216.
348. A Péntek, T Tél, Z Toroczkai, *Fractal tracer patterns in open hydrodynamical flows: the case of leapfrogging vortex pairs,* Fractals 3(1) (1995) 33-53.
349. A Péntek, T Tél, Z Toroczkai, *Transient chaotic mixing in open hydrodynamical flows,* Int J Bif and Chaos 6(12B) (1996) 2619-2625.
350. A M Perelomov, *Integrable systems of classical mechanics and Lie algebras,* M.: Nauka 1990.
351. A E Perry, M S Chong, *A series expansion study of the Navier-Stokes equations with applications to three-dimensional separation patterns,* J Fluid Mech 173 (1986) 207-223.
352. A E Perry, B D Farlie, *Critical points in flow patterns,* Adv Geophysics B 18 (1974) 299-315.
353. Michael Plischke, Birger Bergensen. *Equilibrium Statistical Mechanics.* 2nd Edition. World Scientific, Singapore, 1994.
354. H Poincaré, *Théorie des tourbillous,* Paris, 1893.
355. Y B Pointin, T S Lundgren, *Statistical mechanics of two-dimensional vortices in a bounded container,* Phys Fluids 19(10) (1976) 1459-1470.
356. L M Polvani, D G Dritschel, *Wave and vortex dynamics on the surface of a sphere,* J Fluid Mech 255 (1993) 35-
357. S B Pope, *Turbulent Flows,* Cambridge University Press, Cambridge, Massachusetts, 2000.
358. C Pozrikidis, *Introduction to theoretical and computational fluid dynamics,* Oxford University Press, 1997.
359. E G Puckett, *Vortex methods: An introduction and survey of several research topics,* Incompressible Computational Fluid Dynamics: Trends and Advances, Cambridge Press, Cambridge, 1993.
360. J W S Rayleigh, *On the stability and instability of certain fluid motions,* Proc Lond Math Soc 9, 57-70, 1880.
361. Michael Reed, Barry Simon. *Methods of Modern Mathematical Physics I: Functional Analysis,* Academic Press, New York City, 1980.
362. G M Reznik, *Dynamics of singular vortices on a beta-plane,* J Fluid Mech 240 (1992) 405-
363. P B Rhines, *Waves and turbulence on a beta plane,* J Fluid Mech 69(3) (1975) 417-443.
364. S Richardson, *on the no-slip boundary condition,* J Fluid Mech 59 (1973) 707-719.
365. P Ripa, *General stability conditions for zonal flows in a one-layer model on the beta-plane or on the sphere,* Journal of Fluid Mechanics, 126:463-489, 1983.
366. G F Roach, *Green's Functions,* 2nd Edition, Cambridge Press, Cambridge, 1982.
367. Christian P Robert, *The Bayesian Choice: From Decision-Theoretic Foundations to Computational Implementation.* 2nd Edition. Springer-Verlag, New York City, 2001.
368. P Robert, *A maximum-entropy principle for two-dimensional perfect fluid dynamics,* J Stat Phys 65 (1991) 531-553.
369. R Robert, J Sommeria, *Statistical equilibrium states for two-dimensional flows,* J Fluid Mech 229 (1991) 291-310.

370. B Robertson, *Application of maximum entropy to non-equilibrium statistical mechanics,* in R D Levine, M Tribus, Eds, *The Maximum Entropy Formalism,* MIT Press, Cambridge, 1979.
371. Sheldon Ross, *A First Course in Probability.* 4th Edition. Macmillan College Publishing Company, New York City, 1994.
372. N Rott, *Constrained three and four vortex problems,* Phys Fluids A2(8) (1990) 1477-1480.
373. N Rott, H Aref, *Three-vortex motion with zero total circulation,* J of Appl Math and Phyz (ZAMP) Vol 40 (1989) 473-500.
374. E J Routh, *Some applications of conjugate functions,* Proc Lond Math Soc 12 (1881) 73-89.
375. G S Rushbrooke, *Introduction to statistical thermodynamics,* Clarendon Press, Oxford, 1964.
376. Murry L Salby, *Fundamentals of Atmospheric Physics.* Academic Press, San Diego, 1996.
377. E B Saff, A B J Kuijlars, *Distributing many points on the sphere,* Mathematical Intelligener 19 (1997) 5-11.
378. P G Saffman, *Vortex Dynamics.* Cambridge University Press, 1992.
379. P G Saffman, *The approach of a vortex pair to a plane surface in an inviscid fluid,* J Fluid Mech 92 (1979) 497-503.
380. P G Saffman, *The stability of vortex arrays to two- and three-dimensional disturbances,* Fluid Dynamics Research 3 (1988) 13-21.
381. P G Saffman, J S Sheffield, *Flow over a wing with an attached free vortex,* Stud in App Math 57 (1977) 107-117.
382. Carl Sagan, *The Demon-Haunted World: Science as a Candle in the Dark,* Ballantine Books, New York City, 1996.
383. Takashi Sakajo, *Invariant dynamical systems embedded in the N-vortex problem on a sphere with pole vortices,* Physica D 217 (2006) 142-152.
384. Giovanni Sansone, Johan Gerretsen, *Lectures on the theory of functions of a complex variable,* Wolters-Noordhoff, Groningen, the Netherlands, 1969.
385. E N Selivanova, *Topology of problem of three dot vortices,* Works Math Inst RAN Vol 205 (1994) 141-149.
386. J Serrin, *Mathematical principles of classical fluid dynamics,* Handbuch der Physik 8, 125-263, Springer-Verlag, New York, 1959.
387. S Schochet, *The weak vorticity formulation of the 2D Euler equations and concentration-cancellation,* Comm Part Diff Eqn 49(5) 911-965.
388. S Schochet, *The point-vortex method for periodic weak solutions of the 2D Euler equations,* Comm Part Diff Eqn 20(5-6) 107-1104.
389. Erwin Schrödinger, *What is Life?* Cambridge University Press, 1944, 1992.
390. J A Sethian, *A brief overview of vortex methods,* in *Vortex methods and vortex motion,* Eds K E Gustafson, J A Sethian, Society for Industrial and Applied Math, Philadelphia, 1991.
391. J S Sheffield, *Trajectories of an ideal vortex pair near an orifice,* Phys Fluids 20 (1977) 543-545.
392. T G Shepherd, *Applications of Hamiltonian theory to GFD,* 1993 GFD Summer School Lecture Notes, Woods Hole Oceanographic Institution Technical Report WHOI-94-12, pp. 113-152, 1994.
393. T G Shepherd, *Non-ergodicity of inviscid two-dimensional flow on a beta-plane and on the surface of a rotating sphere,* J Fluid Mech, 184 (1987) 289-302.

394. T G Shepherd, *Symmetries, conservation laws, and Hamiltonian structure in geophysical fluid dynamics.* Adv Geophys 32 287-338.
395. J Shi, *The role of higher vorticity moments in a variational formulation of Barotropic flows on a rotating sphere,* submitted 2006.
396. Barry Simon, *The Statistical Mechanics of Lattice Gases I.* Princeton University Press, 1993.
397. Stephanie Frank Singer, *Symmetry in Mechanics: A Gentle, Modern Introduction.* Birkhäuser, Boston, 2001.
398. K R Singh, *Path of a vortex round the rectangular bend of a channel with uniform flow,* Z Angew Math Mech 34 (1954) 432-435.
399. Lawrence Sirovich, Chjan C Lim. *Wave propagation on the von Kármán vortex trail,* Physics Fluids 29 (1986) 3910-3911.
400. S Smeil, *Topology and mechanics,* Uspekhi mat nauk Vol 27 Num 2 (1972) 77-133.
401. D Smith, *Variational Methods in Optimization,* Dover
402. R A Smith, *Phase transitions behaviour in negative temperatures guiding center plasma,* Phys Rev Lett 63 (1990) 1479-
403. V I Smirnov and N A Lebedev, *Functions of a complex variable: Constructive theory,* MIT Press, Cambridge, Massachusetts, 1968.
404. R A Smith, T M O'Neil, *Nonaxisymmetric thermal equilibria of a cylindrically bounded guiding center plasma or discrete vortex system,* Phys Fluids B 2 (1990) 2961-2975.
405. E B Smith, *Basic chemical thermodynamics,* Oxford University Press, 1990.
406. E R Smith, C J Thompson, *Glass-like behaviour of a spherical with oscillatory long-range potential,* Physica A 135 (1986) 559-
407. A Sommerfield, *Thermodynamics and statistical mechanics,* Academic Press, New York, 1964.
408. K R Sreenivasan, P Kailasnath, *An update on the intermittency exponent in turbulence,* PHys Fluids A 5 (1993) 512-514.
409. I Stakgold, *Green's functions and boundary value problems,* Wiley, New York, 1979.
410. H E Stanley, *Spherical model as the limit of infinite spin dimensionality,* Phys Rev 176 (1968) 718-
411. C Sulem, P L Sulem, *The well-posedness of two-dimensional ideal flow,* Journal de Mécanique Théorique et Appliquée, Special Issue, 217-242, 1983.
412. Daud Sutton, *Platonic and Archimedean Solids,* Walker and Company, New York, 2002.
413. A G Sveshnikov, A N Tikhonov, *The theory of functions of a complex variable,* Mir Publishers, Moscow, 1971.
414. J L Synge, *On the motion of three vortices,* Can J Math 1 (1949) 257-270.
415. J L Synge, C C Lin, *On a statistical model of isotropic turbulence,* Trans R Soc Can 37 (1943) 45-63.
416. A Szeri, P Holmes, *Non-linear stability of axisymmetric swirling flows,* Philos Trans R Soc London Series A 326 (1988) 327-
417. R Takaki, M Utsumi, *Middle-scale structure of point vortex clouds,* Forma 7 (1992) 107-120.
418. J Tavantzis, L Ting, *The dynamics of three vortices revised,* Phys Fluids Vol 31 Num 6 (1988) 1392-1409.
419. C J Thompson, *Classical Equilibrium Statistical Mechanics,* Oxford Science Pub, 1988.

420. C Thompson, *Mathematical Statistical Mechanics,* Princeton University Press, 1972.
421. J J Thompson, *A treatise on the motion of vortex rings,* London: Macmillan, 1883.
422. W Thompson, *On the vortex atoms,* Phil Mag Ser 4 34 Num 227 (1867) 15-24.
423. Lu Ting, Denis Blackmore, *Bifurcation of Motions of three vortices and applications,* Proceedings 2004 ICTAM Conference Warsaw (2004).
424. A C Ting, H H Chen, Y C Lee, *Exact solutions of non-linear boundary value problems: the vortices of the sinh-Poisson equation,* Physica D 26 (1987) 37-
425. Lu Ting, Rupert Klein, *Lecture Notes in Physics 374: Viscous Vortical Flows,* Springer-Verlag, Berlin, New York City, 1991.
426. Howard G Tucker, *A Graduate Course in Probability.* Academic Press, New York City and London, 1967g
427. K K Tung, *Barotropic instability of zonal flows,* J Atmos Sci, 38, 308 - 321, 1981.
428. K K Tung, W T Welch, *The k^{-3} and $k^{-5/3}$ energy spectrum of atmospheric turbulence,* J Atmos Sci (2001)
429. S R Turner, *An introduction to combustion, concepts and applications,* 2nd edition, McGraw-Hill, New York City, 1972.
430. B Turkington, *on steady vortex flows in two dimensions,* Comm Part Diff Eqns 8(9) (1983) 999-1071.
431. B Turkington, *On the evolution of a concentrated vortex in an ideal fluid,* Arch Rat Mech Anal 97 (1987) 57-87.
432. B Turkington, *Statistical equilibrium measures and coherent states in two-dimensional turbulence*, Comm Pure Appl Math LII (1998) 1-29.
433. B Turkington, N Whittaker, *Statistical equilibrium computations of coherent structures in turbulent shear layers,* SIAM J Sci Comp 17 (1996) 1414-
434. J H G M van Geffen, V V Meleshko, G J F van Heijst, *Motion of a two-dimensional monopolar vortex in a bounded rectangular domain,* Phys Fluids 8 (1996) 2393-
435. Washington University Libraries: Edward U Condon, 1902 - 1974. Web page, http:// library.wustl.edu/units/ spec/ exhibits/ crow/ condonbio.html visited May 2005.
436. W K Hastings, Statistician and Developer of the Metropolis-Hastings Algorithm, Web page, http://probability.ca/hastings/index.html visited June 2005.
437. J B Weiss, J C McWilliams, *Nonergodicity of point vortices,* Phys Fluids A 3(5) (1991) 835-844.
438. J B Weiss, J C McWilliams, *Temporal scaling behavior of decaying two-dimensional turbulence,* Phys Fluids A 5 (1993) 608.
439. Eric W Weisstein *Eric Weisstein's World of Scientific Biography,* http:// scienceworld.wolfram.com/ biography/ visited May 2003-October 2005
440. Neil Wells, *The atmosphere and ocean: a physical introduction.* Taylor and Francis Ltd, London, 1986.
441. F M White, *Viscous Flows* 2nd Edition, McGraw-Hill, New York City, 1974.
442. J A Whitehead, *Thermohaline Ocean Processes and Models,* Ann. Rev. Fluid Mech. 27 (1995) 89.
443. E T Whittaker, *A treatise on the Analytical Dynamics* 3rd Ed. Cambridge University Press, 1927
444. G P Williams, *Planetary circulations: I,* J Atmos Sci, 35 (1978) 1399-

445. J H WIlliamson, *Statistical mechanics of a guiding center plasma*, J Plasma Physics, 17 (1977) 85-92.
446. J Wilks, *The Third Law of Thermodynamics*, Clarendon Press, Oxford, 1961.
447. A Wintner, *The analytical foundation of celestial mechanics*, Princeton University Press, 1941.
448. Carl Wunsch, Biographical Memoirs, Henry Stommel, September 27, 1920 - january 17, 1992. http://fermat.nap.edu/ html/ biomems/ hstommel.html visited May 2006.
449. S Yoden and M Yamada, *A numerical experiment on 2D decaying turbulence on a rotating sphere*, J Atmos Sci, 50, 631, 1993.
450. W R Young, *Some interactions between small numbers of baroclinic geostrophic vorticities*, Geo Astro Fluid Dyn 33 (1985) 35-61.
451. L Zannetti, P Franzese, *Advection by a point vortex in a closed domain*, Eur J Mech B/Fluids 12 (1994) 43-
452. L Zannetti, P Franzese, *The non-integrability of the restricted problem of two vortices in closed domains*, Physica D 76 (1994) 99-
453. G Zaslavsky, *Chaotic dynamics and the origin or statistical laws*, Physics Today (1999) 39-45.
454. R K Zeytounian, *Meteorological Fluid Dynamics: Asymptotic modelling, stability, and chaotic atmospheric motion*. Springer-Verlag, Berlin, 1991.

Index